ALSO BY PAUL BLOOM

The Sweet Spot: The Pleasures of Suffering and the Search for Meaning

Against Empathy: The Case for Rational Compassion

Just Babies: The Origins of Good and Evil

How Pleasure Works: The New Science of Why We Like What We Like

es' Baby: How the Science of Child Development Explains What Makes Us Human

How Children Learn the Meanings of Words

Psych

Psych

The Story of the Human Mind

PAUL BLOOM

ecco

An Imprint of HarperCollins*Publishers*

HarperCollins books may be purchased for educational, business, or sales promotional use. For information, please email the Special Markets Department at SPsales@harpercollins.com.

Ecco® and HarperCollins® are trademarks of HarperCollins Publishers.

Images on pages 17, 26: Hendri Maulana. Pages 44, 171, 172, 176: Christina Starmans. Page 162: © Fibonacci, licensed under the Creative Commons and the GNU Free Documentation License, found at commons.wikimedia.org/wiki /File:Café_wall.svg. Page 169: Edward Adelson (with his permission). Page 312: public domain via Wikimedia Commons.

FIRST EDITION

Designed by Angie Boutin

Library of Congress Cataloging-in-Publication Data has been applied for.

ISBN 978-0-06-309635-6

23 24 25 26 27 LBC 5 4 3 2 1

For my boys—Max and Zachary

CONTENTS

Prologue   1

FOUNDATIONS

1. "Brain Makes Thought"   7
2. Consciousness   35
3. Freud and the Unconscious   55
4. The Skinnerian Revolution   71

THINKING

5. Piaget's Project   97
6. The Ape That Speaks   125
7. The World in Your Head   159
8. The Rational Animal   191

APPETITES

9. Hearts and Minds   215

RELATIONS

10. A Brief Note on a Crisis   255
11. Social Butterflies   263
12. Is Everyone a Little Bit Racist?   283

DIFFERENCES

13. Uniquely You   307
14. Suffering Minds   341
15. The Good Life   367

Acknowledgments   387
Notes   391
Index   433

Psych

Prologue

One summer many years ago, I was having a rough Sunday morning. It came after a string of bad days when I struggled with work and relationships, and I was quiet and tense. My eight-year-old son was invited to a birthday party about an hour's drive away, and as we left the house, I snatched a slim hardcover book from a pile near the door. Zachary was tired, and he napped on the way, quietly snoring in the back seat, and this was fine by me. When we got to the party, held in the backyard of a grand house in suburban Connecticut, I made small talk with the adults and then slipped away and sat under a tree and took out the book and started to read.

It was *The Origin of the Universe*, written by John Barrow, a theoretical physicist.[1] It began by describing Edwin Hubble's discovery that the universe was expanding, and then went over the evidence for the "Big Bang" theory of how everything started.

As I read, my heart began to beat faster. It was so exciting that we could know about all this, that I could be reading about events that happened fourteen billion years ago. Perhaps it's what people of faith feel like when reading Scripture—the experience of great truths being revealed. Learning about the universe, I felt insignificant, tiny in space and time. But I also felt proud of our species—that we could know so much about the incredibly long ago and incredibly far away, that we could make real progress on the most fundamental of all questions. And when the birthday party was over and I got up to get my son, the world was full of light.

Driving back, I talked to Zachary about what I learned, and as we spoke, I played with the fantasy of quitting my job as professor of psychology, getting a new degree, and becoming a cosmologist. But I was where I belonged. The tombstone of the philosopher Immanuel Kant has a quote from his *Critique of Pure Reason*: "Two things fill the mind with ever new and increasing admiration and awe, the more often and more steadily we reflect upon them: the starry heavens above me and the moral law within me." I had spent the morning being thrilled by the starry heavens; years later my research would turn to morality and moral psychology, and there again I would experience the same "admiration and awe" as Kant.

Honestly, though, just about all of psychology gives me this buzz. It's about the most interesting topic there is—*us*. It's about our feelings, experiences, plans, goals, fantasies, the most intimate aspects of our being.

The book you are holding is built from an Introduction to Psychology course that I've taught for many years as a professor at Yale University. This is one of the most popular courses at Yale, and I have taught thousands of undergraduates, sometimes as their very first course at a university. Based on these lectures, I created an online course that has had an enrollment, so far, of about a million students.[2]

I love teaching Introduction to Psychology. But there is a limit to how much one can convey in a series of lectures, and there is so much material to cover. And so I decided to write *Psych*. The scope here is broad, and if you choose to read this from cover to cover, you'll have a grounding in every major aspect of the science of psychology. Among other things, *Psych* will put forth the best answers we have to the following questions:

How does the brain—a three-pound lump of bloody meat—give rise to intelligence and conscious experience?
What did Freud get right about human nature?
What did Skinner get right about human nature?
Where does knowledge come from?
How does the mind of a child differ from that of an adult?
What is the relationship between language and thought?
How do our biases affect how we see and remember the world?
Are we rational beings?

What motivates us—and what is the purpose of feelings such as
   fear, disgust, and compassion?
How do we think of other people—including those from other
   social and ethnic groups?
How (and why) do we differ in personality, intelligence, and other
   traits?
What is the cause and treatment of different mental illnesses?
What makes people happy?

Each chapter of this book can be read as a stand-alone piece. It's fine
if you decide to dive in and read about Freud, or language, or mental illness. Or even jump to the end, to the part on happiness. Nobody is judging you. But there is a flow to this book; there are themes and ideas that
stretch across these disparate chapters, and there is a satisfaction to seeing
the story unfold in its proper order.

Some parts of this story make people uncomfortable. We'll see that
modern psychology accepts a mechanistic conception of mental life, one
that is *materialist* (seeing the mind as a physical thing), *evolutionary* (seeing
our psychologies as the product of biological evolution, shaped to a large
extent by natural selection), and *causal* (seeing our thoughts and actions
as the product of the forces of genes, culture, and individual experience).

You might worry that there is something missing here. This conception of mental life might seem to clash with commonsense notions of free
choice and moral responsibility. It might seem to clash as well with the
notion that humans have a transcendent or spiritual nature. The tension
here is nicely illustrated by John Updike in his *Rabbit at Rest*, when Harry
"Rabbit" Angstrom talks to his friend Charlie about Charlie's recent
surgery:

> "Pig valves." Rabbit tries to hide his revulsion. "Was it terrible?
> They split your chest open and ran your blood through a machine?"
>   "Piece of cake. You're knocked out cold. What's wrong with
> running your blood through a machine? What else you think you
> are, champ?"
>   A God-made one-of-a-kind with an immortal soul breathed

in. A vehicle of grace. A battlefield of good and evil. An apprentice angel . . .

"You're just a soft machine," Charlie maintains.[3]

There are different ways to react to all this. I know philosophers and psychologists who confidently assert there's no such thing as free will or moral responsibility. And I've met others who reject the science, who worry that such an approach to the mind takes the specialness away from people, it diminishes us somehow. It's too reductionist, too crude. It reduces us to computers or lumps of cells or lab rats. They reason: "If psychology is going to tell me that I'm just a machine, that the most intimate aspects of my being are nothing more than neural firings, well, so much for psychology."

My own view is that we can find a middle ground here. I think the scientific perspective at the core of modern psychology is fully compatible with the existence of choice and morality and responsibility. Yes, we are, in the end, soft machines—but not *just* soft machines.

I want to end this prologue with a note of humility. We know so much about the physical world and so little about mental life. This isn't because physicists are smart and psychologists are stupid. It's because my chosen domain of study is so much harder than Barrow's. The mysteries of space and time turn out to be easier for our minds to grasp than those of consciousness and choice. In the pages that follow, I'll be honest about the limitations of our young science and critical of some colleagues who think we've solved it all.

But there's a real joy to being part of a young science. I find the study of psychology to be just as exhilarating as this study of the cosmos, and I hope you come to see it this way as well. We have made exciting progress in the field and I can't wait to talk about it. My fondest hope for this book is that the theories and discoveries reviewed here will give rise to a sort of awe in the reader, something akin to what I experienced when I read about the origins of the universe under that tree many years ago.

FOUNDATIONS

1

"Brain Makes Thought"

The Astonishing Hypothesis

In the late afternoon of September 13, 1848, something miraculous and terrible happened to the young foreman of a construction gang working in Cavendish, Vermont. Phineas Gage was preparing the roadbed for the laying of railway tracks, and he had a routine. He would bore holes into rocks, place explosive powder and a fuse inside the holes, and then pile sand and dirt over them. Then he would use a tamping iron—a piece of iron that looked like a javelin, about three and a half feet long and thirteen and a half pounds in weight—to pack it all down, making a plug over the explosive device. Later, the fuses would be lit, and the explosions would clear away the rocks.

Nobody knows what went wrong—perhaps something distracted him—but Gage slammed his tamping iron into a hole before he had poured the sand, and the blasting powder exploded. The iron shot upward with tremendous force. It entered the left side of Gage's jaw, passing behind the left eye through the left side of his brain and continuing out the top of his skull, landing yards away from him.

Gage lost consciousness—but just for a moment. His gang helped him onto an oxcart and took him to the Cavendish Inn, where he was renting a room. He sat on the veranda and told bystanders the story of what had just happened. When a medical expert finally arrived, Gage said, "Doctor, here is business enough for you."

It was touch and go for a while. Gage had an infection and required considerable treatment. But months later, he was seemingly recovered. He wasn't blind; he wasn't paralyzed; he retained the ability to speak and understand language; he didn't lose his intellectual capacities in any obvious way. You might think he was very lucky indeed.

But Gage wasn't lucky at all. As his doctor, John Martyn Harlow, wrote, Gage used to be: "the most efficient and capable man, a man of temperate habits, considerable energy of character, a sharp shrewd businessman." But afterward: "Gage was no longer Gage. He was fitful, irreverent, indulging at times in the grossest profanity manifesting but little deference for his fellows." He was, according to Harlow, "a child intellectually" with "the animal passions of a strong man."

Unable to return to his job as foreman, Gage held a string of jobs in the years that followed, including working as a stagecoach driver in Chile and being an attraction at Barnum's American Museum in New York, showing off his tamping iron and telling his story. Eleven years after his accident, he began to have seizures, and he died a few months later in his mother's home.

The story of Phineas Gage is a vivid illustration of how damage to the brain (and more specifically, damage to the frontal lobe, the part right behind the forehead) can have a profound influence on some of the most important aspects of who we are—our inhibitions, how we treat others, our character.

There has long been controversy over the details of what happened to Gage, and his story has become more and more extravagant as time has gone by.[1] But the account above is as accurate as I could make it. And anyway, the world contains thousands of Gages. There are many unfortunate cases of brain damage profoundly changing a person's nature.

So here's another one, with a different outcome. This is the story of Greg F., described by the neuroscientist Oliver Sacks, in an article called "The Last Hippie."[2] As a teenager, Greg was restless and rebellious. He dropped out of school and became a Hare Krishna, moving to a temple in New Orleans. After he spent some time there, the spiritual leader grew impressed with Greg, calling him a Holy Man. Then Greg slowly began to go blind. This was seen not as something to be treated, but as a spiritual event.

He was "an illuminate," they told him; it was the "inner light" growing. . . . And indeed, he seemed to be becoming more spiritual by the day—an amazing new serenity had taken hold of him. He no longer showed his previous impatience or appetites, and he was sometimes found in a sort of daze, with a strange (many said "transcendental") smile on his face. It is beatitude, said his swami: he is becoming a saint.

After four years, the temple permitted his parents to visit Greg, and when they did,

they were filled with horror: their lean, hairy son had become fat and hairless; he wore a continual "stupid" smile on his face (this at least was his father's word for it); he kept bursting into bits of song and verse, and making "idiotic" comments, while showing little deep emotion of any kind ("like he was scooped out, hollow inside," his father said); he had lost interest in everything "current"; he was disoriented—and he was totally blind.

It turned out that Greg had a tumor in his brain the size of an orange. It destroyed most of the parts of the brain devoted to vision and extended as well into his frontal lobes (up front) and his temporal lobes (on the sides). The tumor was removed but the damage was irreversible. Greg F. was worse off than Phineas Gage. He was not only blind; he lost most of his memory of decades of his life and was unable to form new memories. He was docile and without feeling, unable to survive on his own.

Phineas Gage, Greg F., and so many others illustrate what the Nobel Prize–winning biologist Francis Crick calls the Astonishing Hypothesis:

> You, your joys and your sorrows, your memories and your ambitions, your sense of personal identity and free will, are in fact no more than the behavior of a vast assembly of nerve cells and their associated molecules.[3]

There's a shorter version of this idea in one of Charles Darwin's notebooks: "Brain makes thought."[4]

———

The philosophical term for the position of Darwin and Crick is "materialism" (there is another meaning of this word that has to do with money; ignore this). For the materialist, there is nothing but physical stuff. There are no immaterial souls.

This is an odd and unnatural view.[5] People are more attracted to the doctrine of "dualism," which is that the mind (or the soul) is a fundamentally different kind of thing than the body. We are not one; we are two—bodies and souls. This is an idea that's present in most religions and most philosophical systems (Plato, for instance, was very much a dualist), but the most thoughtful and articulate defender of dualism was the philosopher René Descartes. In his honor, the idea that minds and bodies are distinct is often described as "Cartesian dualism."

Written in the early 1600s, one of Descartes' arguments for dualism had to do with the limitations of physical things. It might surprise you to hear this, but Descartes was familiar with robots. He had visited the French Royal Gardens, which the philosopher Owen Flanagan describes as "a veritable seventeenth-century Disneyland,"[6] and was impressed by the automata driven by hydraulic force:

> You may have seen in the grottoes and fountains which are in our royal gardens that the simple force with which water moves in issuing from its source is sufficient to put into motion various machines

and even to set various instruments playing or to make them pronounce words accordingly to the varied disposition of the tubes which convey the water. . . . In entering [strangers] necessarily tread on certain tiles or places, which are so disposed that if they approach a bathing Diana, they cause her to hide in the rosebuds, and if they try to follow her, they cause Neptune to come forward to meet them threatening them with his trident.[7]

There is an analogy here to the human body; the springs and motors in the robots correspond to muscles and tendons; the tubes in the robots correspond to nerves. So, one can wonder: Are we nothing more than complicated machines?

Descartes said no. This analogy, he argued, was correct for nonhuman animals. Their actions are solely the product of their physical constitution. They are nothing more than *bêtes machines*—beast machines. There are chilling stories, perhaps apocryphal, of Descartes participating in operations on live dogs—vivisection—presumably believing that their shrieks of agony were akin to the noises that broken machines sometimes make. Without souls, after all, they are incapable of feeling.

But humans are different. We see this in the unpredictable nature of our actions. The doctor taps your knee and you kick your lower leg, and, yes, this can be done by your body alone, under the same principles that govern the motions of robot Diana and robot Neptune. But you can also *choose* to kick out your leg right now, just for the hell of it. This is the sort of willful action that Descartes believed a physical thing could never do. And so he concluded: We are not physical things.

Descartes' other main argument for dualism is better known. He starts with the question "What can we know for sure?" and answers: Not much. You believe you were born in such and so place, but maybe you were lied to. Perhaps, as some children fantasize, you are of royal blood, and it's only due to some misadventure that you got stuck with the disappointing family of commoners who raised you. Or, to get really weird, maybe the universe was created five seconds ago, and your memories are all false. This is unlikely, but it's possible.

You might believe that you are now in a certain physical environment, sitting on a chair with your loyal hound by your side, my book in one hand and a cigar in another (or whatever). But Descartes observed that we often believe such things when we are dreaming. You might protest that you are not dreaming right now, but most dreamers don't know they are dreaming.

You can be wrong that you have a body. Philosophers have long worried that our experience might be an illusion created by the devil, and a modern version of this concern is played out in my favorite movie, *The Matrix*, which imagines a world in which everyday human experience is an illusion created by a malevolent computer. Some philosophers take this even further and argue that we are parts of computer simulations—essentially video-game characters. You might agree with me that this seems batty, but how can we know for sure?

We can't. But Descartes notes there's one thing that cannot be doubted—our own existence as thinking beings. The famous line is *Cogito ergo sum*—I think, therefore I am. You might not be sure you have a body, but you can be sure that there is a you who is asking the question. Drawing on this distinction between how we think about minds and how we think about bodies, Descartes concludes,

> I knew that I was a substance the whole essence or nature of which is to think, and that for its existence there is no need of any place, nor does it depend on any material thing. . . . That is to say, the soul by which I am what I am, is entirely distinct from body.[8]

This feels right. Our gut intuition is that we are not our bodies; we inhabit these bodies. We are *Ghosts in the Shell*, in the evocative phrase of manga artist Masamune Shirow. This is why we so easily create and understand fictions where bodies and souls come apart. Think about Franz Kafka's *Metamorphosis*, which begins with: "As Gregor Samsa awoke one morning from uneasy dreams he found himself transformed in his bed into a gigantic insect." Or the scene from *The Odyssey*, in which the goddess Circe transforms Odysseus' men into pigs: "They had the head, and voice, and bristles, and body of swine; but their minds remained unchanged as before. So they were penned there, weeping." Or countless other tales of possession, body swaps, frightening or lovable ghosts, and the like.[9]

Dualism has an appealing real-world consequence. If you are not your body, you can survive its destruction. Maybe you'll end up in some spirit world, or ascend to heaven, or occupy some other body. Now, there are clever ways in which materialists can also arrive at some sort of afterlife belief—perhaps God could somehow reanimate your corpse, repairing it as one would a broken watch. But for the most part materialism is a grim doctrine, tying your survival to the fate of your all-too-fragile flesh.

———

With all the arguments in favor of dualism, then, and all its attractions, why are modern-day psychologists so confident that its opposite—materialism—is correct?

Let's go back to Descartes' arguments. He was correct about the limitations of material things—hundreds of years ago. But now we have an expanded understanding of what such things are capable of. For Descartes, the idea that a machine can do something as complicated as playing chess would be ludicrous. This requires rational deliberation; it's not a matter of reflex. But of course, there are now machines that play chess, better than any human. One might reasonably wonder about other limits of physical things—can computers *feel?*—and we will get to these doubts later on. But the point here is that Descartes' argument no longer flies. The complexity of our actions is not proof of dualism.

As for what Descartes could and could not imagine, many philosophers have pointed out that he was too quick to assume that such a conceptual exercise could tell us about how things really are. Yes, you can doubt that you have a body and can imagine yourself without one. But this doesn't mean that this is possible. After all, I can imagine a spaceship moving faster than light—there are many in science fiction. Descartes' method reflects how we think about minds, not what's true about minds.

Consider all the problems with dualism and all the evidence against it. Talking about an immaterial Cartesian soul, the psychologist Steven Pinker writes, "How does the spook interact with solid matter? How does an ethereal nothing respond to flashes, pokes, and beeps and get arms and legs to move?"[10] This is an old complaint. In 1643, Elizabeth Stuart, the former queen of Bohemia, wrote to Descartes to complain how hard it is to take seriously the idea that "an immaterial thing could move and be moved by a body."[11]

To be fair, the other option, that brains make thoughts, can be equally hard to stomach. Here's Gottfried Leibniz in 1712: "In imagining that there is a machine whose structure would enable it to think, feel, and have perception, one could think of it as enlarged yet preserving its same proportions, so that one could enter into it as one does a mill. If we did this, we should find nothing within but parts which push upon each another, we should never see anything which would explain a perception."[12] More than a few modern neuroscientists have become tempted by dualism toward the end of their careers, often with a similar argument to that of Leibniz—they've spent their lives studying the brain, and they found no physical sign of consciousness residing there, and so perhaps it's in the spirit realm after all.

In the end, what decides the issue is all the evidence that the brain *is* implicated in thought, though not in a way that's apparent to someone peering at an opened skull. The change in Phineas Gage's character was caused by a very physical tamping iron going through his very physical head. And of course, you didn't have to wait until 1848 to appreciate that a blow to your skull can affect your consciousness and your memory, and can obliterate them permanently if the blow is hard enough. Everybody knew that dementia could rob you of your rationality or that coffee and alcohol can, in different ways, inflame the passions. (As Pinker puts it: "The supposedly immaterial soul, we now know, can be bisected with a knife, altered by chemicals, started or stopped by electricity, and extinguished by a sharp blow or by insufficient oxygen."[13])

What is new is that we can now observe the brain at work. You can put someone in a brain scanner, for instance, and tell from the parts of the brain that are active whether they are thinking about their favorite song or the layout of their apartment or a mathematical problem. We may not be too far from the point where we can look at the brain of a sleeping person and know what they are dreaming about.[14]

Is there any hope for Descartes' position? There are important distinctions between mental events and physical events, and some contemporary philosophers defend what they describe as mild forms of dualism.[15] These views are worth discussing, though not here. But almost nobody defends the sort of hard-core dualism maintained by Descartes, so-called substance dualism, where the mind is a different kind of stuff than the brain, where the process of thinking occurs in an immaterial realm, separate from the laws of nature. This theory is dead as a theory can be.

Sentient Meat

Okay, then, what is the physical seat of thought? What is the source of our emotions, decision making, our passions, our pains, and everything else? Even a dualist must address some version of the question; the soul must connect to some part of our physical being to make the body act and receive its sensory information. (Descartes believed the conduit to be the pineal gland.)

For most of history, people thought that the answer was the heart. This was apparently the belief of diverse populations around the world, including the Maya, the Aztecs, the Inuit, the Hopi, the Jews, the Egyptians, the Indians, and the Chinese. It is a view at the foundation of Western philosophy; Aristotle wrote, "And of course, the brain is not responsible for any of the sensations at all. The correct view [is] that the seat and source of sensation is the region of the heart. . . . The motions of pleasure and pain, and generally all sensation plainly have their source in the heart." After all, the heart responds to feelings; it pounds when you are lustful or angry; it is still when you are calm.[16]

But the brain is also a serious contender. There are psychological experiments suggesting that the commonsense view is that our consciousness is located above the neck.[17] In work that I've done with my colleague and wife, Christina Starmans, we find that even young children, when asked in various ways to locate "where" a person is, tend to answer that their real location isn't the chest; it's right between the eyes.[18]

Head or heart? Over history, this has been the focus of much debate, nicely captured by a line from *The Merchant of Venice*, written in the late 1500s:

> Tell me where is fancy bred,
> Or in the heart or in the head?[19]

As you probably have heard, we now know the answer—it's the head. The brain is a mere one fiftieth of our body weight but consumes about a quarter of the calories we burn off when we are at rest—it's an energy hog. The human brain is also humongous. Baby heads are bowling balls, which is one reason why human females, relative to females of other species, have such a prolonged and painful childbirth.

If you've never seen a brain, you might imagine that it would look impressive. It's often described as the most complicated thing in the known universe, after all. Perhaps it would glow, maybe there would be flashing colored lights or something. But no, it's just meat. One can eat brain—I've had it with cream sauce (not human brain, mind you—you shouldn't eat human brain; you can get this terrible disease, kuru, which is much like mad cow disease, and it's one reason not to be a cannibal). When you take it out of the head, the brain is dull gray; inside the head it is bright red because of all the blood.

There is a science fiction short story by Terry Bisson that nicely captures just how strange this is.[20] The story is in the form of a dialogue by a pair of hyperintelligent aliens traveling through the universe to find sentient beings, and then they find us:

> "Meat. They're made of meat."
> "Meat?"
> "There's no doubt about it. We picked up several from different parts of the planet, took them aboard our recon vessels, and probed them all the way through. They're completely meat."
> "That's impossible. What about the radio signals? The messages to the stars?"
> "They use the radio waves to talk, but the signals don't come from them. The signals come from machines."

"So who made the machines? That's who we want to contact."

"They made the machines. That's what I'm trying to tell you. Meat made the machines."

"That's ridiculous. How can meat make a machine? You're asking me to believe in sentient meat."

The aliens agree to erase the records and report that our solar system is unoccupied.

———

To explore the mystery of "sentient meat," we will start small, with neurons, and work our way up. By weight, most of the brain is fat and blood, and there are cells in the brain other than neurons (about half the brain is composed of glial cells, which support, clean up after, and nourish neurons). But the story of mental life is fundamentally the story of neurons, which is why the study of the biological basis of thought is called *neuroscience*.[21] Below are the parts of a neuron.

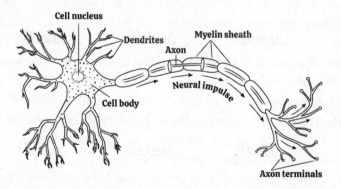

Like other cells, neurons have a cell body that keeps the cell alive and that houses a nucleus that contains the chromosomes that are made up of DNA. The cell body also coordinates the inputs from other neurons or from the senses. It gets this information through dendrites, which protrude from the cell body like tree branches—"dendrite" comes from the

Greek word meaning "tree." If the cell body receives the right sort of input from these dendrites, it causes the neuron to fire, and then an electrical signal goes down a long part of the neuron known as the axon. While dendrites are tiny, axons are long; there are single axons that run from your spinal cord all the way to your big toe. Axons have myelin sheaths—coatings of fatty tissue—that work like insulation on a wire, making the communication in the neuron run more efficiently. Diseases such as multiple sclerosis involve damage to the myelin sheath, which causes problems with action, perception, and thought.

The neuron then communicates with other neurons or, less frequently, to organs and muscles.

Summing up, then, here's how information typically flows:

dendrites > cell body > axon > the dendrites of other neurons

Some neurons are sensory neurons and take in information from the external world; some are motor neurons and go out to the external world. If you touch something hot, and you feel pain, that is because of sensory neurons: If you reach for something, that's due to the working of motor neurons. Other neurons—interneurons—don't connect directly to the world; they connect to one another, and this is how the thinking happens.

There's a puzzle here: When neurons talk to these other neurons or connect to the world, their communication is all or nothing. Neurons fire or they don't fire. It's like a gun—the bullet doesn't go faster if you pull the trigger with all your strength. But perception and action are graded. You can feel the difference between touching a warm plate and a hot stove; you can poke someone gently or really hard.

The solution to the puzzle is that assemblies of neurons have certain ways to represent the intensity of experience and action. One is the number of neurons that fire. If N neurons correspond to a mild experience, then N × 100 neurons may correspond to an intense experience. There is also the frequency of firing of individual neurons; an individual neuron might denote a mild sensation with fire . . . fire . . . fire . . . fire; and an intense sensation with firefirefirefirefirefire. Similar means of coding explain how motor neurons can code for intensity, allowing you to choose to pound the wall with your fist or gently stroke your newborn's cheek.

One major finding about neurons was discovered by the neuroscientist Santiago Ramón y Cajal in the 1800s. I've said that neurons talk to one another when the axon of one communicates with the dendrite of another. But neurons don't touch. There is a tiny gap between the axon of one neuron and the dendrite of another—typically about 20 to 40 nanometers. This gap is known as a *synapse*.

One of the great scientific disputes of the last century was over how the message got across this passage. This was known as the War of the Soups and the Sparks; the options were chemical (soup) or electrical (spark).[22] Long story short, the Soups won. As Cajal discovered, when neurons fire, axons release chemicals that we now call neurotransmitters; these cross the synapses to act on the dendrites of other neurons.

I said before that the cell body decides whether to fire based on the inputs of the dendrites impinging on it. Now I can spell this out a bit. The effect of these incoming neurotransmitters can either be excitatory, which is that they increase the likelihood of a neuron firing, or inhibitory, so that they bring down the likelihood of a neuron firing. The cell bodies put this all together, calculating whether all the increases and decreases add up to enough of a sum to fire.

Neurotransmitters are a large part of the story of how the brain works, and they have considerable practical import. We have invented drugs that interact in different ways with the workings of the neurotransmitters, and these can treat diseases, enhance pleasure, or increase focus.

Or kill. For an example of some deadly interaction, take curare, a drug used by some Indigenous people in South America while hunting—it's put on the tip of a dart or arrow. Curare is an antagonist, which means that it makes neurotransmitters less available for use. More specifically, it inhibits sensitivity to a neurotransmitter called acetylcholine, which is how motor neurons communicate with muscles. This is how curare paralyzes the prey. In large enough doses, it kills, because motor neurons also keep an animal breathing. Conveniently, curare is safe to eat, so you can dine on an animal that you felled with a curare-tipped dart.

Other drugs are agonists; they increase the availability of neurotransmitters for the brain to use. More specifically, they work on neurotransmitters such as norepinephrine that are involved with arousal, increasing euphoria, wakefulness, and control of attention. This is how (in different ways, and to different extents) drugs like speed and Ritalin and cocaine work.

———

This is what thinking is, then: neurons talking to other neurons via neurotransmitters. By one estimate, the brain of an adult human contains about eighty-six billion neurons, each connecting to thousands or tens of thousands of other neurons, leading to hundreds of trillions of connections, a combinatorial explosion that's mind boggling.[23]

But how does this give rise to experience? How does fire, fire, [not fire], fire, [not fire], etc., give rise to laughing at an excellent tweet or grieving the death of someone you love? And what about action? Our brains are physical things, but they are wired up so that they guide us to act in ways that seemingly transcend the laws of physics. William James puts it like this:

> If some iron filings be sprinkled on a table and a magnet brought near them, they will fly through the air for a certain distance and stick to its surface. . . . But let a card cover the poles of the magnet, and the filings will press forever against its surface without its ever occurring to them to pass around its sides. . . . If now we pass from such actions as these to those of living things, we notice a striking difference. Romeo wants Juliet as the filings want the magnet; and if no obstacles intervene he moves towards her by as straight a line as they. But Romeo and Juliet, if a wall be built between them, do not remain idiotically pressing their faces against its opposite sides like the magnet and the filings with the card. Romeo soon finds a circuitous way, by scaling the wall or otherwise, of touching Juliet's lips directly. With the filings the path is fixed; whether it reaches

the end depends on accidents. With the lover it is the end which is fixed, the path may be modified indefinitely.[24]

Other creatures with brains have similar capacities for feelings and for rational action. A chimpanzee might shake with fear or bellow with rage. A cheetah chasing an antelope who darts behind a tree won't run into the tree but will move around it. How do brains do all this?

People often get caught up in the seemingly paradoxical nature of this inquiry—isn't it weird that we're using our brains to understand our brains? One physicist, Emerson M. Pugh, wrote, "If the human brain were so simple that we could understand it, we would be so simple that we couldn't." The comedian Emo Phillips says, "I used to think that the brain was the most wonderful organ in my body. Then I realized who was telling me this."

Charles Darwin added a twist to this, implying that the human brain was the *second* most interesting thing in nature. What could be more marvelous? Well, look at the ground below you:

It is certain that there may be extraordinary mental activity with an extremely small absolute mass of nervous matter: thus the wonderfully diversified instincts, mental powers, and affections of ants are notorious, yet their cerebral ganglia are not so large as the quarter of a small pin's head. Under this point of view, the brain of an ant is one of the most marvellous atoms of matter in the world, perhaps more so than the brain of a man.[25]

There is reason for optimism, though, when it comes to understanding how brains make us—and ants—smart. When I talked about Descartes' dualism, I noted that computers prove that brute physical things are capable of capacities we associate with intelligence. One can take this further now and consider that computers work through simple processes that do stupid things like turn 0 to 1 or 1 to 0. If you have enough processes of this sort and they are put together in the right way, then chess playing, mathematical ability, language parsing, and all the rest arise. And here we meet up with neuroscience, as these binary operations look intriguingly like the basic dichotomy expressed by neurons inside the brain—fire versus not fire. This is progress, then: Computers suggest that the project of neuroscience is feasible, that intelligence can arise from the proper interaction of components that are themselves entirely unintelligent. As the polymath Alan Turing speculated in the 1940s, the human mind may be a computing machine.[26]

The mind as a computer? There is a dismissive reaction you sometimes get here: We used to see brains as hydraulic machines or clocks; then as telegraphic networks; then as telephone exchanges; and now, finally, we see them as computers. Perhaps this is yet another metaphor, a way of talking that will one day be supplanted by something else.

I agree that seeing the brain as akin to a Mac or a PC is just a metaphor, and not a very good one. Neurons are much slower to communicate than parts of computers are, and the brain is "wired up" differently than the computer I'm using to write these words. Much of the brain operates simultaneously—in parallel—whereas computers are largely serial.

And there are a thousand more specific ways in which computers function differently than brains. To take a case we'll explore later in our discussion of memory, when you ask someone a question about a past experience, the question itself can alter their recollection of the scene. If you show someone a movie and later ask, "Did you see the children getting on the school bus?" the person is more likely to remember, later on, a school bus in the scene, even if there wasn't one.[27] Indeed, repeated questioning can lead to the creation of false memories. Computers don't work this way. You can search for "school bus" a hundred times; this will not create a file with "school bus" on your hard drive. Human memory and computer memory work in very different ways.

But there is another sense in which the brain really is a computer. The brain processes information—it *computes*. Not long ago, at the time that Alan Turing did his pioneering work that led to the foundation of artificial intelligence, "computer" referred to a sort of person, someone who worked at the job of computing. To call the brain a computer in this sense isn't a metaphor. It is an interesting claim. It means that it carries out mathematical and logical calculations, it manipulates symbols. Calculating that one plus one is two is computation, and so is reasoning that if all men are mortal and Socrates is a man, then Socrates is mortal. The idea that brains are computers in this sense has shaped psychological theories of mental life, and we'll return to it when we talk about capacities such as language and perception.

Thinking about the brain as a computer has an interesting implication: Just as the study of computation can inform us about psychology, studies of the mind can help us build better computers. If you want to build machines that can walk in a straight line, recognize faces, and understand language, it's sensible enough to check out how people do it, in the same way that Leonardo da Vinci studied the wings of birds to figure out how to make a flying machine.

―――――

The brain is not just a large bowl of porridge. It contains parts that do different things. These parts sometimes get called areas, systems, modules, or faculties—the linguist Noam Chomsky called them "mental organs" to emphasize how they can be as different from one another as the organs below the neck, like the kidney or the spleen.[28]

Actually, the idea that mental life has parts has been popular since before anyone even knew about the brain. Plato, for instance, talked of a trinity—a "spirit" that lives in the chest and is involved in righteous anger, the "appetite" located in the stomach and related to desires, and "reason," in the head (at last!), which oversees the other two.

One attempt to subdivide the brain came from Franz Josef Gall, who founded the school of phrenology. Gall had a very good idea and a very bad idea. The good idea was that the different parts of the brain were specialized for different things, many of which would make a contemporary neuroscientist nod with agreement, such as number, time, and language. Gall's ideas were popular in the early 1800s and left us with these lovely diagrams in which the skull is depicted with dotted lines segmenting different parts, like the drawings of cows that one sometimes sees in steakhouses—marking off chuck, sirloin, round, and so on—except that here the parts are mental traits and capacities.

The bad idea—phrenology—was that these brain areas bloat up the more they are used, and this causes bumps on the skull. Someone skilled in the tools of phrenology, Gall claimed, can put their hands on people's heads and learn about their characters. Phrenology used to be quite the thing. Karl Marx was a convert, and he would sometimes rub the heads of people he met. Queen Victoria was similarly entranced and hired phrenologists to test out the skulls of her children.[29]

I don't need to tell you that this is a quite goofy notion. But in Gall's insistence that different parts of the brain have different functions, and not just general functions like reason or appetite, but specific ones, like language, he was a scientist ahead of his time.

If the brain is composed of parts, then we can learn about how it works by taking it apart. This idea was nicely expressed in 1669 by the anatomist Nicolaus Steno:

> The brain being indeed a machine, we must not hope to find its artifice through other ways than those which are used to find the artifice of the other machines. It thus remains to do what we would do for any other machine; I mean to dismantle it piece by piece and to consider what these can do separately and together.[30]

Neuroscience can be said to have properly begun when scholars started to put this strategy into practice, by looking at those sad cases where natural causes did the dismantling. In 1861, a French physician named Paul Broca discovered a patient who was intelligent and could fully understand what was said to him, but who could only produce one word, "tan," which he said no matter what was said to him, usually twice in a row—"tan, tan." After he died an autopsy found brain damage in part of his frontal lobe, now known as Broca's area.

Years later, the neurologist Carl Wernicke discovered a patient with a different language disorder—she had problems understanding speech and could talk rapidly and fluently, but what she said was gibberish. This was related to another part of the brain, located in the back of the temporal lobe, usually on the left side of the brain, which has come to be known as Wernicke's area. (Note that finding the precise locations of these areas is of practical value; when doctors are cutting into brains during surgery, they want to avoid hitting areas that serve valuable functions.)

Language provides one illustration of how different parts of the brain have different capacities. Let's look at others, taking a brief tour of the brain.

The cortex is the part on the surface, right under the skull, and many parts of the brain that are highly relevant to our mental lives are subcortical, meaning that they lie just below the cortex. A pretty metaphor I've heard envisions the brain as a peach—the skin is the cortex, and the subcortical structures are parts of the stone. (The flesh of the peach is the white matter, largely composed of glial cells.) Such subcortical structures include:

The medulla, which controls automatic functions like heart rate, blood pressure, and swallowing.

The cerebellum, which is involved in movement, posture, motor learning, and certain aspects of language. (To get a sense of how complex these systems are, note that the cerebellum contains about thirty billion neurons.)

The hypothalamus, which is involved in sleep and wakefulness and hunger and thirst and sex. (This corresponds best to what Plato was talking about when he speculated about the appetitive part of the soul, although he located it in the stomach.)

The limbic system, involved with the emotions.

The hippocampus, involved in long-term memory storage and memory of locations and things in space.

The pituitary gland, which secretes hormones involved in sex and reproduction and other things, and is interesting to historians of science and philosophy because, according to Descartes, it served as the conduit between body and soul.

Now let's turn to the skin of the peach. The first thing you notice when you look at a brain is that it's all wrinkly. This is because it's crumpled up. If you were to take a brain, pull out the cortex, and smooth it out, it's about two feet square.

The cortex breaks down into different lobes. You have the frontal lobe (conveniently enough on the front), the parietal lobe, the occipital lobe, and the temporal (next to the temple!) lobe.

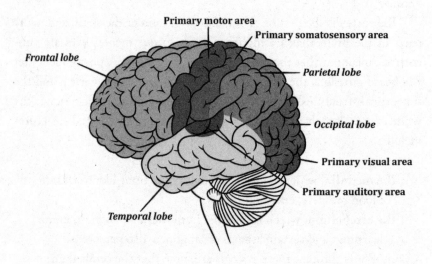

Some parts of the cortex contain "maps"—areas of the brain that correspond to parts of the body. If you give mild electrical shocks to neurons in the primary motor area, associated parts of the body twitch accordingly, while shocks to the primary somatosensory (soma = body) area lead to sensations in the corresponding areas. They are called maps because they are isomorphic with the body. For instance, the part of the brain representing the right index finger is close to the part of the brain representing the right thumb, which is close to the part of the brain representing the right wrist.

While the organization of these motor and sensory maps matches the organization of the body, the size does not—rather, brain size corresponds to the extent to which there is a lot of motor control or sensory discrimination. For instance, the part of the brain corresponding to the hand is bigger than the part corresponding to the chest because there is so much more sensation going on in your hand than in your chest, and so it gets more brain.

Aside from these maps, much of the rest of the cortex is involved in higher-order functions, such as language, reasoning, and moral judgment. Fish don't have any cerebral cortex, reptiles and birds have a little bit, mammals have more, and primates, including humans, have a lot.

How do we know which parts of the cortex do what? We've already mentioned studies where electrical impulses are applied to parts of the brain, but these are unusual, typically done with people who are having some sort of brain surgery. Far more common are those methods that look at the real-time brain activity of healthy people with intact heads. One popular technique is fMRI, which uses a strong magnetic field to look at the distribution of blood flow to the brain, seeing which parts are active when people think about different things. It is this that has the possibility of, in an almost literal sense, reading minds.

More techniques are emerging all the time. One method doesn't scan the brain at all—it *influences* it. This is TMS—transcranial magnetic stimulation, which uses magnetic fields to stimulate cells in the brain. Apply TMS to one part of the brain and it can impair language; apply it to other areas and it can cause the body to move without the person's volition. (I got to experience this once myself while I was visiting a laboratory in Kyoto—it felt weird to have my fingers twitch for no reason.)

We also know a lot about the brain from so-called natural experiments when people have tumors or strokes or accidents, as with unfortunate individuals such as Phineas Gage or Greg F. From these tragic cases, we can learn about which parts of the brain correspond to which functions, helping us to understand the correspondences between mind and brain.

As one example, some types of brain damage lead to agnosia—disorders of perception. Those with agnosia can see just fine, but often fail to recognize objects. When shown a picture, they can often describe the parts, but can't recognize how these parts make up the whole. More specifically, there is prosopagnosia, where one can't recognize faces. Oliver

Sacks wrote a classic book many years ago called *The Man Who Mistook His Wife for a Hat*.[31] This was a series of profiles of people who had surprising neurological disorders, including a man whose prosopagnosia was so bad that, as it says in the title, he couldn't distinguish his wife's face from a hat. In more common and milder forms, someone suffering from prosopagnosia can recognize faces as faces, but they can't recognize whose faces they are. This all illustrates the distinction between sensation and perception, something we'll get to later.

If you just look at a brain—if you remove it from somebody's head and put it on a table—it looks symmetrical. But it isn't. The asymmetry of the brain is manifested in handedness. Some people are right-handed, and others are left-handed, and since motor control comes from the brain, this suggests that the brain itself is asymmetrical. It has a right side and a left side, and they are not identical.

The difference between the sides is often exaggerated in popular articles. There is no such thing as "right-brain" people and "left-brain" people. Most of the functions of the brain are on both sides of the brain.

Still, there are differences. The left side of the brain is usually more involved with language and with the capacity for reason and logic, while the right side of the brain is more involved with social processes, imagination, and music. Some of these right-left differences are inborn; others are produced by experience. As one striking case of how culture shapes our brains, learning to read—a relatively recent human invention—reconfigures the brain, creating a region that is active when looking at words (called a "letterbox") in the left hemisphere and shifting the processing of faces more to the right hemisphere.[32]

The halves of the brain connect to the world in accord with a principle of contralateral organization, which means that due to a quirk of evolutionary history that's not entirely understood, your right brain sees the left side of the world and your left brain sees the right side of the world; your right hemisphere controls the left side of the body and your left hemisphere controls the right side of the body.

Suppose, then, that a psychologist presented a picture very quickly to your left visual field, too fast for your eyes to turn to see it head-on, and asked you to name the image. The information would go to the right side of your brain. Since typically the left side of the brain is involved in the processing of language, there would be a fraction of a second delay for you to name the picture, because the information has to make it to the left side, where the names are stored. If it was flashed to the right visual field, you would be just a little bit faster.

Information goes from one half of the brain to the other mostly through the corpus callosum—a network of neurons in the middle of your skull. If you think of the two halves of the brain as a city bisected by a river, like Budapest with the river Danube running down the middle, the corpus callosum is like thousands of small bridges uniting the city.

What if you were to cut the corpus callosum? This used to be done as a last resort for extreme cases of epilepsy. Epilepsy could be viewed as an electrical storm in the brain; the idea of this radical surgery was to isolate and shrink the electrical storms. And it really did help with the seizures. It also meant that the two halves of the brain couldn't readily communicate with each other, and this had some serious consequences.

In one case, for instance, a split-brain patient would find herself putting on clothes with her right hand and removing them with her left; in another, a patient who was shopping would put something in the shopping cart with one hand and take it out with the other. The left hand of another patient would suddenly strike his wife; another patient's left hand tried to choke him. This sort of behavior is so common in split-brain patients that it has a name—alien hand syndrome.[33] Separated, the two parts of the brain no longer act in unison; they can be seen as two individuals occupying, and sometimes fighting over, the same body.

Some scientists and philosophers draw a disturbing conclusion from the split-brain cases. They argue that for all of us, including those with intact corpus callosa, each half of our brain can be seen as a separate person. There is the language-using you, the one who is mostly in charge, and this is who is reading these words. But there is another you, a silent partner, also conscious, sitting next to the language-using self. When you split the brain, you liberate this silent self from the dominant "you," and the two selves may fight it out for control. But this radical conclusion is controversial, and there's no consensus as to what's really going on in the mind (or minds) of someone with a split brain.[34]

––––––

This brief tour of the brain has come to an end. We will go on to explore, throughout the rest of the book, psychological processes such as decision making, memory, and emotional experience. We know that all of these are the consequences of activity in brains—neurons working in concert with other neurons. This is an amazing discovery.

But sometimes people get *too* interested in the brain, neglecting the mind. Occasionally you bump into a neuroscientist who says that theirs is the real science. Sure, you can talk about ideas, emotions, short-term memory, and so on, but when you really get down to it, the serious theories will be about brain areas, neurons, and neurotransmitters. This is what matters. Neuroscience makes psychology irrelevant.

This attack on psychology is based on a confusion about how scientific explanation works. Just because we know about molecular biology doesn't mean that we stopped talking about hearts, kidneys, respiration, and the like. The sciences of anatomy and physiology did not disappear. Cars are made of atoms but understanding how a car works requires appealing to higher-level structures such as engines, transmissions, and brakes, which is why physics will never replace auto mechanics. Or to take an analogy closer to psychology, you can best understand the strategies that a computer uses to play chess by looking at the program it implements, not the material stuff the computer is made of. The same chess program can run on a 1980s mainframe computer, a 1990s desktop, a 2000 laptop, or a present-day smartphone. The physical structure of the hardware changes with each generation, but the program can stay constant.

If your neuroscientist is skeptical, ask how they would respond to a physicist telling them that the real science of the mind is ultimately about atoms and molecules, which are themselves composed of elementary particles—so why are they wasting their time talking about neurons and glial cells and the hippocampus, and the like? The neuroscientist would promptly protest that certain important scientific findings, such as the discovery that the hippocampus is involved in memory storage or that lack of dopamine is implicated in Parkinson's disease, can't be captured in the language of physics. And this would be a good response. Well, the psychologist can tell the neuroscientist much the same thing.

Indeed, it turns out that one can do psychology without studying the brain, even though the mind *is* the brain. Some of the most astonishing findings in our field have been done by scholars who couldn't tell a neuron from a nematode. And while we're at it, one can do psychology without studying evolution, even though the brain has evolved, and one can do psychology without studying child development, even though we were all once children. Many routes understanding has, as Yoda would put it.

I think part of the enthusiasm about brains reflects the commonsense dualism we talked about earlier. Occasionally when boasting about the effects of some sort of therapeutic or educational intervention, people claim, "It changes the brain!" But *everything* changes the brain. Reading this sentence just changed your brain, because you're thinking about it and thinking takes place in the brain. Indeed, reading this sentence creates long-lasting changes in your brain, because you're going to remember a bit of it tomorrow (I promise you), and this means that the structure of your brain has been modified by this experience. If there was some mental activity that *didn't* change the brain, it would prove Cartesian dualism and would be one of the most amazing discoveries of our time. But this will never happen because Cartesian dualism is mistaken.

By the same token, while the details of mind-brain relationships can be interesting, the fact that the brain is involved in mental life should be seen as obvious—and it sometimes isn't. There was a *New York Times* article, in the Science section, many years ago, titled: "In Pain and Joy of Envy, the Brain May Play a Role." And my reaction was: *May?* Where else is the pain and joy of envy going to be, the big toe?

———

Some think that neuroscience tells us little about psychology. They are particularly uninterested in findings of localization. The philosopher Jerry

Fodor wrote: "If the mind happens in space at all, it happens somewhere north of the neck. What exactly turns on knowing how far north?"[35] An astute contemporary observer, Matthew Cobb, chimes in with a similar point: "A map—and at their best that is what fMRI data are—does not tell you how something works. Where is not how. The next time you read a claim that a particular ability, or emotion, or concept has been localized to a particular region of the human brain using fMRI, ask yourself, 'So what?'"[36]

I'm not quite so skeptical. While I think that the relevance of neuroscience is often overblown, some of the results really do matter for psychological theory.

Just as one example, in research by Naomi Eisenberger and her colleagues, subjects have their brains scanned while they play a virtual ball-tossing game that they believe is with two other people.[37] Actually, it's a computer program, and it's designed to give them the feeling of being excluded, by having the other characters toss the ball to one another, and leaving the human out.

This hurts. Being shunned is painful, and this study was designed to explore the theory that the pain of rejection shares deep commonalities with actual physical pain. And this is what the brain scanning found: Relative to those subjects who didn't get socially shunned, people in the social exclusion condition had increased activation in parts of the brain such as the dorsal anterior cingulate cortex and anterior insula, the same parts of the brain that are activated when feeling physical pain. This finding has a surprising (though controversial) consequence, which is that interventions that reduce one sort of pain should reduce the other, and indeed there is some evidence that drugs like Tylenol, designed to work on physical aches and pains, can also diminish the hurt of loneliness.[38]

Psychology doesn't reduce to neuroscience, but neuroscience really can tell us interesting things about how the mind works.

———

We've seen the case for materialism, the theory that the brain is the source of the mental life. We know a lot about how specific activities of the brain correspond to experience and can observe this through tools such as fMRI. Like Santa Claus in the Bing Crosby song, an attentive neuroscientist can tell when you are sleeping and when you are awake—and perhaps isn't that far from telling whether you are naughty or nice, or at least thinking naughty thoughts or nice thoughts.

But there remains what the philosopher David Chalmers has called "the hard problem" of consciousness.[39] How is it that the activities of the brain correspond to conscious experiences? We know that they do; we know that the workings of physical stuff—meat—give rise to what it's like to slam your hand in a car door, or eat scrambled eggs with hot sauce, or kiss someone for the first time. But how does this happen? It seems like magic.

To put the problem in a different way, let's return to computers. My laptop can do smart things like playing chess. This is a rebuke to Descartes, who doubted that mere physical objects are capable of complex action. But as far as we know, my laptop cannot feel the pang of loneliness, the heat of anger, and so on. When it wins or loses, it feels nothing. If I drop it down the stairs, nobody besides me is going to suffer; if I decide to salvage it for parts, this isn't like murder, not even a little bit. So, what's missing? What does one have to add to a machine to give it the capacities to feel? Or can this never happen—must a conscious being be flesh and blood?

There are many avenues to pursue here, and we'll get to some in the next chapter. But I think the honest response, right now at least, is that nobody yet knows.

2

Consciousness

The topic of this chapter can be seen as a quirky diversion. We'll see that behaviorists like B. F. Skinner believed that an adequate science of psychology would say nothing about conscious experience. After all, we don't discuss consciousness when talking about rats, and we're no different from rats. The cognitive psychologists who followed Skinner rejected just about all his views—except for that one. After all, we don't discuss consciousness when talking about computers, and we're no different from computers.

I did my own graduate studies at MIT, which was ground zero for cognitive psychology. My dissertation was on children's language learning, and when struggling with questions like how children figure out what words mean, it never occurred to me to wonder what the experience of learning language felt like for a child. My fellow graduate students and professors studied language, perception, attention, memory, and reasoning, and like me, they thought of these capacities in terms of brain processes and computational mechanisms. Consciousness wasn't relevant. We had philosophers on the other side of campus, in an old World War II timber structure designated as Building 20—let them worry about it.

If we were asked to defend our dismissal of consciousness, we would point out that intelligence does not require sentience. A calculator can do math, a GPS can compute directions, a thermostat can adjust temperature—all without the slightest spark of awareness. It's not just machines that are smart without feeling, it's also other creatures in the natural world. Check out this excerpt from a review of *Entangled Life*, a recent book on fungi:

> Fungi are used to searching out food by exploring complex three-dimensional environments such as soil, so maybe it's no surprise that fungal mycelium solves maze puzzles so accurately. It is also very good at finding the most economical route between points of interest. The mycologist Lynne Boddy once made a scale model of Britain out of soil, placing blocks of fungus-colonised wood at the points of the major cities; the blocks were sized proportionately to the places they represented. Mycelial networks quickly grew between the blocks: the web they created reproduced the pattern of the UK's motorways ("You could see the M5, M4, M1, M6"). Other researchers have set slime mold loose on tiny scale-models of Tokyo with food placed at the major hubs (in a single day they reproduced the form of the subway system) and on maps of Ikea (they found the exit, more efficiently than the scientists who set the task). Slime molds are so good at this kind of puzzle that researchers are now using them to plan urban transport networks and fire-escape routes for large buildings.[1]

That looks plenty smart, and since slime molds are not conscious (as far as we know), this suggests that you can have intelligence without consciousness. And so why worry about consciousness if we're interested in how humans come to do intelligent things?

———

Times have changed since my graduate student days. Consciousness is now front and center in the science of the mind, as it should be. Regardless of whether it's essential for intelligence, no theory of psychology could be complete without it. It is, after all, the most certain thing there is. As Descartes pointed out, I might question almost everything, but my conscious experience is beyond doubt. I am more confident that I am experiencing, right now, a twinge in my neck than I am about anything else in the world.

Questions of consciousness are also of moral importance. I'm comfortable swinging an ax against a tree because I am confident that it doesn't feel anything. If I were to discover otherwise, I would probably stop. As the philosopher Jeremy Bentham put it, when it comes to moral issues, the relevant question isn't whether something can reason or speak; it is whether it can suffer.[2] And suffering requires consciousness.

The immediate problem that arises with trying to understand consciousness is that it appears to be a fundamentally first-person event. While I am sure that I am an experiencing being, the internal life of other people (sorry, but this includes you, Dear Reader) is more in doubt.

Perhaps there's a period in the life of every reflective person when they start to worry about this, when they get a little bit solipsistic. They wonder if they are surrounded by empty suits, creatures that go "ooh" and "aah" and smile and scream and sob, but really don't have any experiences. My virtual AI assistant—my Alexa—has thanked me, told me she's sorry, and said she wants to help me, but I'd be a fool to believe her. She isn't thankful, is not sorry, and doesn't want anything. Perhaps other people are like that too, only they put on more of a show.

If you worry about this, you're on your way to becoming a philosopher. Now, most of us—and most philosophers too, when they're not doing philosophy—put aside these skeptical concerns and assume that others are conscious. Humans look the same, have the same sorts of brains, and act the same, so it makes sense that our phenomenology would also be shared. And as we'll see soon, we might well be naturally constituted to assume that others have minds, that they possess beliefs and desires and experiences. To take too seriously the notion that those around us are automata would be a form of madness.

There remains, however, a more reasonable concern. We can accept that others are conscious while still worrying about how well their experiences match with our own. We know there are differences. Genetic variation causes some people, usually men, to become color-blind, unable to distinguish a red apple from a green one. Other genes influence taste—for some unfortunate individuals, cilantro tastes like soap. Some people have little or no power to conjure up pictures in their heads; others have vivid mental imagery. There are a few who even have synesthesia, in which sensory input is experienced in an unusual way; later in this book, we'll meet someone who could taste colors.

Other differences in experience might be hard to notice, perhaps even impossible. To take a classic philosophical case developed by John Locke, what if what I see as one color, you see as another?[3] We both use "green" for grass and "purple" for eggplant, because these are the words we've come to associate with our experiences, but what if these experiences are flipped, so that what you describe as "green" I see as purple, and vice versa? How could we ever tell?

It's hard to know what's in someone else's head. Marcel Proust writes, "A real person, profoundly as we may sympathize with him, is in a great measure perceptible only through our senses, that is to say, remains opaque, presents a dead weight which our sensibilities have not the strength to lift."[4]

Now, we can't be total failures at this. Communication involves making educated guesses about the minds of others, and we often succeed at this. "Nice restaurant you have there," says the mobster, "shame if something were to happen to it," and though it's not explicit in the sentence, the owner gets the threat loud and clear. Sometimes we infer the mental states of others through their expressions, as when we look at someone's face and realize that they are angry or bored or afraid. And sometimes we succeed using our own experience as a benchmark. When I cook pasta for others, I taste the sauce, and this simple act reflects the tacit belief that what it tastes like for me is pretty much what it tastes like for them. We can use our own experience as a stand-in for the experience of others.

But sometimes we fail. Sometimes we rely too much on our own experience and don't appreciate how different others are. In my Introduction to Psychology class, my joke about Ross Perot falls flat, because my students aren't as familiar as I am with the 1992 presidential election. I enjoy garlic but my friend does not, and so he barely touches the pasta puttanesca that I made for our dinner. I once thought that *Fight Club* was an appropriate date movie and was quite mistaken, but, you see, the trailer looked great to *me*.

The exercise of getting into the heads of other people, often known as empathy, really falters when we try to make sense of the consciousness of those who are very distinct from us. Can you imagine what it's like to be Attila the Hun? Someone suffering from paranoid schizophrenia? A monkey?

Or what about a bat? In Thomas Nagel's classic article "What Is It Like to Be a Bat?" he concedes that we can imagine ourselves flying and seeing the world out of a bat's eyeholes.[5] But still, we wouldn't be bats, we'd be ourselves in the bodies of bats. To know what it's like to be a bat you must be a bat.

Sometimes, we can make educated guesses. Consider what it's like to be a baby. William James famously suggested that mental life starts off as chaos: "The baby, assailed by eyes, ears, nose, skin, and entrails at once, feels it all as one great blooming, buzzing confusion."[6] The developmental psychologist Alison Gopnik makes a case for a different view.[7] Adult attention can be willful (we will discuss this more later in the book), and while it can be captured by external events—we will notice if someone calls our name—we also have control over what to attend to. But the part of the brain responsible for inhibition and control, the prefrontal cortex, is among the last to develop, and Gopnik suggests that babies lack this capacity for willful control; they are, for the most part, at the mercy of their environments. They are trapped in the here and now. No wonder they so often scream.[8]

Much of our discussion about consciousness so far draws from philosophy, literature, and everyday observation, but when we get to the baby example, we begin to see how scientific research, such as studies of brain structure, can give us some clues to the nature of consciousness.

One example of this concerns the subjective experience of the passage of time. Think of the experience of tossing an apple into the sky and watching it hit the ground. You can imagine one creature for whom this whole event would be excruciatingly slow and another who would see the act as instantaneous. If we ever encounter sentient alien life, successful communication requires that our subjective experiences of time are roughly similar.[9] If a second for us would feel like a year to them, or vice versa, it's going to be tough to interact.

Forget extraterrestrial life—how does our temporal experience correspond to those of other creatures on our own planet? You might think this can't be answered, but there are efforts, admittedly tentative and controversial, to compare the subjective speed of experience across different species.[10] One method is to look at "critical flicker-fusion frequency" (CFF). This is the point at which a rapidly flickering light no longer seems to flicker but just glows, where the mind doesn't work fast enough to consciously experience each flash as a distinct event. The human CFF is about sixty flashes per second. But other species have different numbers, and for dogs, it's about eighty. According to this test, then, their consciousness is speedier than ours; perhaps to them, we move in slow motion.

We know quite a bit about the neural expression of consciousness. We know that how conscious one is—awake, asleep, attentive, distracted—is reflected in processes that involve most of the brain. An EEG (electroencephalogram) records oscillations in electrical activity in the cortex; the frequency of the oscillations reveals how conscious you are. If this activity is in Alpha, between 8 and 15hz (spikes/second), this corresponds to relaxation; if it is in Theta, below 4hz, it corresponds to sleep; if it is higher, in Beta, between 16 and 31hz, it corresponds to agitation.

We also know that certain conscious experiences correspond to activity in specific areas in the brain. If you were to lie in an fMRI machine with separate screens in front of each eye, and a neuroscientist showed a picture of a house to the left eye and a picture of a face to the right eye, you would either see the house or the face depending on what you were focusing on. The conscious experience flits back and forth between the two images.[11] Different brain areas are activated for houses and for faces, and so a neuroscientist analyzing the data from the fMRI machine can identify which one you are conscious of.

Such findings aren't just of theoretical interest. One research team used this method to test people who were believed to be in a persistent vegetative state.[12] These individuals are "locked in"—they cannot speak or move, but researchers are able to use fMRI methods to test for brain activity. It turns out that if you imagine walking around your house, it causes a spike of activity in the parahippocampal gyrus in the frontal lobe, while if you imagine playing tennis, this will activate the premotor cortex. In a fascinating study, researchers told one such patient—known as Patient 23—that they were going to ask him questions and he could signal "yes" by imagining playing tennis or "no" by imagining walking about his house. This was their first dialogue, initiated by Martin Monti, a neuroscientist:

> "Is your father's name Alexander?"
> The man's premotor cortex lit up. He was thinking about tennis—
> yes.
> "Is your father's name Thomas?"
> Activity in the parahippocampal gyrus. He was imagining walking
> around his house—no.
> "Do you have any brothers?"
> Tennis—yes.
> "Do you have any sisters?"
> House—no.
> "Before your injury, was your last vacation in the United States?"
> Tennis—yes.

The answers were correct. Astonished, Monti called [his collaborator] Owen, who was away at a conference. Owen thought that they should ask more questions. The group ran through some possibilities. "Do you like pizza?" was dismissed as being too imprecise. They decided to probe more deeply. Monti turned the intercom back on.

> "Do you want to die?"

For the first time that night, there was no clear answer.[13]

———

One proposal of how consciousness works in the brain is called the Global Workspace Theory.[14] The core idea here is that sensory areas of the brain are activated in response to information from the world (such as when you are at a party and someone near you is talking) and this is initially unconscious. But when this information is attended to (such as when you focus on what they are saying), the prefrontal and parietal areas of the brain are activated, and the information becomes conscious. This information is then broadcast to other parts of the brain for further conscious processing (such as when you mull over what they just said). The Global Workspace Theory makes testable predictions, and there is a vibrant and productive debate over the merits of this account and its competitors.[15]

But there is a larger question that is a lot harder to get a handle on, which is how to think about consciousness more generally.[16]

Some philosophers argue that consciousness is the product of certain computational systems. The physical stuff underlying the computations doesn't matter. In the brain, it happens through neurons, but in other creatures, consciousness might arise through another medium—perhaps, say, in the sort of silicon-based life-form described in science fiction. It doesn't even have to be biological. Indeed, under some versions of this view, someone can scan your brain and upload all the information into a robot, and then that robot would be you—or a duplicate of you if the original sticks around. This notion of consciousness makes immortality possible.

Others see consciousness as a biological phenomenon, akin to digestion or mitosis. This perspective argues that expecting a computer that is simulating human thought processes to be conscious is as dumb as expecting a simulation of a thunderstorm to get you soaking wet.[17] To talk about "uploading" someone's consciousness onto the cloud is confused; your consciousness is the product of your physical brain; lose the brain, lose the consciousness.

Everyone I know who is involved in this debate believes that one of these views is ridiculous.[18] But they disagree about which one.

If the computational view is correct, it suggests that if you wire up my

MacBook Pro just right—perhaps also adding extra memory and kicking up the processing speed a bit—then it can be as fully conscious and pain-feeling and loving as any person. (And to destroy it would be murder.) This seems like a bizarre conclusion.

On the other hand, if it's the physical stuff of the brain that's essential, it is impossible to program a machine to be conscious. An android could be perfectly indistinguishable from a person in how it acts, talks, emotes, and so on, but if this view is right, the android would truly be no more sentient than a rock. To many, this seems like an arbitrary and ultimately immoral conclusion.

What's frustrating here is that it's hard to know how we would ever find out which view is right. If we did build a machine that could act in complex ways (like the friendly and not-so-friendly robots that thrill us in movies), how would we tell if it's conscious? The same problem that we faced as young sophists has come back to haunt us yet again.

———

We've strayed into the hard problem of consciousness—the question of how physical brains can give rise to sentient experience. So let's step back and ask: What are the *easier* problems?

The philosopher David Chalmers, who introduced the hard/easy distinction, described easy problems as explaining "the performance of all the cognitive and behavioral functions in the vicinity of experience—perceptual discrimination, categorization, internal access, verbal report."[19] Explaining how the brain gives rise to a certain ineffable experience of seeing the color red, that's the hard problem, but figuring out how we categorize something as red, how children learn the word "red," why some people are color-blind—all of these are easy problems. (I should add, by the way, that the "easy" is a bit of a joke here. Chalmers understands they are savagely difficult; his point is that they are *tractable*, solvable by the standard methods of our science.)

Another way to think about all this is the distinction made by the philosopher Ned Block between phenomenal consciousness and access consciousness.[20] Phenomenal consciousness is "what it is like to be in that state" (hard problem); access consciousness is about "availability for use in reasoning and rationally guiding speech and action" (easier problem). If you are conscious of something in the sense that you can tell people about it (such as the words you are reading right now), then it is access consciousness, regardless of what (if anything) it feels like.

Psychologists have made discoveries about access consciousness (I'll just call it *consciousness* in what follows). One is that it is limited. We saw an example of this earlier; when a picture of a house is shown to one eye and a picture of a face is shown to the other, you can experience the house or the face, but not both. As another example, when you perceive the world, you typically see objects as standing out from a background— "figures" separate from a "ground." Usually, it's obvious which is which. But clever psychologists have invented displays that are ambiguous, where there's a reasonable way to see one part of the display as figure and the other as ground, but another reasonable way to see the opposite. Here's a classic example, developed by Edgar Rubin in 1915.[21]

You can see this as two black faces looking at each other with a white background, or you can see this as a white vase with a black background. But again, you can't see both at the same time, and so the images will flit back and forth in your mind.

We see the limits of consciousness in other modalities. One clever method in cognitive psychology is to have people wear headphones and listen to two separate speeches, one in the right ear, one in the left. It would be neat if people could attend to both—imagine following two podcasts simultaneously—but we can't. If someone is forced to focus on one ear—usually by asking them to repeat what they are hearing in that ear, called "shadowing"—the information in the other ear is mostly unattended to.

But not entirely. You might not be access conscious of this other information, but you are *unconsciously* attending to it. If you are shadowing sentences spoken into one ear, and you are hearing a series of words in the other ear, you won't usually know what they are. But if one of those words is your name, it'll pop out to you—you'll notice.[22] It'll capture your attention.

You've probably experienced this in real life, in what's called the "cocktail party effect." If you are at a party intensely involved in conversation with someone, not listening to anyone else, you'll notice if certain words crop up, such as your name. It also works for certain taboo words.[23] Perhaps you've had that experience; there is a hubbub of conversation, and suddenly someone says a remarkably obscene phrase and there is a sudden silence, as everyone stops and focuses their attention on the speaker.

It turns out that only a small fraction of the sensory experience makes its way in; everything else is ignored and lost forever. In one famous study, reported in a paper titled "Gorillas in Our Midst," subjects are shown a video in which people in white shirts and black shirts are standing in a hallway passing basketballs back and forth.[24] The subjects' task is to focus on the white shirts and count the passes they make. People don't find this hard, but it does take all their attention. Here's the twist: In the middle of the video someone dressed as a gorilla walks onto the scene, stops in the middle and pounds his chest, then walks off. About half of the subjects miss it, though the presence of the gorilla is screamingly obvious for anyone who is not told to focus on the passing of the basketballs.

We tend to be ignorant of these limitations. It feels like we are conscious of the world, not just a small sliver of it. It feels like we can attend to multiple things at the same time, rather than being forced to move our attention back and forth. Our limitations are harmless enough if we are writing emails when watching television or listening to a podcast while mowing the lawn. But they can be fatal in cases where something occasionally needs our full attention, such as driving. Talking on the phone, even using a hands-free device, slows our reaction time on the road, to an extent that is roughly the same as being legally intoxicated.[25]

How do we make it through life with such limited powers of conscious focus? It turns out that we have an excellent trick. When we do something repeatedly, it becomes unconscious, which frees up the conscious mind for other things. When you first learn to drive a car, it takes up all your focus, and you can barely do anything else at the same time. But slowly, over a long period, it becomes habitual, leaving your mind to wander off to do other things, like having a conversation or listening to the radio. We go on autopilot. We can walk and chew bubble gum, drive and talk, mow the lawn and listen to a podcast. The habitual can be consciousness-free.

Habit liberates us. Aristotle argues that goodness is ultimately a matter of making the right act natural and instinctual, without thought. Similarly, the Confucians suggested that the way to find *wu wei*, the valuable skill of effortless action, is ritual and repetition.[26] And the best contemporary self-help books will remind you that the trick of any lifestyle change—in diet, in exercise, in dealing with those you love—is to make the desired behavior habitual, and hence easy.

William James, in *The Principles of Psychology*, published in 1890, makes the case for habit with typical eloquence.[27] (If you update the prose a bit, this wouldn't be out of place in a modern-day *New York Times* bestseller.)

The more of the details of our daily life we can hand over to the effortless custody of automatism, the more our higher powers of mind will be set free for their own proper work. There is no more miserable human being than one in whom nothing is habitual but indecision, and for whom the lighting of every cigar, the drinking of every cup, the time of rising and going to bed every day, and the beginning of every bit of work, are subjects of express volitional deliberation. Full half the time of such a man goes to the deciding, or regretting, of matters which ought to be so ingrained in him as practically not to exist for his consciousness at all. If there be such daily duties not yet ingrained in any one of my readers, let him begin this very hour to set the matter right.

You are on your way to getting in shape if your exercise routine is automatic; you're failing if you struggle every day to figure out what to do. The first thing you will hear if you talk to a therapist for insomnia (and I speak from experience) is to go to bed and wake up at regular hours. I advise my graduate students to set aside the same block of time for writing every day, so that they fall into the practice without rumination. Mornings are best for this not just because it's when most people are psychologically best attuned for deep work[28] (though it is) or because it is usually the time with the fewest distractions (though it is), but rather because it's the easiest time to build in a thoughtless routine—get up, go to the bathroom, make your double espresso, and walk to your desk and get to work. (And indeed, this is how I've been writing this book—one hour every morning, when I wake up.)

We saw earlier that one way in which our consciousness differs from that of babies is that we can control it. In *Mastermind*, her book on how to think like Sherlock Holmes, Maria Konnikova begins with a quote by W. H. Auden: "Choice of attention—to pay attention to this and ignore that—is to the inner life what choice of action is to the outer. In both cases man is responsible for his choices and must accept the consequences."[29]

But this control isn't always there. There is a bit of the baby in us; we too have minds that can be taken hostage by the outside world, as in the cocktail party effect. If someone is standing next to you telling you something that you don't want to hear, your best bet is to plug your ears, because once the auditory neurons start firing in response to the words, you're stuck; you can't choose not to hear it.[30] This is true too for reading—once you learn to read and it becomes automatic, you can't help but read. No matter how much you don't want to read the words on the following page, once your eyes fall on them, you will:

THERE ARE MONKEYS IN YOUR HAIR!

This fact about reading led to a 1935 discovery known as the Stroop Effect, named after its discoverer, John Ridley Stroop.[31] Suppose you see a list of words written in different colors. Perhaps the word "cup" is red, the word "game" is green, the word "square" is yellow. Your task is to ignore what the word is and just name the colors. That's not so hard; you'll quickly say: red, green, yellow.

But now imagine that the words themselves are color names—now, the word "green" is red, the word "blue" is green, the word "red" is yellow. The task is the same: name the color of the words. But this is tough, and you'll be slower. The problem here is that as a practiced reader, you read "green" and you want to give the practiced response "green." You can override this and give the correct answer of "red," but it will slow you down.

I can't resist telling you my idea about how to use the Stroop Effect to uncover spies. There was a television show that I loved called *The Americans*, which was about a couple of Russian agents living in America in the 1980s. Imagine yourself as an FBI agent interrogator interviewing them, and you ask them if they can read Russian. They deny it, of course, they tell you that they've lived in the States for their entire lives and have no interest in the Soviet Union. You suspect they are lying, but how can you tell?

Simple. Give them a long list of foreign words in different colors and ask them to say the colors as quickly as they can. After getting them to practice with words in Hindi, Korean, and Finnish, you show them this.

красный (in green)
зеленый (in purple)
синий (in yellow)

For me, as a non-Russian speaker, this is easy: I'd say: green, purple, yellow. But the trick, and I bet you figured it out already, is that these are Russian words—the green one is "yellow," the purple one is "red," the yellow one is "blue." If you know Russian, you can't help but see them as color names, and so, by the logic of the Stroop Effect, the spies will read these words slower than the rest.

———

Consciousness is a topic that spans all of psychology. Neuroscientists study its embodiment in the brain. Cognitive psychologists explore access consciousness in sensation, perception, judgment, and choice. Developmental psychologists study how consciousness arises in children, including the emergence of so-called self-conscious emotions such as embarrassment and pride. And clinical psychologists study disorders of consciousness, such as when people feel as if their bodies are not their own.

Some of the most interesting findings come from the research of social psychologists, who ask about the nature of our conscious experience as we go about everyday lives. In one study, people were given an app on their iPhones that went off at random times during the day.[32] When it went off, they had to report what they were feeling, what they were doing, and whether their minds were wandering ("Are you thinking of something other than what you're currently doing?"). This "experience sampling" study found that our minds wander *a lot*—almost half of the time. It also turns out, by the way, that our mental states when we are mind-wandering are reported as less pleasant than when we are focused on the here and now; the paper was titled: "A Wandering Mind Is an Unhappy Mind."

When we are not mind-wandering, we often think about what others are thinking—and we conclude that they are thinking about us. In 2000, the psychologist Thomas Gilovich and his colleagues reported a study where they asked undergraduates to wear a T-shirt with a photograph on it that they would find embarrassing—this turned out to be the deeply uncool singer-songwriter Barry Manilow.[33] The researchers got the students to go to class while wearing these T-shirts and, afterward, asked them how many people noticed them. The answers tended to be overestimates. In another study, participants were allowed to choose a T-shirt with a cool person on it, this time with a picture of Bob Marley, Martin

Luther King Jr., or Jerry Seinfeld—and again, people radically overestimated how much others noticed it.

This is the "spotlight effect": We overestimate the extent to which others notice us, in both negative and positive ways.

The spotlight effect arises from the first-person nature of consciousness. Gilovich and colleagues quote David Foster Wallace here: "There is no experience you've had that you were not at the absolute center of." I'm very conscious of my own appearance, so I assume that others are too, but I don't realize that, like me, they're conscious of *their* appearance.

I like telling undergraduates about these findings. Many of my students are just coming out of the teenage years, which is a period where one dominant feeling is that all eyes are on you, judging how you look, act, and mess up. It's reassuring to realize that people don't notice your embarrassing moments as much as you might think they do.

These findings are relevant to the rest of us as well. Gilovich and colleagues connect this work to findings about regret.[34] When you talk to people who are soon to die, their regrets are, for the most part, about what they didn't do, and one reason why people don't do these things—didn't take certain risks and make certain choices—is that they were worried about appearing foolish in the eyes of others. The finding that people for the most part aren't looking is, again, liberating.

———

What can move us away from this focus on ourselves? One answer is being in a crowd. As the social psychologist Gustave Le Bon noted in 1895, being in a group can make consciousness of the self partially disappear—"An individual in a crowd is a grain of sand amid other grains of sand, which the wind stirs up at will."[35]

This loss of self-focus is a mixed bag. It can be a source of great pleasure, as when synchronized movement with others—in a rave, a concert, or a rockin' Bar Mitzvah—blurs the boundaries of self and other. Lose yourself, as Eminem so nicely put it.

But being in a crowd can also make people do stupid and cruel things;

as consciousness goes, so does conscience. And it makes us less likely to help those in need, something called bystander apathy.[36] The more people there are around you, the less you care. One explanation for this is the diffusion of individual responsibility. If I'm walking alone through the woods and see a crying child who is lost, there is a compulsion on my part to help, but if I'm part of a crowd, I can tell myself it's someone else's problem, and that my inaction is less of a big deal.

Another way to turn down the dial of consciousness is as close as your liquor cabinet. Edward Slingerland, in his book *Drunk*, is mindful of the many problems that alcohol causes, but makes a case for inebriation as a tool for releasing us from the stress of being conscious of ourselves.[37] He cites Tao Yuanming here, from his "Drinking Wine" series.

> Old friends appreciate my tastes;
> So they arrive carrying a jug of wine to share.
> We spread our mats and sit under a pine tree,
> And after a few cups are thoroughly drunk.
> Just some old guys rambling from topic to topic
> Losing track of whose cup should be filled next.
> Completely free of self-consciousness.

And finally, the right sort of pain, under the right circumstances, can get you out of your head.[38] There is this line from a dominatrix that everyone who studies masochism loves to quote: "A whip is a great way to get someone to be here now. They can't look away from it, and they can't think of anything else."[39]

———

There are advantages to access consciousness, to having information in one part of our psyche available to other parts of our psyche. One advantage has to do with language. To communicate, to convey some information from one head to another—the fruit tree is over there, I have some fish to trade, your friend has been injured—it's not enough to possess this information, we must be *conscious* of having this information.

Not everything in our heads is accessible in this way. People aren't directly conscious of their blood pressure and heart rate, though they can sometimes infer this information in indirect ways. We will see in later chapters that we are not conscious of how the brain turns visual sensation into visual perception or transforms sounds and signs into language. Sometimes we are not conscious of specific ideas we have, or biases, or even moods. There's a children's song that begins with the command "If you're happy and you know it, clap your hands," and it must have been written by someone savvy about the mind, someone who appreciated the possibility that some children are happy *but don't know it*.

Why isn't everything conscious? Perhaps our system can handle just so much. You can see the vase or the face—but never both. A smarter creature, or perhaps a human with some futuristic neural enhancement, would do better.

But another consideration is that we might do better having some information hidden from ourselves. The next chapter is on Freud, and we'll see that he argued that some information is too disturbing to be fully appreciated by the conscious mind, and so it is repressed. A boy might hate his father, but this is shameful and dangerous, and so this hatred is hidden from consciousness, though perhaps revealed in subtle ways like slips of the tongue or dreams. If asked, the boy would honestly insist that he loves his father. A healthy mind might keep some thoughts under wraps.

You might be skeptical about Freud's theory (I am), but here's a related proposal, developed by the evolutionary biologist Robert Trivers.[40] He speculated that the advantage to keeping information unconscious is that it helps you deceive other people.

To illustrate this, imagine that you're playing poker but you are an expressive sort, and your reactions to your cards are obvious to everyone. If there was some way to play your hand without consciously being aware of what it was, you'd have the perfect poker face. We can't do this for poker, but Trivers argues that this goes on in other avenues of life, including romantic overtures and aggressive confrontations.[41] Want to trick someone into believing that you are in love with them? Consciously believe that you *are* in love with them; keep your deception hidden from your consciousness. Want to convince an adversary that you will never, ever back down? Consciously believe that it's true—even if it isn't.

For a simpler example of the advantage of ignorance, when a hare is being chased by a dog, it moves in a random pattern, working to be as unpredictable as possible in its zigzagging plan to escape.[42] Suppose it knew which direction it was going to jump next. Then it could give away cues in its posture and the dog might suss it out. Best for it not to know up until the instant it moves. Lack of self-awareness can be the trick to survival.

———

Sometimes we want to obliterate consciousness altogether. The most common form of this is trying to fall asleep. It's odd how difficult this can be. Our minds don't have an off switch; we must carefully engineer our environments to calm them down. The advertising executive Rory Sutherland nicely sums up the strangeness here:[43]

> Imagine an alien species with the power to fall asleep at will—they would regard human bedtime behaviour as essentially ridiculous. "Rather than just going to sleep, they go through a strange religious ritual," an alien anthropologist would remark. "They turn off lights, reduce all noise to a minimum and then remove the seven decorative cushions which for no apparent reason are placed at the head of the bed. Then they lie in silence and darkness, in the hope that sleep descends upon them. And rather than simply waking up when they wish, they program a strange machine which sounds a bell at an appointed time, to nudge them back into consciousness. This seems ridiculous."

As a lifetime insomniac, I take solace in knowing that it could be worse. There is a German myth about a nymph named Ondine, who falls in love and marries a human, who is then unfaithful to her. She punishes him by taking something that we all do unconsciously—breathing—and making it conscious. If her husband falls asleep, relinquishing conscious control of his breathing, he will die.

It turns out, tragically, that there is a rare condition much like this—Ondine's curse, or central hyperventilation syndrome.[44] Those afflicted with this syndrome don't maintain regular breathing while sleeping. This often results in death, though if it is diagnosed early such people can be kept alive through respirators. Consciousness, so often a blessing, can be a terrible burden.

3

Freud and the Unconscious

I f there's one psychologist you've heard of, it's Sigmund Freud. He was born in 1856, spent most of his life in Vienna, and died in London in 1939, soon after retreating there once the Nazis had taken over Austria. For most of his adult life, he was famous—a good comparison, in influence and popularity, is to Albert Einstein. But while Einstein just told us about space and time and energy, Freud revealed the secrets of the soul.

The psychiatrist Peter Kramer sums up Freud as seen by his acolytes, then and now.

> He appeared to possess special powers of observation that allowed him to turn his work with patients into innovative science. Using methods he had himself developed, Freud had discovered and mapped the unconscious. He had named the components of the mind and explored the principles by which they operated. He had charted the sequence of human psychological growth, from infancy to mature adulthood. He had identified the causes of most mental illnesses and invented a method for treating them.[1]

And his discoveries were extraordinary:

Beneath apparent rationality, Freud had discerned dark impulses and contradictory yearnings that coalesced into predictable patterns he called complexes. He had demonstrated that, in the culture and in the lives of individuals, hidden symbols abound; our customs and behaviours simultaneously hide and reveal sexual and aggressive drives incompatible with the requirements of civilized society. Freud's theories seemed to update ancient philosophies, casting our lives as tragic dramas of a distinctively modern sort. It was as if, before Freud, we had never known ourselves.[2]

This is one view of Freud, and there are many who would still defend it. But it is no longer popular. Unlike the wild-haired and irreverent Einstein, destined to be forever loved by all, Freud is dismissed and despised by many.

Part of this is due to revelations about the man himself. As Kramer notes, documents were discovered showing that Freud "had altered fact to fit theory, conducted therapies in ways that bore scant relationship to his precepts, and claimed success in treatments that had failed."[3] He took credit for others' accomplishments and blamed others for his failures. He once cajoled one of his patients (also his colleague) into marrying a wealthy woman—who was also the colleague's patient!—and admitted that he did so in the hopes that she would give money to support Freud's psychoanalytic program. And we're just getting started. One reviewer, discussing a biography of Freud by Frederick Crews, summarizes Crews' view of the great man as: "liar, cheat, incestuous child molester, woman hater, money-worshiper, chronic plagiarizer and all-around nasty nut job."[4]

And then there are his theories. Students are often surprised that psychology departments seldom have courses on Freud; actually, you can get an undergraduate degree in psychology at a major university without hearing his name ever spoken. (There is more interest in the humanities—you're more likely to find an English professor talking about Freud than a psychologist.) When I was starting out as an assistant professor at the University of Arizona, I taught a seminar called Darwin, Freud, and Turing: Three Perspectives on the Mind, and more than one of my senior colleagues told me that it would be a better course if I took out the man in the middle. Many psychologists would concede that the psychodynamic movement founded by Freud is an essential part of the history of our field—but they see it as an embarrassment, like a pharmaceutical company that got its start by selling meth.

And to be fair, a lot of Freud's ideas really are strange. He insisted on the importance of penis envy (the trauma resulting from a girl's discovery of her lack of a penis), castration anxiety (a boy's concern that he will lose his), and the "primal scene" (a term that he coined for children witnessing their parents having sex). The effects on a boy if the mother dies and the father raises him? Homosexuality, due to "exaggerated castration anxiety." A young girl in a therapy session who pulls the hem of her skirt over her exposed ankle? Tendencies toward exhibitionism. A young man who smooths the crease of his trousers in place before lying down in his first session? Freud writes that this man "reveals himself as an erstwhile coprophiliac of the highest refinement"—where a coprophiliac is someone with an erotic fascination with feces![5]

There's nothing wrong with surprising claims. The best psychologists come up with unintuitive and discomforting theories, and this is a good thing, a sign of a mature science that goes beyond common sense, that tells us what we didn't already know. Perhaps sex really is more important than many think. But Freud had few good arguments for his fantastical claims—often he had no real arguments at all—and few of his radical proposals have survived the passage of time.

Despite all this, I believe that there is a lot of value to Freudian thought. I think we can retain some of the core insights of his theory and jettison the silly stuff. This is what many of Freud's own followers did. Alfred Adler, for instance, put more emphasis on self-esteem (he coined the term "inferiority complex"), while Karen Horney focused on the importance of warmth and affection in the early years.

For me, the best idea that remains is the notion of an unconscious mind at war with itself. For Freud and his followers, we are not singular deliberating beings. Rather, each of us is driven by internal turmoil, much of which is hidden. Freud was hardly the first to focus on unconscious motivation and conflict, but the amount of thought he gave to it was unprecedented. His science of the mind was at root a science of the unruly unconscious.

————

The primacy of the unconscious is a disturbing idea, perhaps even more so than the notion of materialism outlined in the earlier discussion of the brain. If it's tough to give up on an immortal and immaterial soul, it's even harder to give up on the idea that we are in full control of our own lives.

Suppose you decide to get married, and somebody was to ask why. You might say something like, "Well, I'm ready to get hitched. It's the right time of my life. I really love this person and don't want to live without them." But a Freudian might insist that you're wrong. Perhaps you really want to marry John because he reminds you of your father, or you want to marry Laura to get back at your mother for betraying you, or some other bizarre reason. Now your response to these alternatives might be angry denial, but this wouldn't deter a Freudian. The Freudian might think he or she knows you better than you know yourself. (Indeed, the Freudian might think that your anger is evidence that these theories are on the right track.)

The marriage case is extreme, but there are a lot of simpler examples of how we seem to be influenced by forces outside of our conscious awareness. Have you ever had a powerful attraction to a person, or a powerful dislike, but didn't know why? Have you ever forgotten someone's name at exactly the wrong time? Have you ever missed an important appointment even though you seemingly had every intention of being there?

The unconscious, according to Freud, bleeds out into much of our everyday life, showing up, among other places, in dreams, jokes, and certain speech errors—or what are now commonly called *Freudian slips*. As he puts it, "He that has eyes to see and ears to hear may convince himself that no mortal can keep a secret. If his lips are silent, he chatters with his finger-tips; betrayal oozes out of him at every pore."[6]

Donald Trump, talking about his wife, Melania, said, "She's got a son," seemingly forgetting that the son, Barron, is also his. A neuroscientist, lecturing to a room of hundreds, describes a set of studies with people with autism, and says, "There is similar research done with humans," implying he thinks that people with autism are not humans. In an episode of the show *Friends* ("The One with Ross's Wedding"), Ross's wedding is ruined when, during his vows to Emily, he says, "I, Ross, take thee, Rachel," inadvertently referencing the woman he truly loves.

Now, for each of these examples, one might be skeptical whether they really carry any significance. But there are cases where a Freudian approach really does seem to tell us something we didn't already know, and we're going to see some examples of this in the discussion that follows.

———

The primacy of the unconscious wouldn't be such a problem if the unconscious mind were a rational computer mindful of your best interests. But for Freud, it's a mess—a contentious trinity, with three distinct processes that are in violent conflict.

The first process is the *id*, which is present at birth. This is the animal part of the self. The id wants to eat, and drink, and excrete, and get sensual pleasure. It works in accord with what Freud called the *pleasure principle*—it wants immediate gratification.

Sadly, the world doesn't work that way. The desires of even the most pampered baby are rarely satisfied right away. The baby might want milk, but the mother's breast isn't there; the baby needs a cuddle, but the cuddler is in the other room.

This failure of the world to give you what you want leads to a second system called the *ego*. This is where consciousness emerges—your ego is *you*. With the emergence of the ego, there is some understanding of reality; the ego enables you to either pragmatically satisfy your desires or to suppress them. Here we see the *reality principle* at work.

Later in development, the trinity is complete—the *superego* emerges. This is the part of the mind that has internalized a moral code, first from the parents, and then from society more generally. An enraged baby might want to hit his father in the face and, being nothing but id, will do just that. Later on, possessing an ego, the toddler might reason that this is going to have a bad outcome—father's anger—and so holds back. But much later on, possessing a superego, a child might decide to restrain herself just because it's wrong. At a certain point, one's desires are inhibited not merely by fear of consequences but by some sort of moral code.

For Freud, then, the ego serves two masters. It's stuck between raging animal desires on the one hand, the id, and a conscience on the other, the superego. Now, it's tempting to think, as I'm framing it this way, that the id is dumb and animalistic and the superego is advanced and civilized, corresponding to the picture of a person (ego) with a devil over one shoulder (id) and an angel over the other (superego). But this isn't quite right. A lot of the prohibitions set up in the superego are grounded on the prejudices and beliefs of society, and may not reflect a clearheaded moral understanding. It's possible that you might believe, intellectually, that some act, perhaps something sexual, that you engage in is fine—nobody is harmed—but your superego may scream at you that it is disgusting and wrong. And this can limit your happiness and flourishing.

———

These are the rudiments of Freud's developmental theory—we start off as an id, develop an ego, and then develop a superego. You'll see that some of his other ideas about development are familiar; invented over a hundred years ago in Vienna, they have seeped into popular culture all over the world.

Freud proposed that there were five main developmental stages, each associated with a different part of the body. If you get into a problem at a certain stage and don't adequately resolve it, you could end up stuck there—*fixated*, as Freud put it. You subsequently will try to achieve pleasure in ways that correspond to this immature period of your life.

The first stage, which lasts throughout the first year of life, is the *oral stage*, where the mouth is associated with pleasure. For Freud, weaning a child incorrectly could lead to oral fixation in adulthood. In the literal sense, that could mean eating too much or chewing gum or smoking. In a more metaphorical sense, the person could be dependent or needy; oral fixation relates to problems revolving around trust and envy.

The next is the *anal stage*, running from roughly the first year to about age three. Now it's the anus that is associated with pleasure. The key challenge involves toilet training. Adults who struggled through this stage of development might be compulsive, clean, and stingy, because they are (to put it metaphorically) unwilling to part with their feces. This conception has ended up in language, so we might say, "Oh, he's so anal," to refer to someone who is obsessively concerned with getting things exactly right.

Then there's the *phallic stage*, between around age three and five, where the focus of pleasure shifts to the genitals. For Freud, there is an important drama that occurs at this point, which he called the Oedipus complex. This is based on the play *Oedipus Rex*, written by Sophocles in 439 BC, about a man who unwittingly kills his father and marries his mother, Jocasta.

Freud's idea is that an analogous event happens for every boy. In the phallic stage, he is focusing on his penis, and he seeks an external object of affection. Who is the woman he loves and who loves him back? Mom. He wants to sleep with Mom. But what about his father? Three's a crowd, and so Freud claimed that the child comes to hate his father and wants him out of the way. Murder Dad, marry Mom.

Freud also believed that the child will believe that his thoughts are public. He worries that his father will discover his plans and retaliate—by castrating him. This is terrifying. The child gives up the murder plot and, instead, allies himself with his father. Success at this stage? According to Freud: Identification as a man, heterosexual desires later in life. Failure? Well, it could be homosexuality; or alternatively, a sort of hypermasculinity, with too much focus on power and authority.

You may have noticed that this is a theory just for boys. Do girls go through their own developmental stage? There was a lot of debate about this among the psychoanalytic community. Some argued for an Electra complex, the term coined not by Freud, but by his famous follower, Carl Jung. (The term comes from another Greek myth, in which Electra competes with her mother for the affections of her father.) Jung's story is complex, involving penis envy, a desire to create a son who would have the missing penis, a shift from the clitoris to the vagina as an expression of mature sexuality, and much else.

After all the turmoil of the phallic stage, there is a respite. This is the *latency stage*, where sexuality is repressed. The child identifies most with the same-sex parent, and focuses on hobbies and school and friendship.

At puberty, sexual feelings reemerge, and healthy adults find pleasure in sexual relationships as well as other pursuits. We are at the *genital phase*. Those unfortunate enough to have had difficulties with earlier stages, such as with breastfeeding or toilet training, will, according to Freud, struggle with associated psychological problems throughout their adult lives.

Even for a reasonably healthy adult, all sorts of challenges remain. One of them is that the id is generating various desires, many of which are forbidden by the superego. It's not merely that you can't act upon them; it's that you shouldn't even be thinking about them. And so they get repressed. But some of this forbidden material makes it out, in jokes, in slips of the tongue—the Freudian slips we talked about earlier—and in dreams.

Freud has a lot to say about dreams. One of his best-known books was *The Interpretation of Dreams*, where he argued that they represent taboo wishes.[7] These are then disguised as they make their way to consciousness. Accordingly, Freud distinguished the *manifest dream*—the dream you remember—from the *latent dream*—the one that really happened.

Then there are what Freud described as defense mechanisms. These mechanisms were further elaborated by Freud's daughter, Anna Freud, who was a great psychodynamic theorist in her own right. Like a lot of Freud, defense mechanisms might be familiar to you, as they have infused our everyday understanding of the mind. Here are some of the major ones:

1. **Sublimation.** Taking desires that are unacceptable and directing them to more valuable activities. For instance, somebody who has strong sexual desires might devote a lot of energy to his or her work or studies.
2. **Rationalization.** Taking desires that are unacceptable and re-

construing them in a more acceptable way. A father who enjoys physically punishing his children might think of his violent acts as being for their own good.

3. **Displacement.** Redirecting shameful thoughts to more appropriate targets. If a boy hates his father and wishes him dead, this defense mechanism might refocus his aggression toward more appropriate targets, making him violently competitive with other boys, for instance.

4. **Projection.** Taking one's own shameful thoughts and attributing them to someone else. Imagine a woman who wants to have sex with other women, but who was raised so that her superego tells her that this desire is unacceptable. She might become unconscious of it—and come to believe that other women are sexually drawn to her.

5. **Reaction formation.** Replacing shameful thoughts and fantasies with their opposites. Consider romantic comedies in which couples who are in vicious disagreement later fall in each other's arms—their apparent dislike masked their true attraction. Or consider the cliché that men who have the most negative feelings toward gay men are themselves wrestling with homosexual desires (a claim, by the way, that there's little evidence for[8]).

For the Freudians, these defense mechanisms are part of normal psychological functioning. But sometimes the mechanisms fail to properly repress the impulses from the id, and real problems might arise. In the late 1800s, the French physicians Jean-Martin Charcot and Pierre Janet examined patients, usually women, who suffered from *hysteria*. They would become amnesic, blind, or paralyzed, even though they seemed to have nothing physically wrong with them.

Freud studied with Charcot and then developed the theory that the symptoms of hysteria are mechanisms through which we keep forbidden aspects of ourselves under lock and key. As a simple example, someone might lose their memory of an event that would be too painful to remember. Initially, Freud linked many cases of hysteria to actual childhood sexual abuse, though later in his career he changed his mind and argued that these events didn't actually occur; they were fantasies of his patients.

Treatment of hysteria brings us to the development of Freud's own brand of therapy—what one famous patient (known as Anna O.) of Freud's colleague Josef Breuer called "the talking cure." Unlike modern approaches such as cognitive behavioral therapy, which deal directly with the everyday difficulties that patients face, Freud saw these problems as mere symptoms. The goal of psychoanalysis was insight—to bring to the patient's consciousness the core issue, often a repressed trauma early in life. Through insight, Freud believed, patients can ultimately be freed from the problems that brought them to therapy in the first place. In his earlier days, he used hypnosis to achieve this understanding, but then he moved to techniques like dream interpretation and free association, where the patient reports what comes to mind, trying not to filter or control it.

Freud spoke of transference, a phenomenon where one projects one's desires and feelings from one person onto another. In treatment, this can mean that the patient may start to think of the therapist in terms of a significant individual in his or her life, treating the therapist as a father or mother, or perhaps as a romantic partner. Transference can be useful, and the patient can be urged to explore this reaction, but it has its perils, and therapists are trained to be wary of this happening in the other direction—*countertransference*, where a therapist might begin to have inappropriate feelings toward the patient.

Freudian therapy, even in its most orthodox fashion, continues today, but it's not common, at least not in North America. For most readers, what I'm talking about here will be most familiar from movies and television (the sessions depicted in the HBO series *The Sopranos*, particularly in the first season, do a good job of illustrating many themes of Freudian therapy).

———

What should we make of Freud's theory? How much of this is true? How much of it is good science?

One concern builds from an idea of the philosopher Karl Popper.[9]

This is the notion of *falsifiability*. Popper argued that one thing that distinguishes science from nonscience is that scientific theories make claims about the world that run the risk of being proven false.

Consider claims like these: Everywhere in the world, men are, on average, more physically violent than women. A high proportion of children's first words are nouns. IQ scores predict life expectancy. Certain drugs cause the symptoms of schizophrenia to diminish. We will discuss all of these later in this book, but the relevant point here is that these are good hypotheses, in part because they can be proven false. This makes them the stuff of science.

Now, few philosophers today believe that falsifiability is the absolute criterion through which you can distinguish science from nonscience.[10] It's more complicated than that; large-scale theories have wiggle room, ways to explain away apparently disconfirming evidence. There is no single observation that could falsify the germ theory of disease, say. But still, there is a real insight here: *We should suspect theories that cannot be proven false.* The physicist Wolfgang Pauli famously derided the work of a colleague by saying, "He's not right. He's not even wrong," and that's a stinging insult, because what scientists aspire to do is to generate theories that are substantive enough to be tested against the world.

Just as one example outside of Freud, I have often heard versions of the claim: "People are hedonists—we only do what gives us pleasure." Now, this could be an interesting claim if it's made sufficiently narrow, so that certain things that people do could conceivably count against it. But often this view is presented in such a way that it's unfalsifiable; any action counts as supporting it. If a skeptic about psychological hedonism (such as myself) points out that people sometimes choose to die for a cause, or suffer to help someone they love, the defender of hedonism can just respond, "Well, they wouldn't have done that if at some level it didn't give them pleasure." If a claim remains true no matter what happens, then it's not interesting from a scientific perspective.

This brings us back to Freud. One of the main criticisms of Freudian theory is that it's unfalsifiable. This is clearest in examples of therapeutic insights. Imagine dealing with a therapist—suppose you are treated by Sigmund Freud himself—and he says that at the root of your problems is hatred of your father. And suppose you are outraged and angrily deny it. Freud could say, "Well, your anger shows this idea is painful for you, which is evidence that I am right." But Freud would also see support for his idea if you had said nothing or if you agreed with him. There isn't really anything that could prove him wrong. Indeed, Freud could have said the opposite—that you have an unhealthy romantic obsession with Dad, you love him too much—and he would have taken the very same responses as support for that view.

If this seems like caricature, consider this example from Freud's *Interpretation of Dreams*. First, here is the dream as it appears to the dreamer, who was a young woman suffering from agoraphobia, the fear of being in public places.

> I am walking in the street in summer; I am wearing a straw hat of peculiar shape, the middle piece of which is bent upwards, while the side pieces hang downwards (here the description hesitates), and in such a fashion that one hangs lower than the other. I am cheerful and in a confident mood, and as I pass a number of young officers I think to myself: You can't do anything to me.[11]

Now Freud goes on to explain what it means. He says to her, "The hat is really a male genital organ, with its raised middle piece and the two downward-hanging side pieces. . . . If, therefore, she had a husband with such splendid genitals she would not have to fear the officers; that is, she would have nothing to wish from them." Her agoraphobia, he concludes, is grounded in temptation, in sexual fantasies about strange men.

Well. This is quite the theory, both about the dream and about the source of the woman's troubles. I'm not saying it is wrong. Who knows? But what could convince Freud to reconsider? What piece of evidence could dissuade him? Nothing—and so it loses all value.

This falsifiability objection is powerful, but a defender of psychoanalytic theory could object that much of it pertains to the practice of Freud and his colleagues, about how they treat their hypotheses. In the hands of an objective scientist, Freud's actual theory seems to make falsifiable predictions. So, for instance, if Freud is right, your experience of breastfeeding and of toilet training should be a reliable predictor of your personality later in life; it will influence whether you are an oral person or an anal person. If Freud is right, whether you're raised by one parent or two, by a man and woman or two men or two women, should have a profound effect.

So we can derive falsifiable hypotheses. Unfortunately for the Freudians, none of these claims has real empirical support. In general, when you test such Freudian hypotheses empirically, they do poorly.[12]

In an illuminating investigation, the data scientist Seth Stephens-Davidowitz used computational analyses to explore other Freudian claims.[13] For instance, he studied the foods mentioned in dream reports. From a Freudian perspective, you would expect for there to be a lot of sexual imagery. Perhaps when dreaming of eating, we should tend to dream of phallus-shaped foods, such as bananas, cucumbers, and hot dogs. Stephens-Davidowitz found that this is false: there is no evidence for an abundance of phallic foods in imagery. Rather, the frequency of foods found in dreams corresponds to tastiness and how often we eat them, so that the top two are pizza and chocolate.

Or take Freudian slips. Stephens-Davidowitz looked at forty thousand typing errors found by Microsoft (defining an error as a misspelling that is made, but then immediately corrected). Some do have a Freudian flavor, such as "penistrian" (instead of pedestrian) and "sexurity" (instead of security). But the majority of errors had no salacious qualities, like "pindows" (instead of windows), and computational analyses reveal that the sex-related errors were no more frequent than one would expect by chance.

Now I'm not sure these analyses would worry a Freudian, who can say, reasonably enough, that the sexual content of dreams and speech errors need not show up in such a reductionist way. (Recall that the penis symbol in the dream quoted above was a hat—not exactly your standard phallic symbol.) But now we're moving back to the original problem, that of a failure to provide falsifiable predictions.

Overall, Freudian theory is either vague or unsupported by the evidence. This is why we no longer study Freud in most psychology departments.

Why, then, did I just spend so much time discussing him?

One answer is that he's an important part of our history, and not just the history of psychology, but the history of Western thought as a whole. Even if all his claims are wrong, he's left his mark. Have you ever said "I'm not your mother" or described someone as having an anal personality? Even if you reject all of Freud, you should know about his ideas, just as atheists should know about the Bible.

Freud is worth reading. He was a wonderful writer. He was not only nominated multiple times for the Nobel Prize in Medicine; he was also nominated once for literature, and he did win the coveted Goethe Prize. His books, such as my favorite, *Civilization and Its Discontents*, brim with clever thoughts, and have sparked all sorts of valuable insights in other writers and thinkers.

Just as a random example, I recently read an article arguing that absent-minded professors are jerks. It claimed that being forgetful and irresponsible isn't a trait; it's a choice, one that lets you treat others poorly and prioritize your own selfish concerns.[14] And it's a choice usually taken by powerful people; there is no such thing as an absent-minded graduate student because graduate students can't get away with such behavior. During this argument, the author noted that Freud got there first, in his book *The Psychopathology of Everyday Life*, and he quotes this passage:

> There are some who are noted as generally forgetful, and we excuse their lapses in the same manner as we excuse those who are short-sighted when they do not greet us in the street. Such persons forget all small promises which they have made; they leave unexecuted all orders which they have received; they prove themselves unreliable in little things; and at the same time demand that we shall not take these slight offenses amiss—that is, they do not want us to attribute these failings to personal characteristics but to refer them to an organic peculiarity. I am not one of these people myself, and have had no opportunity to analyze the actions of such a person in order to discover from the selection of forgetting the motive underlying the same. I cannot forgo, however, the conjecture *per analogiam*, that here the motive is an unusually large amount of

unavowed disregard for others which exploits the constitutional factor for its purpose.[15]

Or take Freud's focus on the complexity and centrality of sexuality. I've been critical about his obsession here, but there's a more sympathetic take, by the essayist George Prochnik:

> By identifying sexual desire as a universal drive with endlessly idiosyncratic objects determined by individual experiences and memories, Freud, more than anyone, not only made it possible to see female desire as a force no less powerful or valid than male desire; he made *all* the variants of sexual proclivity dance along a shared erotic continuum. In doing so, Freud articulated basic conceptual premises that reduced the sway of experts who attributed diverse sexual urges to hereditary degeneration or criminal pathology. His work has allowed many people to feel less isolated and freakish in their deepest cravings and fears.[16]

There is something liberating about Freud's openness to sexuality, particularly to those who are sexually marginalized, seen by many as degenerates. If everyone is a pervert, perhaps nobody is a pervert.

Finally, and most importantly for the purposes here, his grand idea of the primacy of the unconscious lives on. In the chapters that follow, we will discuss all sorts of aspects of the mind—language and prejudice and emotions and much else—and it will be a continuing theme that much of what we're talking about lies beneath the surface, inaccessible to the conscious mind. Freud was wrong about so many things, but he was right about this.

4

The Skinnerian Revolution

You can think of B. F. Skinner as the American Freud. Burrhus Frederick Skinner, born in 1904, died in 1990, was as famous as Freud—appearing on the cover of *Time* magazine, recognized on the street, the topic of op-eds and dorm room debates. If he were around now, he would have a popular YouTube channel and a million followers on Twitter.

Like Freud, Skinner was known for his radical views on psychology. But their theories were opposites. A typical passage from one of Freud's books might detail how a young man's dream of drowning reflects his wish to have sex with his mother, while a passage from one of Skinner's articles would outline the precise conditions required to get pigeons to peck more frequently at a black square. While Freud expanded our conception of mental life, arguing for a rich and dynamic unconscious, Skinner wanted to get rid of the focus on internal psychological processes altogether. He thought that a proper psychology would make no principled distinction between people and animals like rats.

The men also had different relationships to the theories they were identified with. Though he did have his influences, Freud invented much of the psychoanalytic perspective by himself. In contrast, when Skinner entered the field, most of the core ideas of behaviorism were already there, developed by scholars such as Ivan Pavlov, John Watson, and Edward

Thorndike. Freud was the first to talk about psychoanalytic theory, while Watson wrote a classic article, "Psychology as the Behaviorist Views It," when Skinner was just a boy. Freud was an originator, while Skinner took the existing ideas of behaviorism and ran with them, defending them to both the scientific community and the world at large.

And what were these ideas? There are different schools of behaviorism, but they tend to agree on three radical and interesting claims.

The first is the centrality of learning, and the corresponding rejection of innate ideas or innate differences. This idea was nicely summed up by John Watson, who wrote,

> Give me a dozen healthy infants, well-formed, and my own specified world to bring them up in and I'll guarantee to take any one at random, train them to become any specialist I may select—doctor, lawyer, artist, merchant-chief, and, yes, even beggar-man and thief, regardless of his talents, pensions, tendencies, abilities, vocations, and race of his ancestors.[1]

There is something appealing about this view. Writing in 1930, Watson was saying that all men are equal, not just in their rights but in their capacities. Your race, he says, won't hold you back from any destiny you might choose (one would imagine that if Watson was writing now, he would say the same about one's sex). This fits well with a modern egalitarian sensibility.

The second idea is that there are no interesting differences between species. The cognitive psychologist Ulric Neisser wrote that when he was a student at Swarthmore in the early 1950s, the consensus was that "no psychological phenomenon was real unless you could demonstrate it in a rat."[2] There were decades in which doing foundational psychological research was synonymous with wearing a white coat and working in a rat lab. After all, rats are far easier to study than humans. You don't have to pay a rat, or get its consent, or debrief it about the study later.

Every psychologist now would agree that some mental capacities of humans are also present in other animals—many of the most important studies on visual perception were done on cats, for instance, and most psychology departments include researchers who study nonhumans; one of my favorite colleagues is famous for her research on monkeys and dogs. But the behaviorists took this theory to its extreme. I once asked a behaviorist why a rat doesn't grow up to understand language or appreciate logical arguments or create art, just like a person does, and he shrugged and said that we raise rats and children in different environments. Ivan Pavlov—more about him soon—once told a journalist, "That which I see in dogs, I immediately transfer to myself, since, you know, the basics are identical."[3]

The third theme of behaviorism, held by some but not all of its adherents, is what you can call antimentalism. To get a grasp of what that involves and how radical it is, take this passage from the first page of Virginia Woolf's *Mrs. Dalloway*, which relays the internal dialogue of Clarissa Dalloway.

> For having lived in Westminster—how many years now? over twenty,—one feels even in the midst of the traffic, or waking at night, Clarissa was positive, a particular hush, or solemnity; an indescribable pause; a suspense (but that might be her heart, affected, they said, by influenza) before Big Ben strikes. There! Out it boomed. First a warning, musical; then the hour, irrevocable.[4]

This sort of modernist prose is said to reflect how people think, and who could doubt that our minds wander in this way, that we have streams of thought? Remember that Descartes said the one thing that you cannot deny is that you are a thinking self. And this self is *active*. Here is my own shot at describing a graduate student's lazy day:

> You wake up late on Sunday, irritable after a poor night's sleep. You know you should clean your apartment or work on your dissertation, but you lie in bed instead and surf the web, scrolling aimlessly, compulsively, trying not to be annoyed at how it seems that everyone but you is winning prizes. Breakfast is cereal at the kitchen counter. You decide to do a bit of shopping at Ikea, and so you get dressed and leave the apartment, and you feel a jolt of panic

when you step outside and there's an empty space where your car should be. It's been stolen! But then you remember that the street was full when you came home late last night, so your car is parked on the next block. Finding your car cheers you up, and as you get onto the highway, you start thinking about a vacation you're planning with a friend.

Fine, I'm no Woolf, and it's not really modernist prose. But this is mental life as most of us see it. We make choices, consult our memories, get reminded of things, daydream, and so on. We feel boredom, jealousy, surprise, and pleasure. We are conscious, thinking beings.

Nonsense, says the behaviorist. A complete theory of psychology is a theory of behavior, nothing more. "The time seems to have come," wrote John Watson in 1913, "when psychology must discard all reference to consciousness; when it need no longer delude itself into thinking that it is making mental states the object of observation."[5] Behaviorists agree we need to explain why the student stayed in bed, walked to the car, and so on, but these explanations will not appeal to feelings or planning or memories or anything like that. (Some behaviorists would concede, when challenged, that an internal life does exist, it's just not the proper subject of science; others would deny its existence entirely.)

The antimentalism of behaviorism is often seen as a reaction to the excesses of Freud, who posited all sorts of mental constructs—the id, ego, superego, defense mechanisms, and much more—that are revealed in often mysterious ways. In response to this, the behaviorists said, "We want to do real science, and so we need to get away from this Freudian mumbo jumbo—and while we're at it, we'll also get rid of those unscientific things that everyone else talks about, such as desires and goals and memories and emotions."

There is some truth to this historical reconstruction. James Watson, in his book *Behaviorism*, mocks Freud; he calls psychoanalysis "voodooism."[6] But like most origin stories, this oversimplifies. For one thing, as discussed earlier, behaviorism and psychoanalysis shared an important idea—conscious thought doesn't matter as much as we thought it did. (We'll see later in the book that this rejection of the importance of consciousness has reemerged in the twenty-first century in some schools of social psychology.)

Also, the behaviorists were more sympathetic to Freud than you might have expected. Early in his career, Watson praised the psychoanalysts, and Skinner himself was a lifelong admirer of Freud, citing him more than any other person. Skinner even applied to undergo psychoanalysis, which was offered for free by the U.S. government, as part of the GI Bill of Rights. But there were too many applicants, and he wasn't accepted. (Why did he apply? "My motives are complex," Skinner writes, "but first among them is the belief that in extrapolating to human behavior [as I find myself doing more and more], I stand to gain from firsthand experience with the Freudian point of view."[7])

———

Behaviorism is all about stimulus and response; how an animal's experience with its environment shapes its future behavior. To put it another way, it's about learning. The behaviorists focused on two important types of learning that apply to all manner of creatures, including us.

The first is *classical conditioning*, which we can illustrate with an experiment that you've probably heard of, from the lab of physiologist Ivan Pavlov.

Pavlov won his Nobel Prize in 1928 for studies on the physiology of digestion. To do his research, he needed to collect saliva from dogs. He created an apparatus for doing so, strapping the dog in place and giving it meat powder. When the dog salivated, this drool was collected through a cannula put into its mouth.

As he did this work, Pavlov noticed something. The dog didn't need the meat powder to produce saliva. It would salivate when the person carrying the powder approached. Curious, Pavlov began to play around and tested whether he could train dogs to respond to other cues to the arrival of meat, and he found they could. For instance, after repeated pairings of a metronome and the food, the dog would salivate at the sound of the metronome.

(Beyond the importance of this discovery, it's worth marveling at this wonderful case of scientific serendipity, where a researcher looking at one thing is clever enough to notice something entirely unexpected. We should all aspire toward this receptivity. There was also an intermingling of science and commerce in Pavlov's work. Russians at the time believed that a dog's gastric fluids were a good treatment for dyspepsia—or indigestion—and so Pavlov would sell the dogs' drool and use the money to help run his lab.[8])

The process of classical conditioning can be broken into three stages:

STAGE 1: BEFORE ANYTHING HAPPENS

The animal starts with some sort of natural, or unconditioned, response to some stimulus in the world, either innate or learned in the past. In Pavlov's case, it was the food in the mouth (the unconditioned stimulus) and the salivation in response (the unconditioned response). That's what the animal brings to the study before anyone messes with it.

STAGE 2: CONDITIONING

Then the experimenter adds a neutral stimulus, something that doesn't evoke any response, such as a bell. (Pavlov himself actually never used a bell, but it's what most people associate with his studies, and it makes for a fine example.) This neutral stimulus (bell) and the unconditioned stimulus (food) are then presented together. Because of the presence of the unconditioned stimulus, the animal will provide the unconditioned response (salivation). But over time, the animal will come to also associate the neutral stimulus (bell) with the unconditioned stimulus (food) and so . . .

STAGE 3: AFTER CONDITIONING

. . . the bell will change from a neutral stimulus to a conditioned stimulus and give rise to salivation, which, in this case, will be the conditioned response.

As an example, imagine you go to a dentist's office for the first time. The drill is turned on and makes a particular sound, but that means nothing. The sound is a neutral stimulus. When the drill hits your teeth, it's painful and you flinch. The pressure is an unconditioned stimulus and your flinching is an unconditioned response. You are in Stage 1.

As the drill presses against your teeth, you hear its whine. This started off as a neutral stimulus, but now is associated with the unconditioned stimulus, the pressure of the drill. You are in Stage 2. Behaviorists talk about reinforced trials, and this is what happens in Stage 2 when the conditioned stimulus and unconditioned stimulus are brought together. For the dog, the bell and the food go together; for you, the whine of the drill and the pressure of the drill go together, and this increases the connections. This is how learning takes place.

And then, through the magic of classical conditioning, what will happen over time is that the whine of the drill will make you flinch and draw back. You are in Stage 3—you have been conditioned. The whine, which itself is painless, is now a conditioned stimulus giving rise to a conditioned response that used to occur only with the actual drilling.

There can also be unreinforced trials, when the conditioned stimulus happens without the unconditioned stimulus, and this causes the association to diminish. Imagine you were to ding the bell at Pavlov's dog but don't provide food, and you do it again and again and again. Over time, the dog will produce less and less saliva. Similarly, if you were to become a dentist yourself, you will hear the whine of the drill but experience no pain, and if this happens repeatedly, you'll gradually come to respond to it less, until ultimately the response dies out. The logic here is that the world changes, and you should be flexible enough to learn when two things that used to go together no longer do. If X is no longer a signal that Y is coming, you'll want to unlearn the link between X and Y.

Now, for conditioning to be of any use, it must generalize over different stimuli. You never step into the same river twice, says Heraclitus; no two experiences are ever entirely identical. Classical conditioning would be useless if you didn't respond to a slightly different dentist drill, if the dog didn't salivate to a somewhat different sound.

The power of generalization in the context of classical conditioning was illustrated in a classic experiment by John Watson in collaboration with his graduate student Rosalie Rayner.[9] In the paper reporting this study, their subject was described as a normal healthy eleven-month-old who was given the pseudonym of Little Albert. At the start of the experiment, they showed Albert a white rat, a dog, some masks, and burning newspaper(!). Albert was mellow about all of this. Then Albert was allowed to play with the rat, and he was calm about this too—so we can see the rat as a neutral stimulus, as it elicited no response.

After this first interaction, though, whenever Albert reached out to touch the rat again, Watson hit a steel bar, which made a loud noise, and caused Albert to cry and tremble. Sticking with the language of classical conditioning, the noise was an unconditioned stimulus giving rise to an unconditioned response—loud noises are frightening to children. Soon, Albert would recoil in terror from the rat. It became a conditioned stimulus leading to a conditioned response of fear. Watson and Rayner created a phobia in this little child.

Albert also grew to be afraid of a white rabbit, a fur coat, and a Santa Claus mask, just because they were similar to the rat. In other words, Albert's fear generalized.

This experiment is a famous and powerful demonstration, but there's so much wrong with it. For one thing, most experiments these days don't test just a single subject; they test dozens, hundreds, or more, because we want to see if our results hold for people in general. This use of a single subject is a particularly fraught issue here, because some psychologists have speculated that Little Albert, who was described as normal and healthy, was a boy named Douglas Merritte, who was in fact very ill before the experiment began, never learned to walk or talk later in life, and died a few years later from hydrocephalus.[10] Also, and I hope this is obvious to contemporary readers, this is an unethical study. The first thing we tell incoming graduate students is: *You are not allowed to create phobias in young children.*

On a gossipy note, Watson—voted "handsomest professor" in 1919 at Johns Hopkins—and Rayner were having an affair at the time, Watson's wife discovered this, and it led to a bitter divorce, during which Watson's love letters to Rayner were published in newspapers. Watson lost his job and he ended up working in advertising—a fitting job for someone who specialized in influencing human behavior. While in advertising, he was assigned to sell Maxwell House coffee, and he used the principles of classical conditioning to come up with one of the better inventions of the modern workplace—the coffee break.[11]

———

There are problems with the behaviorist account of phobias. Many phobias don't have their origin in bad experiences. Many people are afraid of spiders and snakes, for instance, without ever having had negative encounters with them. We'll see later on that this is more consistent with an evolutionary psychology approach, which posits that some fears emerge as biological adaptations.

Now, you can develop a phobia due to a specific negative experience—an attack by a vicious dog, nearly drowning, a plane crash that leaves you with terrible injuries. Surely this is Pavlovian, explained by classical conditioning? Actually, no. Multiple trials are critical to classical conditioning. You need to have a lengthy history of a connection between the conditioned stimulus and the unconditioned stimulus. Developing a phobia based on a single bad experience isn't consistent with this theory.

Whatever we think of this account of the origin of phobias, classical conditioning is the logic underlying certain treatments. One of these, known as systematic desensitization, involves training people to relax when exposed to what frightens them.[12] The idea here is that over time, the connection with fear will go away.

In the real world, though, nothing is ever so simple. I was walking in Toronto with a friend of mine—also a psychology prof—and he pointed to an apartment and said that he used to live there, on the fifty-fourth floor.

"Wait," I said, remembering an earlier conversation. "Aren't you afraid of heights?"

"Yes," he said proudly, "that's why I chose the place. I figured I'd spend time on the balcony and get rid of my fear. Unreinforced trials, you know."

"Impressive," I said.

"Not really," he told me. "I never went onto the balcony."

The reality of classical conditioning as a method of learning is not in doubt. You can use the methods of classical conditioning to teach things to all sorts of animals, not just dogs, cats, chimps, and horses, and so on, but even seemingly simpler creatures like cockroaches.[13] The discovery of this learning mechanism is one of the great findings of psychology.

But it is limited. It's a passive sort of learning: You just sit there and observe how stimuli interact, and come to learn what signals what, helping you respond appropriately. But how do we learn to do things in the world? What explains *action*?

This question brings us to the second, different type of learning at the core of behaviorism, called instrumental conditioning or *operant conditioning*.

The theoretical foundations for operant conditioning were established by the psychologist Edward Thorndike.[14] He asked: How do animals solve problems? To explore this, he built puzzle boxes in which he trapped cats and left them to discover how to escape, which required hitting a certain lever in the right way.

They succeeded—but not by thinking about it and figuring out the solution. Rather, the typical cat would flail away until it hit the lever, and then would dart out of the cage. Then, when placed back in the cage, it wouldn't immediately hit the same lever; it would again do a range of actions—but it would be just a bit quicker at doing the right thing. It would get gradually better over time until it would hit the lever and escape the cage immediately. Under Thorndike's analysis, the actions that led to the positive result became more frequent and the useless actions became less frequent. Based on these studies and others, Thorndike coined the famous Law of Effect:

Responses that produce a satisfying effect in a particular situation become more likely to occur again in that situation, and responses that produce a discomforting effect become less likely to occur again in that situation.

————

Enter Skinner, an English major at Harvard with ambitions of being a novelist. The shift to psychology happened at the age of twenty-three, when he read an essay by H. G. Wells about Pavlov. He then attended a lecture that Pavlov gave and got a signed picture of the great physiologist, which hung in his office for the rest of his life.[15]

In spite of this, Skinner's major research program was not about classical conditioning. Instead, he explored how operant conditioning could shape the behavior of humans and of other animals.

Suppose, for instance, you want to train a pig to step forward when you clap your hands. Operant conditioning provides the solution. Clap your hands and then reward or punish the pig depending on whatever random activity it does—punish it if it does nothing or steps back, reward it if it moves forward. The behaviorists' word for reward is *reinforcement*, and there are two types—*positive reinforcement*, giving the animal something it wants, and *negative reinforcement*, releasing it from something aversive. (So, for instance, if the pig had a heavy object on its back, a reward would be taking it away.) *Punishment* is anything that would diminish the odds of the pig engaging in the behavior, perhaps poking it in some way that pigs do not enjoy.

Suppose now that you want to train a pig to dance when you clap. Here's a bad idea: just clap, wait for the pig to dance, and reinforce it when it does. This plainly won't work. Complex behaviors like dancing don't emerge spontaneously.

Skinner's solution is a process described as *shaping*. When the pig moves in a certain manner that approximates dancing in even the slightest way—it just tilts to the left—you reward it, and now the pig will be more prone to do that behavior. Then when it moves to the left and then bobs to the right, and the behavior is looking just slightly more like dancing, you reward it again. As the pig gradually approximates the behavior you are looking for, you reward it step-by-step. This is how animal trainers do their work, and yes, through shaping, you really can make a pig do the foxtrot.

There is a story, most likely apocryphal, about a large lecture class where the students decided to shape their professor. When he moved to the right side of the stage, they would act bored, talk to each other, shuffle through papers. But when he moved to the left, they would look attentive, nod at his stories, and laugh at his jokes. By the time the lecture had come to an end he was lecturing from the far left corner of the stage, though, as the story goes, he had no idea how he got there. Through shaping, the students guided him to that position by the careful application of reinforcement and punishment.

What goes on in operant conditioning is that you have random behavior, some of this is reinforced, and over time the behavior reaches some optimal end. "Selection by consequences,"[16] Skinner called it, and this was the title of a 1981 paper he wrote for the prestigious journal *Science*, in which he made the analogy between operant conditioning and natural selection: In both cases, there is random variation, and some of it is favored (for operant conditioning, through reward; for evolution, through reproductive success), and over time the behavior/species becomes increasingly optimized to the environment.

It's a nice analogy: No evolutionary biologist believes that an animal has a single mutation that leads to a fully functioning eye that gives it a reproductive edge over other animals. Rather, animals evolve some limited sensitivity to light, giving them a small reproductive advantage. Then, over the course of generations, this feature of the animal gets increasingly complex; differential reproductive success guides the organism on the path from no eye to a complete eye. Similarly, no behaviorist thinks that a pig just starts to dance. Rather, the environment (the trainer) rewards approximations to dancing, and the behavior becomes increasingly complex over time, going from no dancing to a foxtrot.

Typically, operant conditioning leads an animal to more complex and adaptive behavior, but not always. In an earlier article called "'Superstition' in the Pigeon," Skinner put pigeons in an apparatus that provided them with food at regular times, regardless of what they were doing.[17] He discovered that pigeons developed odd behaviors; they would turn counterclockwise around the stage, or bob around, or thrust their heads into a corner of the cage.

Skinner's explanation was that these were the actions the pigeons were doing when they received their food, and so they were more likely to occur, due to the Law of Effect—even though, in this case, unlike for Thorndike's cats, there is no real relationship between the action and the result. As Skinner often did, he drew an analogy with humans, giving the example of a bowler twisting and turning as the ball goes down the lane. If the pins get knocked down, this behavior is reinforced and might continue, even though, of course, it had no role in the success of the effort.

The bowling example raises an important point about reinforcement. In the animal examples so far, the reinforcements were naturally pleasant, such as food. But in bowling the reinforcement was knocking down the pins. What's so natural about that? Skinner observed that you can reinforce and punish in all sorts of ways that don't correspond to innate reinforcements or punishments. A dog will be rewarded by a pat on the head or saying "Good dog." You don't need a treat. How can this come to be?

Such cases combine operant conditioning and classical conditioning. That is, a behaviorist might use classical conditioning to associate a neutral stimulus with a positive unconditioned stimulus. So, for instance, every time you give your dog a delicious treat, you pat her on the head. Pretty soon, a pat on the head will become rewarding to the dog. The same thing works for punishment; you can train a dog to cringe at the sight of a clicker if you had cruelly paired the clicker with electric shock in the past. In this way, previous neutral stimuli can become rewards and punishments.

In the examples so far, we reinforce every time. But real life doesn't work that way. Some actions are good for an organism, but the payoff isn't inevitable. And this brings us to the *partial reinforcement effect*.

If you want to teach something quickly, reinforce it every time. But if you want it to stick once the teaching phase is over, reinforce it occasionally. This makes intuitive sense. If I get a reward whenever I do something and then the rewards stop, I'll try a bit more, and then give up (This just isn't working, I might think). But if I get a reward only occasionally, and then the rewards stop, I'll keep at it, assuming that the reward is just around the corner. (Of course, "thinking" and "assuming" is not how the behaviorist would describe it.)

Imagine a child who is having serious tantrums, and the parents respond by paying attention to him and giving in to his demands every time, but then they decide *enough is enough* and ignore him. This sort of cold turkey approach, though difficult for everyone at first, is likely to succeed; the child will realize that the tantrums aren't working anymore and eventually will come to stop.

But suppose instead that the parents were initially rewarding the tantrums only some of the time. Once the parents stop capitulating, they have a long wait before the child gives up on this strategy. It is as if—and again this is not how behaviorists would put it—the child reasons: "Well, I haven't received reinforcement for a while, but maybe this is a run of bad luck, and rewards are just around the corner, so I'll keep trying." Basically, the parents are a slot machine that pays off occasionally, and the child is a degenerate gambler.

I once got an email from a teenage girl in China who watched an online lecture of mine in which I talked about the partial reinforcement effect. She told me that this helped her resolve a romantic situation. She was flirting with a boy over text and realized that she was responding with a heart whenever this boy texted her. She then worried that if she ever missed a few of his texts, he'd quickly give up on her. But if she responded with a heart only some of the time (partial reinforcement), he would persist in contacting her and then—and this part goes a bit beyond what I said in the lecture—she would eventually win his love.

———

Classical conditioning is real; operant conditioning is real. What, then, about the broader theoretical premises underlying this approach to psychology—the centrality of learning, the lack of difference between people and rats and pigeons, the idea that you can do psychology without positing mental life? Does this have the making of a good theory of behavior, for both people and nonhuman animals?

Not really. Before considering people, I'll give three illustrations of how animals behave in ways that challenge the behaviorist worldview.

The first comes from the work of Marian and Keller Breland—two former students of Skinner's, who were in the business of training animals for television and movies. The Brelands noted that their everyday experience contradicted the behaviorist claim that all animals learn in the same way. It is very hard, for instance, to use the principles of operant conditioning to get raccoons to drop coins into a box—instead, they would tend to rub coins between their paws. In an article pointedly called "The Misbehavior of Organisms" (ironically echoing the title of Skinner's first book, *The Behavior of Organisms*), the Brelands pointed out that this reflected raccoons' natural evolved behavior.[18] It's no accident that the successes of operant conditioning typically involved getting animals to do what they naturally tend to do—pigeons pecking, or rats running around searching for food. The behaviorists, whether they knew it or not, set up their studies to exploit the natural proclivities of the creatures they were working with.

While the Brelands discovered a way in which creatures are resistant to learning in a way that would surprise the behaviorist, the psychologist John Garcia discovered a way in which animals are far better learners than any behaviorist should expect.[19] He exposed rats to stimuli that have distinctive tastes and smells (like sweetened water) and gave them toxins or radiation that made them sick hours later. As he predicted, these rats came to avoid the water with that flavor—they developed an aversion to it. But if Garcia tried to associate something other than a food with the sickness, like a light or tone, it didn't work; the rat developed no aversion.

This is unexpected from the standpoint of classical conditioning, which, as we've seen, requires multiple trials when the conditioned stimuli and the unconditioned stimuli are presented at the same time—and the type of stimulus doesn't matter. But none of these conditions are met in the Garcia studies. It was one-shot learning, the flavor and the sickness were separated by hours, and it only worked for food. As Garcia noted, though, this sort of learning makes perfect sense from an evolutionary perspective. If you eat something and become violently ill hours later, this is a good sign that the food was poisoned and you should avoid it later on.

(Many people have taste aversions that were created that way. In my teenage years, I got sick from drinking too much ouzo—Greek friends in high school—and for a long time afterward, I would gag if I got a whiff of that distinctive licorice smell. Cancer patients are advised to take care about what they eat before they begin a chemotherapy session, so that they don't develop taste aversion to once-liked foods.)

The third example presents what perhaps is behaviorism's fatal flaw. It has to do with the proposal that we should reject all mention of mental representations.

Suppose, to take an old philosophical scenario, a dog chases a squirrel into a forest and starts barking at a tree. A perfectly sensible construal of this is that the dog believes that the squirrel is up the tree—that is, it has a representation in its head that expresses the proposition "The squirrel is up this tree." This provides the best explanation of what the dog is up to. Or, to go back to an earlier analogy, suppose we had to figure out how a chess machine works. You would say, well, it has this sort of rule built into it, it has this sort of strategy, and so on. (It knows that bishops only move on the diagonal; it likes to get its queen out early.) You would appeal to these internal representations when explaining and predicting its actions, and you'd be right to do so, because things like computers have internal representations. Indeed, we put them there.

How could you convince a skeptic that such representations really exist in humans and animals? Here we get to the clever work of the psychologist Edward Tolman.[20] In his studies, he let rats run around a maze to learn to get to a reward. One group of rats always received a reward at the end of the maze and were trained for two weeks. As expected, they got quicker and quicker at running the maze. Fine: operant conditioning, Law of Effect, business as usual. Another group of rats were given no food at all for the first ten days; they just wandered around. When Tolman added a reward in the last four days, these rats learned the maze just as well as the first group. This means that while they were just wandering around, they had to be learning the contours of the maze. Learning without reinforcement!

Tolman went further and suggested that, in their little rat brains, they were constructing a representation of the maze, now known as a *cognitive map*. To get a sense of this, forget about rats for a second and imagine you live in Manhattan and regularly have to walk from your luxury penthouse apartment on Tenth Street and Sixth Avenue to your job at a hedge fund on Eleventh Street and Fifth Avenue. How would get there? One way is to walk north for a block (Tenth Street to Eleventh Street), and then east for a block (Sixth to Fifth Avenue). A behaviorist would view this as two behaviors . . .

> Walk north for a block
> Turn right and walk for a block

. . . that presumably had somehow been established through your history of reinforcement.

But I bet you're smarter than that. Suppose that the buildings you've been walking around suddenly disappeared, turning into green space. You might then take a shortcut, a diagonal trek through the park. You do that because you have a mental representation of the relationship between your apartment and your office.

Edward Tolman found that rats reasoned much the same way. Suppose they're put in a complicated maze in which the food is located at a 45-degree angle to the starting point—northeast. But there is no direct path; to get to the food, they need to go up and then to the right. They'll learn this—but if you then change the maze so that there is a direct route, they'll take this diagonal path, just like you when you went through that park.

Many years later, neuroscientists studied rats' brains, looking particularly at the hippocampus, a brain structure involved in spatial understanding. It turns out that certain cells—*place cells*—are activated based on the precise location of the rat, regardless of what it's seeing and what it's doing.[21] Based on all this evidence, we can be confident that rats, like people, have mental representations of the world in their heads. And this makes perfect Darwinian sense. In their natural environments, they need to know how to get to their food sources and back to their nests even when there are obstacles and threats in their way. If all they knew was "Go straight, turn left, go for three feet, look for nest entrance," then they would be at a complete loss anytime they had to deviate from their regular route.

———

The behaviorists proposed that the principles of classical conditioning and operant conditioning apply to humans. And they were right. As Pavlov found with dogs, neutral stimuli, like the whine of a dentist drill, can take on significance through association with something that hurts us. And as Skinner found with rats, humans respond to contingencies in the environment. If the food is delicious, we'll return to the restaurant; burned by the stove, we stay away. If you cut me, do I not bleed? If you partially reinforce me, do I not persist?

But are these simple mechanisms of learning even close to adequate when it comes to describing our psychologies? We just saw that they fail to work in many cases for rats and pigeons, and it would seem like they are nonstarters for people. Thorndike's cats get out of the puzzle box by flailing around until they find an action that gets some reinforcement, but people can try out different alternatives in their head until they come to the right one. People can complete mazes just by thinking hard.

People can also learn by observing others. "Let me show you," says a mother, as she watches her son ineptly try to throw a Frisbee. "No, sweetie, don't throw it like a ball," and she mimes snapping her wrist while he watches intently, and now he gets the idea. We even learn when we don't want to; many moviegoers became afraid to go into the water after watching the movie *Jaws*, and the dental torture scenes in *Marathon Man* made me delay my next teeth cleaning appointment for a long time.

And spatial maps? Everyone admires Tolman's clever experiments with rats and those neuroscience studies finding that place cells ping when you're in a certain location, but you don't need anything like this to make the case that we have mental representations of the world. Most likely, you can draw a crude floor map of your house or apartment or dorm room, and it's hard to think of any way to explain this other than saying that you have a mental representation of the space (a cognitive map) in your head. Quite a while ago, my wife and I were doing a *New*

York Times crossword puzzle and the question was "Yemen-to-Zimbabwe dir." Without hesitation, she typed in SSW (south-southwest) because she plays online geography quizzes where she is tested over and over about the locations of the countries of the world, and wow, it finally came in handy.

Then there's Skinner's broader claim that we are under "stimulus control"—that everything we do is determined by the environment. Whatever one might say about rats, this is an audacious claim about people. Surely what I write now, sitting by my computer on a cold Canadian winter morning, is driven by what I'm thinking about, not whatever "stimulus" is currently impinging on my senses.

Skinner would disagree. In fact, it was a disagreement over this issue that motivated Skinner at age thirty to write what he believed to be his finest book, *Verbal Behavior*. Here's how he tells the story:

> In 1934, while dining at the Harvard Society of Fellows, I found myself seated next to Professor Alfred North Whitehead. We dropped into a discussion of behaviorism, which was then still very much an "ism," and of which I was a zealous devotee. Here was an opportunity which I could not overlook to strike a blow for the cause, and I began to set forth the principal arguments of behaviorism with enthusiasm. Professor Whitehead was equally in earnest—not in defending his own position, but in trying to understand what I was saying and (I suppose) to discover how I could possibly bring myself to say it. . . . He brought the discussion to a close with a friendly challenge: "Let me see you," he said, "account for my behavior as I sit here saying, 'No black scorpion is falling upon this table.'"[22]

The next morning, Skinner started to outline *Verbal Behavior*, which was published over twenty years later, in 1957. It's an ambitious book, much of it forbiddingly technical (introducing new terminology such as "mand" and "tact") and was much praised by behaviorists as a successful attempt to extend Skinnerian principles to the richest of human behaviors.

The book's publication also played a role in the end of behaviorism as a dominant perspective in psychology. In 1959, a young linguist named

Noam Chomsky wrote a scathing review of Skinner's book, and it's not an exaggeration to say that this was one of the most important book reviews in modern intellectual history.[23]

I recommend reading the review, though it does have its flaws. Many behaviorists argued afterward, with some justification, that it was unfair, and mistakenly blurred together Skinner's ideas with those of other behaviorists.[24] Nevertheless, it is a powerful and forceful argument, one that is often misunderstood by those who have heard of it but never read it. Chomsky later became known for his views that language is an inborn capacity—more on this later—but despite how the review is often described, he hardly talks about his own nativist position. Chomsky would also later argue that there is something morally suspect about behaviorism, but that wasn't in the review either. Nor does he discuss any new experimental evidence or take issue with any of the previous laboratory studies.

Rather, Chomsky's main argument was this: In *Verbal Behavior*, Skinner goes over countless examples of what people say, and for each, he gives a behaviorist reconstruction in which this verbal behavior is under the control of the stimulus, caused by the environment. For instance, Skinner points out that you might respond to certain music by saying "Mozart" or a certain painting by saying "Dutch," and this is elicited by "extremely subtle properties of the stimuli." You might talk to yourself, or give bad news to an enemy, or imitate someone, and in all these cases this is due to some reinforcing consequence of the action. By going through such examples, Skinner purports to be demonstrating that our verbal behavior can be explained in terms of the behaviorist notions of stimulus and response.

Chomsky's concern is that this is not so much false as empty. He considers the example of someone calling a painting "Dutch" and points out that the person might have also said,

Clashes with the wallpaper
I thought you liked abstract work
Never saw it before
Tilted
Hanging too low

Beautiful

Hideous

Remember our camping trip last summer?

Is that a problem for Skinner? In a sense, not at all; he could say that each of these different responses is under the control of different aspects of the environment. But Chomsky's point was that if you can explain everything, you explain nothing. As Chomsky puts it, "It is clear from such examples, which abound, that the talk of stimulus control simply disguises a complete retreat to mentalistic psychology. We cannot predict verbal behavior in terms of the stimuli in the speaker's environment, since we do not know what the current stimuli are until he responds."[25] In other words, if you can take anything someone says—including a comment that has nothing to do with the current environment, like the remark about the camping trip—and say that it's under the control of some stimulus, then the notion of stimulus control has no explanatory power and no predictive power. It's just a fancy way of saying that what we talk about makes sense to us at the time.

Or take the notion of reinforcement. People sometimes talk to themselves or make music in private, and children sometimes like to imitate the sounds of cars and airplanes. Skinner explained all of this in terms of self-reinforcement; we do these things because we find it reinforcing to do so. Again, Chomsky argues that this can "explain" everything because it doesn't really say anything:

When we read that a person plays what music he likes . . . thinks what he likes, reads what books he likes, etc., BECAUSE he finds it reinforcing to do so, or that we write books or inform others of facts BECAUSE we are reinforced by what we hope will be the ultimate behavior of reader or listener, we can only conclude that the term *reinforcement* has a purely ritual function. The phrase "X is reinforced by Y (stimulus, state of affairs, event, etc.)" is being used as a cover term for "X wants Y," "X likes Y," "X wishes that Y were the case," etc. Invoking the term *reinforcement* has no explanatory force, and any idea that this paraphrase introduces any new clarity or objectivity into the description of wishing, liking, etc., is a serious delusion.[26]

Now, notions such as "stimulus" and "reinforcement" *do* have clear technical meanings when Skinner is talking about his experiments with rats. But these strict definitions don't apply in Skinner's examples; we aren't reinforced for what we say in any literal sense, or for singing to ourselves, and so on. The gist of Chomsky's critique is that behaviorism applied to human behavior is either false (when the terms are taken literally) or is trivially and boringly true (when they are taken metaphorically).

We see an example of what Chomsky worries about near the end of *Verbal Behavior*, when Skinner goes back to the challenge that Whitehead posed, to explain: "No black scorpion is falling upon this table." Skinner says, sensibly enough, that Whitehead said this (or as Skinner puts it, "the response was emitted") to make a point. But why didn't he say "autumn leaf" or "snowflake"? Skinner concludes that "black scorpion was a metaphoric response to the topic under discussion. The black scorpion was behaviorism." Now, for all I know, this could be right. But this appeal to metaphor illustrates how far we've gotten from the science of behaviorism as it applies to nonhuman creatures. And notice that Skinner is starting to sound a little bit like Freud, and not in a good way.

———

As he discusses Whitehead's remark, Skinner makes a plaintive comment: "The stimulus may not have been much, but in a determined system it must have been something."[27] Now in a sense, this seems right. Everything has *some* cause. I assume that Chomsky would agree with this; Freud certainly would. But if this was all that Skinner was saying, it would be uninteresting. His interesting claim was that the cause for Whitehead's remark had to be a *stimulus*.

This just isn't true. In some cases, the environment is connected to our behavior, but this is mediated through a lot of thinking. I was talking to friends recently and someone mentioned that Obama's favorite television character was Omar, from *The Wire*, and while the conversation went on,

I remembered that there was a guy called Bill Rawls in the series, which made me think of the philosopher John Rawls, which reminded me that I wanted to add something about impartiality to a lecture that I was giving in a month, and I wrote a note about this several hours later when I was at my desk. In some sense, my writing of that note (and my lecture, a month later) was *influenced* by the "stimulus" of what my friend mentioned, but it certainly wasn't caused in the same sense as Pavlov's metronome caused the dog to drool. Rather, and here a psychologist might use the same terms as anyone else, it got a chain of thoughts going. But you can't describe what happened unless you grant the reality of thoughts—and more generally, the reality of mental life.

The mind is far more complicated than Skinner's framework allowed for. If the problem of Freud is that his theories were too all-encompassing and too vague and too ungrounded in empirical effect to ultimately become a successful account of the mind, the problem with Skinner is that his theories just fail to explain the richness of human psychology. As this book proceeds, we'll talk about all sorts of aspects of the mind that the ideas of behaviorism are simply not sufficient to address.

But I'll end by pointing out that, just like Freud, Skinner's ideas live on, and deservedly so.

For one thing, he has given us a better understanding of some important learning mechanisms.

For another, regardless of what you think of Skinner's theoretical edifice, it has practical implications. For instance, Skinnerian ideas play a role in some successful treatments of phobias and are used to deal with behavioral problems in animals.

Finally, sometimes, just as with Freud, insights from Skinner become relevant to everyday life. I often think of the power of partial reinforcement, of how a diet of rare and random rewards can make a behavior difficult to extinguish. I don't currently have to deal with toddlers who throw tantrums and I've never been tempted by slot machines. But I often find myself lost online, staring at my phone, numbly clicking on links, watching videos, doing the drag-down-to-refresh gesture in the hopes of seeing something that makes me feel good, and when I do all this, I am reminded of a rat in the behaviorist's cage.

THINKING

5

Piaget's Project

All of psychology is fascinating, but for me the study of children carries a special wonder. I started to do research in developmental psychology many years ago, when I was an undergraduate at McGill University, and my fascination with the minds of children has never wavered.

I bet you're interested in the topic as well. Who hasn't been amused by the crazy things that children sometimes say? Who hasn't marveled at how quickly they come to learn about the world, but also at how stupid they can seem? Who hasn't looked into the eyes of an infant and wondered what (if anything) is looking back?

I'll be up-front here and admit that developmental psychology doesn't get the same respect as other areas of psychology. I'm not quite sure why. Some of this may be due to sexism. Taking care of children is traditionally women's work, and perhaps people feel the same about the study of children's minds and so take it less seriously than they should. Another point against the developmentalists is that the field is usually pretty low-tech. Developmental psychologists tend to do their work in testing rooms and day cares and schools; with some notable exceptions, we don't produce those colorful brain scans that so impress granting agencies, university administrators, and media outlets.

Regardless of its cause, this lack of respect is a mistake. The study of children has the potential not just of satisfying our curiosity about the small creatures that live among us, but of answering broader questions that people have wondered about for a very long time.

Just as one example, take the question of destiny. How much of our fate is sealed early in life, set in stone by our genes and our early experience?

Nobody can perfectly predict how someone will turn out based on how they are as a baby or as a child. But consider a study conducted by Avshalom Caspi.[1] He got hold of some data from 1975, where psychologists did ninety-minute interviews with three-year-olds in Dunedin, New Zealand. The original investigators used the children's responses to classify them as falling into one of three categories:

1. **Undercontrolled:** impulsive and restless, emotionally labile
2. **Inhibited:** slow to warm, fearful, easily upset by strangers
3. **Well adjusted:** just right, capable of self-control, relatively confident and calm

These individuals were given a series of tests and questionnaires eighteen years later, and Caspi was able to ask about the relationship between this early assessment and how the children turned out. He found that those who were "undercontrolled" at age three tended, at the age of twenty-one, to report more employment problems, alcohol abuse, and brushes with the law. Those who were categorized as "inhibited" were more likely, as adults, to be depressed and to report receiving lower levels of support. And those (lucky) children who were described as "well adjusted" were less likely to suffer from any of these problems.

Now, these are not enormous effects—there are plenty of undercontrolled and uninhibited three-year-olds who grew up just fine and plenty of well-adjusted three-year-olds who have problems later on. Still, there is a regular relationship. Accordingly, Caspi uses a quote from Wordsworth to title his paper: "The child is father of the man."

There are many potential explanations for this stability over time. It could be genetic—we keep the same genes over time. It could be environmental—we tend to stick to our environments, particularly when young, going to the same schools, living in the same neighborhoods, surrounded by pretty much the same people. Or it could be both. As Caspi notes, one's genes and one's environment, nature and nurture, often reinforce each other. We tend to fall into groups, for instance, that best fit our natural propensities, and these in turn can cause these propensities to flourish. An aggressive youngster might choose a peer group in which aggression is rewarded—a rough crowd, as my grandma would have called it—and so this aggression might stick. Similarly, a bookish and argumentative boy might choose a community and then a profession that rewards these traits, and who knows, he might even become a psychology professor and write a book that you're now reading.

We'll return to the topic of destiny later in the book. This chapter focuses on a set of questions that are perhaps even more fundamental: Where does knowledge come from? How much of what we know as adults is inborn and how much of it is learned? And more generally, what's the difference between the minds of children and the minds of adults?

———

Babies don't look smart and don't act smart. One striking demonstration of their limitations was reported in a national newspaper about twenty years ago. Researchers at the UCLA Institute of Child Development tested babies' brainpower by putting them in a series of difficult situations in which they had to act intelligently to survive. The babies had to escape from rooms filled with cyanide gas, stave off starvation by using can openers to get food, and master the ability to scuba dive so as to breathe underwater. The babies failed—and all died. The scientists concluded: "Human babies, long thought by psychologists to be highly inquisitive and adaptable, are actually extraordinarily stupid."

The story was from the satirical newspaper *The Onion*, and the target of the satire was the babies-are-smart research discussed in the popular press, research that we are soon going to turn to. But the babies-are-stupid moral of the made-up study would have resonated with many scholars throughout history. We saw in a previous chapter that, in 1890, William James described the mental life of a baby as "a blooming, buzzing confusion."[2] (Given my last name, here's hoping that nobody uses that as a title of a review for this book.) A century earlier, Jean-Jacques Rousseau made this point in even harsher terms, saying that if a child was born in an adult body, "such a child-man would be a perfect idiot, an automaton, a statue without motion and almost without feeling; he would see and hear nothing, he would recognize no one."[3]

This view—that knowledge comes through exposure to the environment—is known as *empiricism*. This resembles the view of the behaviorists we read about earlier, such as John Watson and B. F. Skinner, who thought that the environment was all-important. But it's not clear that behaviorists would count as empiricists since they eschewed all talk of knowledge. (You can just imagine Skinner sneering at the very idea of *knowledge in the rat*.)

Rather, the best early representatives of this view were a group of philosophers collectively known as the British Empiricists. This included scholars such as John Stuart Mill, David Hume, and John Locke. The classic expression of their view, often attributed to Locke, is that the mind is a blank slate: "Let us then suppose the mind to be . . . white paper; void of any characters; without any ideas; how comes it to be furnished? To this I answer in one word: experience."[4] Metaphors change; a modern-day empiricist might say that a baby's mind is like an iPhone without any apps.

The opposing view is *nativism*, which proposes that much of our knowledge and capacities are part of our natural endowment. We are born with them. Going back to the iPhone analogy, when you buy a new one it comes with several apps, some preinstalled contacts, maps, a dictionary, and so on. Nativism is the view that our brains are like that. Early philosophers such as Plato explained the existence of innate ideas as the result of souls recollecting knowledge learned in past lives; modern-day nativists think of this as the product of our evolutionary history, encoded in our genes.

Some are skeptical about a nature-nurture dichotomy. One excellent Introduction to Psychology textbook tells us that to ask which of the two is more important is like asking "which is more important in defining the area of a rectangle: the length or the width? Development requires both and, in fact, nature and nurture require each other."[5]

There is a deep truth here. There is no learning (nurture) without initial machinery (nature). Even Skinner would agree that rats learn to run a maze and rocks do not, and this is because only rats have brains capable of operant conditioning (and legs capable of running). And this initial machinery (nature) typically requires some contact with the world (nurture) to work. Even mental systems like color vision, which we are tempted to see as pretty much hardwired and built in, won't operate without some experience—kittens raised in darkness go blind. There are some horrific cases where children are isolated from social stimulation by cruel or deranged parents, and this lack of the normal environment does terrible damage to both body and soul.[6]

One nice example of how innate machinery is sensitive to the environment comes from the sleep-wake cycle. We have evolved a twenty-four-hour circadian rhythm in response to a long-standing fact about the environment—the duration of the Earth's rotation. It's hardwired. But if you take away the environmental clues of light and dark by putting someone in darkness, our internal clock goes out of sync by a few hours. As the neuroscientist David Eagleman puts it, "This exposes the brain's simple solution: build a non-exact clock and then calibrate it to the sun's cycle. With this elegant trick, there is no need to genetically code a perfectly wound clock. The world does the winding."[7]

A more general way in which nature and nurture mesh together is that the brains of humans are evolved to acquire new information and new capacities. We are, it is often said, a cultural species, and our learning capacities far exceed any other mind on the planet, natural or artificial. Language learning, one of the topics of the next chapter, is a paradigm case of this: Some universals of language are already present in babies' brains (nature), and yet the specific language they learn is influenced by the environment they find themselves in (nurture).

So, no nurture without nature and no nature without nurture. Still, there are cases where it makes sense to say that something is learned and other cases where it makes sense to say that something is innate. I know the plots of all the John Wick movies and not even the most nativist of nativists would say that this is inborn. Other thoughts and actions are different. The same textbook authors who warn us not to pit nature and nurture against each other have a section a few pages later called "I was born ready: Early capabilities in the newborn." They point out that if you put your finger in a baby's palm, the baby will grasp it, instinctively. The authors then say, sensibly enough, that this is "a reflex likely inherited from our apelike ancestors who needed to cling to their mothers right from birth as they were carried."[8] It sure seems like they think it's innate.

We should be mindful of the interaction between nature and nurture, knowing you can't have one without the other, but this doesn't make the nature-nurture debate go away.

————

I said in an earlier chapter that if you've heard of one psychologist, it's going to be Sigmund Freud. If you've heard of two, they're likely to be Freud and Skinner. But if you've heard of one developmental psychologist, it's probably the Swiss polymath Jean Piaget, born in 1896, died in 1980.

Piaget was by any standard a genius—at age ten, as a high school student, he wrote his first scientific article, on the albino sparrow. He was bored in classes, unhappy at home (his mother was strictly religious— Piaget was not—and she was probably also mentally ill), but was an engaged scholar from the start. One observer, talking about Piaget's early life, wrote, "It's probably not fair to call Piaget a nerd,"[9] which is often what someone writes about someone he thinks is a nerd. When Piaget graduated from high school he was a renowned expert on mollusks and had published twenty scientific papers on this topic. He got his doctorate in the natural sciences at the age of twenty-one.

Piaget was a productive researcher and writer through his sixty-year career, publishing more than fifty books and five hundred papers. He was known as "Le Patron" (the boss) not just by his graduate students and research assistants but often by people who never met him. It was an affectionate term, more Bruce Springsteen than Tony Soprano. His students adored him; unlike his more famous Austrian counterpart, he had no reputation as a jerk.

Actually, Piaget and Freud did meet once. Piaget gave a lecture at the International Psychoanalytic Congress in 1922. This was an anxious experience for him, since he disagreed with Freud on critical matters and was worried that the psychoanalysts would tear him apart. It went well, but Piaget described his encounter with Freud like this:

> Freud was seated to my right in an armchair smoking his cigars, and I was addressing the public, but the public never glanced at their lecturer. They looked only at Freud, to find out whether or not he was happy with what was being said. When Freud smiles, everyone in the room smiles, when Freud looked serious, everyone in the room looked serious.[10]

Piaget's ultimate interest was not child development. This was just a means to an end. Rather, he was interested in the development of knowledge in the human species—*genetic epistemology*, as he called it, where "genetic" refers to origins and "epistemology" refers to knowledge. He just believed that the best way to pursue that interest was to look at the intellectual development of individual humans. The logic underlying his strategy is summed up in a very snooty phrase coined by the zoologist Ernst Haeckel: ontogeny recapitulates phylogeny. This means that the development of the species (phylogeny) is repeated—recapitulated—in the development of every individual (ontogeny). Interested in the earliest stages of humanity? Focus your attention on what goes on in playpens, day cares, and elementary schools.

With regard to the innateness-empiricism debate, Piaget's sympathies were empiricist. At the core of his theory was the idea that sustained interaction with the physical and social environment is how we come to think like adults. Unlike the British Empiricists, though, Piaget didn't see learning as merely the recording of associations. And unlike the behaviorists, he didn't think in terms of reinforcement and punishment.

Piaget proposed that mental life includes complex cognitive structures, which he called *schemas*, and he posited two psychological mechanisms that lead to the transformation of these schemas and the creation of new ones. The first, *assimilation*, is the process of using already-existing schemas to deal with new situations. Take a baby who has a simple schema—it knows how to suck on its mother's breast. Assimilation is when the baby sucks on a rattle or sucks on its own toes. But to do so successfully, the baby must modify the behavior in certain ways, and this is *accommodation*, a process through which existing schemas are changed or new schemas are created to fit the new information and new experience.

This is a concrete example, but Piaget argued that assimilation and accommodation applied to more abstract attainments. There are schemas having to do with objects and numbers and people, and the processes of extension and modification—assimilation and accommodation—enable the child to reach ever greater intellectual heights. Like Skinner, Piaget was tempted by the analogy between learning and biological evolution, and he sometimes said that through these two learning processes, the developing child can be seen as "adapting" to the environment, working so that there is an equilibrium—a proper balance—between mind and world.

Piaget also proposed that the child goes through a series of stages. Each stage corresponds to a distinct style of thinking, a different way of making sense of how the world works. (Stages also have substages—at times, Piagetian theory reaches a Borgesian level of complexity—but we are going to ignore these here.) We've already seen a stage theory in Freud, but his was psychosexual, focusing on appetites and desires. Piaget's stages are G-rated, having to do with different ways of thinking.

The newborn begins at the *sensorimotor stage*. For about the first two of years of life, the baby is a purely sensory creature. At the start of this stage, it perceives and manipulates, but doesn't reason. There's no sense of time, no differentiation between itself and other people, and critically, no object permanence.

This last claim has long fascinated psychologists—because it's so audacious. When a ball rolls behind the dresser, of course you know it's still there. What could be more obvious? But Piaget's claim was that before about six months of age, babies don't understand that objects exist independently of their actions or perceptions of them. Out of sight, out of mind—literally. This might be why they are so amused by the game of peekaboo. You cover up your face, and then reveal it, and then babies crack up or gasp, and it's because when you covered up your face, they thought you were *gone*.

Piaget was not the first to hold this view about our initial natures. George Berkeley, one of the British Empiricists, believed that an understanding of objects arises only once babies begin to move through space and manipulate the world. Visual experience is two dimensional, after all—the light that hits our eyes is akin to patterns of paint splashing onto two canvases—and so, Berkeley argued, touch is necessary for us to come to appreciate that we live in a three-dimensional world.[11]

Piaget's second stage, the *preoperational stage*, runs from about age two and lasts until about age seven. The baby is now a child and starts to reason. Children at this age can think, they can differentiate themselves from others, they have a rudimentary understanding of time, and they understand that objects continue to exist when out of sight. But they have certain interesting limitations.

One of these is what Piaget called *egocentrism*. He didn't mean this in the same way I do when I say that one of my colleagues is egocentric because he spends all his time boasting about his accomplishments and never notices when I get a new haircut. Piaget meant that children literally can't understand that the world is seen and understood differently by others; they can't take other people's perspectives.

One classic demonstration of this is the Three Mountains task, which can be done with three- and four-year-olds. Show a child a diorama with mountains of different heights—say, from left to right, they are small, medium, and large—and ask the child to draw them. Now, you ask the child how someone else, on the opposite side of the mountains, would draw them. An adult would answer by flipping the orientation, and say that the other person would draw them, from left to right, large, medium, and small. But children can't do this. They just draw the mountains as they see them. Piaget's conclusion is that children believe that everyone else sees the world as they do.

Anyone who spends time with children will notice their egocentrism. It lingers, in a milder form, long past the preoperational stage. My niece recently FaceTimed me to tell me about the presents that she got for her sixth birthday, and she was excited, talking with breakneck speed about this toy and this doll and this bag of candy, and what Sophie got her and Sheri got her. I didn't know anyone she talked about, and she didn't care;

she ignored all the rules of conversation, in her charming way, and it was like I wasn't there at all. This was adorable for a six-year-old but would be bizarre for an adult.

Another limitation, according to Piaget, is that children in this stage fail to appreciate that certain operations on the world change some properties but not others—that some properties are *conserved*. If I have a bunch of candies and I slide some of them to make them take up more space and then ask, "Are there more candies now?" you will laugh and say, "Of course not, they're just spread out." But young children usually say there *are* more candies now. Similarly, if you pour water from a short, fat container into a long, tall container, adults know that the volume of water is conserved through the transformation, but young children will say that there is now more water, because it is taller.

A third limitation we find in the preoperational stage is a failure to distinguish appearance from reality—to appreciate that something might appear to be one thing but be another. As I am writing this, there is a popular set of videos where an object looks like a book or a shoe or a laptop, and then a knife descends, cuts into the object, and, wow, it's cake! This is funny, but if Piaget is right, children should be befuddled: If it looks like a book, it's a book.

A vivid demonstration of children's limitations in this stage comes from a classic study from the 1960s.[12] Rheta De Vries brought a cat named Maynard into a lab and let children of various ages play with him. They all agreed that Maynard was a cat. Then, while the children were watching, she put a mask of a fierce-looking dog on Maynard's head. Now many of the three-year-olds (but not the six-year-olds) said that they thought that Maynard had become a dog. When asked what was under Maynard's skin, they said that he would have a dog's bones and a dog's stomach. He looks like a dog, so he must be a dog.

The sensorimotor and preoperational stages are the focus of most of Piaget's research and the research of his proponents and critics. There are two more stages, which I'll breeze through for the sake of completeness. Upon reaching the *concrete operational stage*, starting at roughly seven years of age, the child is sophisticated but has some difficulties with abstract and hypothetical reasoning. Full cognitive maturity awaits the *formal operational stage*, said to be achieved around age eleven, though Piaget noted that "not all persons in all cultures reach formal operations, and most people do not use formal operations in all aspects of their lives."[13]

When outlining Piaget's theory, I gave rough ages, but Piaget was mellow about the precise points when children reached and surpassed these stages. He was amused by the question he often got: "How can we make children go through these stages faster?"; his answer to what he called "the American question" was: Why would we ever want to?[14]

———

Piaget was hardly a humble theorist, but unlike Freud or Skinner, he never aspired to a grand theory of human nature. He wasn't the sort to write a psychohistory of Leonardo da Vinci (Freud) or try to reform the criminal justice system (Skinner). He stayed in his lane.

But within this lane, he was a bold theoretician. Indeed, one enthusiastic scholar writes that "the total of Piaget's work amounted to perhaps the single most intensive, coherent, sweeping theoretical integration of all the life sciences the world had yet seen."[15] This seems like overkill to me—what about Darwin?—but I agree with its general spirit. There is a richness to Piaget's ideas that rewards a deep dive. If you are tempted to do so, I'd point you to secondary sources. Le Patron was a terrible writer, both in English translation and (I've been told) in the original French.

Generations of developmental psychologists followed up on Piaget's work, replicating and extending his findings, trying to explain them—and often trying to explain them away. In our earlier discussion, we complained about how the theories of Freud and Skinner couldn't be falsified. The problem with saying that all dreams are wish fulfillment or that everything you do is under the control of the stimulus isn't that they are wrong, it's that they are so vague and slippery that they don't even get to be wrong. This is not a problem for Piaget. He made interesting and falsifiable claims. If you discover, for instance, that one-year-olds do have object permanence or that three-year-olds can pass a conservation task, that'd be really damaging to Piaget's theory, and he knew it.

Many of Piaget's observations have proven robust. Just as one example, you really can demonstrate failures of conservation in a cooperative young child. Put a small number of candies in front of you and then the same number of candies in front of the child and match them up one to one. Ask who has more and the child will say it's the same. Then take one line of candies and spread them out and now ask the same question, and you'll typically find that the child says that the longer line has more candies, just as Piaget found. These are surprising discoveries about the minds of children.

There is a lot to this theory, then. But science develops, and there are few scholars today who fully accept Piaget's account of child development.

One limitation is theoretical. Piaget talked about development in terms of the processes of assimilation and accommodation, but it's not clear how to make these notions explicit. He often drew upon biological ideas, but they never rose above the status of metaphor. Neither Piaget nor his followers explained how this transformation and restructuring works at a neurological level; there are no broadly accepted computational models for how such a change takes place.

Another concern is methodological. Piaget developed his theory largely based on his own interactions with babies and children, drawing on his substantial powers of observation. When he was with children who were old enough to be able to speak, he would interview them, asking them to solve problems and explain what they were thinking as they did so. He clearly got insights in this way, but there are limitations to these methods. Children are notoriously inarticulate and—like adults but only more so—have trouble explaining what's going on in their own minds. They are also quite sensitive to what psychologists call *task demands* or *experimenter demands*, sometimes trying hard to give the "right answer" instead of saying what they believe.

Finally, Piaget and his colleagues were sometimes uncharitable when it came to interpreting why a child thinks and acts in such and so a way, too quick to attribute something to cognitive immaturity when there might be another explanation. Let's return to the claim that children don't distinguish appearance from reality, and go back to the De Vries study, where a dog mask was put on Maynard the cat and the younger children said that it was now a dog. But is this really a mistake? Let's shift the example a bit with this made-up dialogue between a father and a four-year-old:

Father: Who am I?

Child: Daddy

Father: [puts on Darth Vader mask] Who am I now?

Child: Darth Vader!

Father: [to himself] Wow, quite the appearance-reality confusion. I guess Piaget was right. She'll grow out of it in a few years.

This is unfair. The child might not be confused at all. After all, if she really thought her father turned into Darth Vader, she'd be terrified, not delighted. Perhaps what develops more generally isn't some deep understanding of reality, but the capacity to suss out contexts in which questions are intended to be taken literally. An older child, knowing that she's being tested, would be careful to answer that he's not *really* Darth Vader; he is just pretending that he is. But it's not really a mistake to call him "Darth Vader," or for the children studied by De Vries to call a cat a dog. (Why else did she put that dog mask on the cat, except for the child to realize that it was, for the moment, "a dog"?)

Progress in psychology doesn't just come in terms of new theories or new findings; there is also the development of methods, such as the dream interpretation of the psychoanalysts, the brain-imaging techniques of neuroscientists, and the reaction-time studies of cognitive psychologists. Similarly, developmental psychologists have invented clever ways to tap the minds of babies and children. When you use these methods, you find that babies and children are far smarter than Piaget claimed.

———

What do humans know from the very start? How can we address the great debate, running from Plato to Locke to modern-day psychologists and cognitive scientists, over how much knowledge is inborn? One way is to look at babies.

We've already seen that there are many ways to learn about the mental life of other creatures besides talking to them, as when the behaviorists got rats to run mazes. Young babies, though, present us with special challenges. They don't talk; they don't walk; they can't peck or pull levers; older ones can crawl, but not far, not fast, and not always in a straight line. Most of the time, they just sit there, being adorable, crying or gurgling or excreting.

Can't we just study what their brains are doing? This would seem to be the most elegant way of exploring the mental life of a creature who can't yet tell you what he or she is thinking. And indeed, there is a lot of developmental work along these lines. There are studies that use fMRI, the same method used for adults, seeing which parts of the brains of babies are most active when they are looking at different kinds of displays. A more recent development is a method called fNIRS—functional near-infrared spectroscopy—which studies mental processing by shining infrared light into the head. (This method works best with bald heads and thin skulls, making it perfect with infants.) And there's long been research using methods such as ERP (event-related potential), which looks at patterns of electrical activity, better for telling the time course of mental activity rather than the location of the activity.

These are exciting methods, and we've learned from them something about the brain areas involved in actions such as listening to speech and attending to numbers.[16] But, so far at least, the most interesting discoveries about the minds of babies have come about through simpler methods. These make use of the few things that babies are successful at, such as sucking on a pacifier and moving their eyes. This might not sound like much to go on, but in the hands of clever researchers, they reveal the secrets of the infant soul.

Suppose you wanted to know whether babies enjoy listening to the sound of their mothers' voices more than the voices of strangers. One method is to stick a pacifier in a baby's mouth and put headphones on the baby's ears, setting it up so that if the baby sucks on the pacifier, she gets to hear sounds from the headphones. On some occasions, the sound is Mom's voice; on others, it's the stranger's voice. Remember the Law of Effect? Babies, just like rats and pigeons, should increase a behavior if it's rewarding. And so by comparing the sucking behavior across different contexts, you can see who the baby prefers to listen to. You can also ask about preferences for languages: Does a baby born to English-speaking parents prefer English over Russian? (For the answer, wait for the next chapter on language.)

Another method is *habituation*. This is defined as a declining tendency to respond to stimuli that are familiar due to repeated exposure, which is just a fancy way of saying that we get used to things. If in the middle of a lecture I stop talking into the microphone and scream BOO!!!, the students would jump. But, if I go BOO!!! a minute later and then did it again and again, it won't surprise them anymore. They become used to it. They habituate.

Habituation turns out to be an essential psychological mechanism because it keeps us focused on novelty. Something that's been in an environment for a while shouldn't capture our attention as much. We've already registered it and dealt with it accordingly. But something that is new could be important, possibly dangerous, and worth attending to. When I was a teenager, I worked in a sheet metal factory over one summer and it was incredibly noisy, with sparks flying and machines moving, and at the beginning, all of this captured my attention (and my anxiety), but quickly I grew used to it. I habituated. But if something new happened, a new noise, or a machine moving in a different way, it would capture my attention. As it should.

It turns out that habituation is a good way to study what babies know. Imagine you were a baby and were shown a series of pictures:

A red circle
Another red circle
Another red circle . . .

You would get bored. You would look around, seeking novelty. But now you see:

A green circle

and you would perk up—Hey, new color! A psychologist can tell from your perking up (maybe reflected in you looking longer at the screen; maybe in your increased heartbeat) that you perceive the difference between red and green. If your color vision wasn't up to snuff, you wouldn't perk up; you would remain bored.

Or imagine you were seeing:

Two cats
Two cats
Two cats
Two cats . . .

Again, you would gradually become bored. Then you see

Three cats

If this captured your attention, this shows you could tell the difference between two cats and three cats.

Does this mean that you can tell twoness from threeness—that you know about numbers? Not so fast. After all, three cats typically take up more space than two cats, so maybe you are responding to the size difference. So, fine, in the next study, the psychologist can make the three cats smaller, so they take up the same amount as the two cats. But now babies might just be distinguishing between big cats and small cats. How do you design a study that tests for sensitivity to number while avoiding the possibility that other cues are driving the distinction? There are solutions (I think), but such experiments get exquisitely complicated. See the endnotes if you're interested in the details.[17]

There's the Law of Effect, there's habituation, and then there's surprise. Babies, like the rest of us, look longer at what they don't expect. Imagine you were staring at a man on the street, and he gradually placed a hat on top of his head. Ho-hum. But now imagine you were looking at a man and he held a hat in his hand, and the hat gradually floated upward and landed on his head. You'd stare. Your staring says something about how your mind works, which is that what you've seen violates your expectations. Babies also look longer at what they find strange, and psychologists have exploited this fact to make some discoveries about their understanding of objects.

In one classic study, babies faced a block on a table in front of them, with a screen lying flat in front of the block, and then the screen rotates up to obscure the block, and then it keeps rotating.[18] Now, suppose it's true that babies believe, as Piaget said, out of sight, out of mind. Then babies should expect that screen to continue moving all the way down because there's no block to get in its way. But when five-month-olds see this actually happen (because you removed the block through a trap door), they look longer. It is as if they are reasoning, "Hey, that's so weird—there should have been a block there to stop the movement of the screen."

Or how about this: You see an empty stage. A hand places a single Mickey Mouse doll on the stage. A screen is placed in front of the doll to hide it from view. Then the hand brings out another Mickey Mouse doll and places it out of sight behind the screen. Then the screen is removed. As an adult, you know that one Mickey plus another Mickey equals two Mickeys, so there should be two, not one or three. Five-month-olds know this too; they are surprised when the screen is removed and there is one Mickey or three Mickeys. And this suggests that they appreciate that once a Mickey goes behind the screen, it continues to exist.[19]

Based on these studies and many more, the developmental psychologist Elizabeth Spelke has proposed that there is an inborn system for reasoning about objects, one present in babies as young as researchers are able to test (and also present in species other than humans, such as newborn baby chicks).[20] She argues that babies have an initial understanding of the physical world that includes the following principles:

1. **Objects are cohesive:** they are connected masses; if you pull on one part of an object, the rest will come with it.
2. **Objects are solid:** they are not permeable; if you touch an object with your finger, your finger won't go through.
3. **Objects move on continuous paths:** they don't disappear from one location and reappear in another.
4. **Objects move on contact:** they won't move spontaneously.

(Unlike the other principles, this only applies to certain objects, not to animate creatures, like people and dogs, and also certain complex artifacts, such as robots and cars.)

The extensive body of research suggesting that these principles show up before children's first birthday is a refutation of Piaget's claims about the limitations of babies. But more than that, this research has led to one of the great discoveries of modern psychology—evidence for an innate understanding of the physical world.

———

We've focused so far on what babies know about objects, so now let's look at somewhat older children, and explore evidence for a different domain of understanding, having to do with how we make sense of certain categories in the world. Let's consider the claim that we are all natural-born essentialists.[21]

Essentialism is usually thought of as a metaphysical doctrine, a claim about how the world is. It comes in different strengths, but the main idea is that certain categories in the world don't reduce to their surface qualities, to what you can see or touch. Rather they possess deeper properties that make them what they are. As John Locke put it, "The real internal, but generally . . . unknown constitution of things, whereon their discoverable qualities depend, may be called their *essence*."

Gold, for instance, is typically a metal of a certain color, but we know that something might be gold and not look like it. Or it can look like gold but not be gold, as with pyrite—fool's gold. What makes a cat a cat, similarly, isn't just what it looks like; it has to do with deeper properties that the animal possesses. This is why adults, at least, appreciate that Maynard did not cease to be a cat once it wore a dog mask.

We often have some idea what the essences are—most scientifically literate people would assume it has to do with atomic structure for water and gold, and DNA for cats—but you can be an essentialist without this knowledge. Long before modern science, people appreciated that all that glitters is not gold.

People, then, know that there are two ways to apprehend the world. There is appearance, what things seem to be superficially. And then there is their real nature. These sometimes pull apart, and this is why science, which is in the business of discovering the essential nature of things, can tell us surprising things, such as that hummingbirds, ostriches, and falcons, however different they appear, are all birds—but bats, which look like birds, are not. We rely on these discoveries in everyday life. When you go to a doctor with a rash, you'll be unsatisfied if you're told that it looks like sunburn; you want to know that it *really* is—and this is what procedures such as blood tests and biopsies are for.

Essentialism isn't just a modern Western belief. It shows up to some extent or another in every culture that has been studied. Everywhere, people understand that something might look like an X but really be a Y; they know that one can always ask, "But what is it *really*?" People everywhere seem to believe in essences, though what these essences are depends on the belief system of specific cultures—when talking about species, for instance, Yoruba farmers don't talk about genes, but rather "structure from heaven."[22]

This universality raises the possibility that essentialism reflects a natural way of thinking about the world. This nativist view would have been anathema to Piaget, who argued that children start off with a superficial orientation toward the world, limited to what they can see, hear, and touch. There is evidence, though, that Piaget was wrong, that some essentialism shows up in the preschool years.[23]

In one study, three-year-olds are shown a picture of a robin and told that it has a hidden property, such as a certain chemical in its blood. Then they are shown two other pictures: one of an animal that looks similar but belongs to a different category, such as a bat; the other of an animal that looks different but belongs to the same category, like a flamingo. Which one has the same hidden property? Children tend to generalize based on category, choosing the flamingo. This doesn't show that they are fully essentialist, but it does show that they are sensitive to something deeper than appearance. Other studies using modified procedures have found the same effect with children before their second birthday.

Other experiments find that young children believe that if you remove the insides of a dog (its blood and bones), it isn't a dog anymore, but if you remove its outside features, it still is. These studies find as well that children are more likely to give a common name to things that share deep properties ("have the same sort of stuff inside") versus those that share superficial properties ("live in the same kind of zoo and the same kind of cage").

My colleague at Yale, Frank Keil, came up with some of the most striking demonstrations of child essentialism.[24] He showed children pictures of a series of transformations, such as a porcupine surgically transformed so as to look like a cactus, and a real dog made to look like a toy. His main finding is that children rejected such radical transformations as changing the category—regardless of what it looks like, it is still a porcupine or a dog. Only when the children were told that the transformations occurred on the inside—that the innards of these creatures were changed—could they be persuaded that these transformations led to a real change in category.

In these examples, children get things right; their intuitions match those of adults, including adult scientists. They appreciate the depth of categories. But essentialism can go overboard and make children, and adults, a bit stupid. One example of this is when we essentialize groups of people, and I'll get to that later in the book. Another example involves what we think about organ transplants. It turns out that young children, and some adults, hold the belief that transplants of organs involve a transfer of deeper properties of the animal. For instance, four-year-olds tend to think that the recipient of a pig heart, say, would become more like a pig.[25] Essentialism run amok!

———

We've talked about knowledge of the physical world and about essentialism. Now we come to a third research program. This is about what's sometimes known as *naive psychology* or *theory of mind*. It's about our understanding of other people.

There is evidence that we are social creatures from the get-go. Just a few minutes after birth, babies will move their eyes to track face-like shapes (with two eyes above a nose above a mouth) more so than non-face-like shapes where the features are scrambled.[26] They prefer to look at faces that are staring straight at them rather than those that are averted to the side, suggesting an interest in social communication.[27] Later, they expect faces to behave differently from other objects. Babies lose interest if a moving object becomes still. But if they interact with a person, and then the person's face becomes still and stays that way, they often get upset.[28]

Babies expect people to have goals. Imagine a table with two different objects on it, and a hand reaching for one of them. Where would you expect the hand to later reach if the objects were to switch locations? Well, for adults, we know that hands are attached to people, and people have goals, and a reasonable goal for a person is to reach for a particular object, not to go to a specific location. (If I was reaching for the corn chips, and you switched the positions of the corn chip bowl and a vase of flowers, my next reach will probably go to the bowl, not the flowers.) Six-month-olds have the same expectation.[29]

Babies also have expectations about how individuals will respond to those who support or thwart their goals. In some research I was involved in many years ago, Valerie Kuhlmeier, Karen Wynn, and I did a series of studies in which babies would watch as a ball goes up a hill and one character helps it up and the other pushes it down. Then we showed the babies two other scenes, one in which the ball approaches the character who helped it and the other in which the ball approaches the character who hindered it. As predicted, babies' looking-time patterns suggest that, like adults, they assume that someone will approach an individual who had previously helped it achieve a goal, not one who has hindered the goal.[30] Later research, led by Kiley Hamlin, found that somewhat older children preferred characters who helped others over those who hindered others.[31]

Other aspects of babies' social understanding are there for anyone to see—you don't need a baby lab. Early in life, usually before their first birthday, babies act to call the attention of adults to something in the world. They gesture, wave, grunt, and eventually are able to point to things. Sometimes they do this to get certain desired results; a baby in a high chair might stare at you and grunt while reaching for a sippy cup, with the message being: "Dude, hand over the cup!" Sometimes the goal is simply shared attention; a baby who might point to a drawing, trying to get you to look at it: "Hey, check that out!"

Such examples reveal something deep about the babies' minds. It suggests that they understand others are thinking beings whose attention can be drawn to things in the world. This appreciation seems to be uniquely human. Mature chimpanzees are much more capable than human babies in so many regards, but in the wild, they do not show, offer, or point to objects.[32]

Once they start to speak, children's developing appreciation of their own and other people's minds becomes obvious.[33] Here is two-year-old Eve, showing some understanding of what she likes:

Adult: Would you like to have a cookie?
Eve: I want some cookie. Cookies, that make me happy.

Here is three-year-old Abe, seeming to appreciate that other people can have views about the world that conflict with his own:

Abe: Some people don't like hawks. They think they have . . . they are slimy.
Mother: What do you think?
Abe: I think they are good animals.

Now three-year-old Adam, explicitly recognizing that his past self can have a view that he no longer holds. He had been eating glue:

Adam: I don't like it.
Adult: Why would you put that in your mouth?
Adam: I thought it was good.

Finally, Abe cracks me up with this one, which I think was a joke.

Abe: I painted on them [his hands].
Adult: Why did you?
Abe: Because I thought my hands are paper.

The attentive reader will notice a tension here. I began with Piaget's findings about the severe limitations of older children, including (as in the case of egocentrism) limitations in understanding other minds. And I said that these findings are real and replicable. But now I'm telling you about how socially savvy babies are, and giving examples of two- and three-year-olds reflecting on their own minds and how their thoughts relate to those of others and of their past selves.

So which is it—are children dumb about how minds work or are they smart?

It's complicated, and this question gets us to one of the most interesting debates in developmental psychology. Check out this story:

Sally and Anne are together in a room that contains a basket and a box. Sally places her marble inside a basket, and leaves the room. While she is gone, naughty Anne moves her marble to the box. Sally returns. Where will Sally look for her marble?[34]

This probably isn't tough for you. The marble is really in the box, but Sally doesn't know that; she thinks it's in the basket, and so this is where she will look. She has a false belief, and this will dictate her behavior.

This "false-belief task" has a neat history. It was first thought up by the philosopher Daniel Dennett during a discussion over whether chimpanzees—not children—can reason about someone else's mental states.[35] Dennett pointed out that if someone has a correct belief about the world, you can anticipate their behavior without getting into their heads. Suppose the marble was never moved. If children, or chimps, predicted that Sally would later go to the basket to get it, would that show that they are reasoning about Sally's mind? Not really—they might just make their prediction based on facts about the world (the marble is in the basket). But when you are asked to predict Sally's behavior based on her *false* beliefs, thinking about the actual location of the marble isn't enough. The only way to give the right answer in a false-belief task is to reason about Sally's mental state.

And this is apparently difficult. When the story is acted out in front of them, five-year-olds (and four-year-olds when it's made especially simple) typically pass this test, but younger children typically do not. They answer that Sally will look in the box, where the marble actually is. (Chimps typically fail as well, by the way, even when the task is simplified so that they should understand it.[36])

We see the same problems with false beliefs outside of the lab. Young children are terrible liars, the sort to insist that they didn't eat the cookies when they have chocolate smeared over their faces. They are also terrible at hide-and-go-seek, which again involves messing with the minds of others, hiding the truth from them. Young children just don't seem to get it; often they hide in the same place where someone just caught them, or hide with their legs or butts sticking out, or they tell the adult to hide in a particular place. One gets the impression that they enjoy the game not due to the deception involved, but just because of the theater, as when parents act confused about their location ("Where did Timmie go? I simply can't find him!") and then shocked at the discovery ("You were under the blanket the whole time? I am stunned!").

Some psychologists defend a Piagetian theory of these limitations. Children just don't understand minds. They lack a full appreciation of mental states and are confused about the uneven relationship between beliefs about the world and the world itself. Perhaps this is because they lack the relevant experience in reasoning about other people, or maybe it's because the brain regions that are involved in reasoning about other people's beliefs just take time to develop.

Other psychologists, though, propose that children do understand other minds. After all, in some real-world contexts, they *do* seem sophisticated about beliefs, including false beliefs. (Recall two-year-old Abe: "I thought my hands are paper.") Furthermore, there are other studies finding that babies around their first birthday—far younger than the age at which they succeed at the false-belief task—show signs that they comprehend false beliefs: Their looking-time patterns suggests that they are surprised when someone who should believe the marble is in one location (a false belief) reaches for where the marble actually is.[37]

How do proponents of this children-are-smart perspective explain failure at the false-belief task? Many of them—including myself, in a paper written long ago—have pointed out that this is a difficult task for reasons that have little to do with false belief.[38] Success requires simultaneously holding in your mind two conflicting pictures of the world—the world as it really is (the marble is in the box) and the world as it is in someone else's mind (the marble is in the basket). This is hard, even for adults, because it involves a form of double bookkeeping.[39] With limited cognitive powers, children find this difficult.

Also, reasoning about false belief involves overriding what's sometimes called *the curse of knowledge*: a pervasive assumption that others have the same knowledge that you do. Several studies with adults find that once someone knows something, such as the answer to a question or whether someone is lying, they will tend to assume that others have the same knowledge.[40] (This is one reason why teaching is so difficult.) And so the question of where Sally will look for the marble poses a challenge in part because the child has to override the default assumption that her own knowledge of the marble's location is shared with Sally.

In a sense, though, these complaints about the difficulty of the false-belief task just push the question back. Why would children have more problems with challenging tasks than adults? The answer here might bring us back to the brain. The brain of a young child is not as developed as that of an older child or adult. Recall that neurons have a fatty sheath around them—myelin—that allows for more efficient transfer of information. It turns out that the process of developing the sheath— myelinization—is a slow one. Hence the brain does not work as efficiently in the young; in particular, the frontal lobes take considerable time to develop and are not fully mature even in adolescence. Every child, then, is something of a Phineas Gage—not dumb, but lacking impulse control. This shows up in tantrums or giggles that just won't go away—but also causes problems with psychological tasks that involve overriding a tempting response, like the false-belief task, where the impulse is to say where the marble really is.

I've laid out here two different views. One is that children don't understand false beliefs; the other is that they do, but suffer from other problems. But it seems that neither view can be exactly right. If children don't understand false beliefs, how do you explain the successes of babies, as well as the occasional glimmerings of knowledge, as shown by children like Abe? If children do understand false beliefs, why do they have such problems with them in the lab and in the real world? Some investigators are tempted by a theory in which there is both a simple innate system that can do some reasoning about false beliefs (and explains the baby findings) as well as a more elaborate, flexible, and conscious system that takes years to develop (explaining the children's failures).[41] But all of this is a matter of ongoing debate.

———

Let's close by returning to a foundational question of developmental psychology: Do children and adults think in different ways?

Some say no. Empiricists believe that children and adults (of every species) learn through the same general principles. And many nativists would also say no—for them, the mind of a four-year-old contains innate knowledge, specialized modules, and dedicated learning mechanisms, and these are no different in kind from what you find in the mind of a forty-year-old.

Others say yes. As we've seen, Piaget was a great proponent of the idea of radical cognitive change over the course of development. And many contemporary psychologists, though they are less prone to talk in terms of stages, agree with Le Patron here.

In my view, the most promising version of the radical change proposal was developed by Susan Carey, my own graduate adviser many years ago. Carey argues that conceptual change in children should be thought of as akin to changes in thought processes by adult scientists—the child-as-scientist theory.[42] She was influenced in her work by Thomas Kuhn's classic book, *The Structure of Scientific Revolutions*, which describes the difference between successive scientific worldviews, such as Newtonian physics and Einsteinian physics, in terms of a paradigm shift.[43] For Kuhn, it's not just that the physicists in the later tradition know more than their predecessors; rather they see the world in profoundly different ways.

Carey proposes that we start life with intuitive theories of the world, and that these theories develop over childhood, sometimes transforming in much the same way as more formal scientific theories do. There are now many elegant analyses in which developmental psychologists explore "theory change" in different domains of understanding, such as how children think about other minds, about biology, about death, and about God.

We've seen a different sort of evidence for the child-as-scientist view earlier when we talked about naive essentialism. Several studies suggest that, just like scientists, children believe that there is a deeper nature of things.

The child-as-scientist view has also led to a shift in how we think about learning, moving toward an insight that many parents would already have told you, which is that children are not merely soaking up information, but rather they are doing experiments and testing hypotheses. Indeed, one view of the Terrible Twos is that children are so awful to their parents because they are exploring how adult minds work, seeing what surprises, amuses, and annoys them.[44]

The early emergence of this scientific mode of thought is nicely illustrated by a clever experiment with eleven-month-olds.[45] The researchers showed the babies various events that violate physical laws, such as an object moving through a wall or floating in space. We've already seen that infants of this age will look longer at such violations. But this new study discovered more: The researchers found that babies would play with these strange objects more than ones that behaved in normal ways. And they would play with the objects in such a way that corresponded to their weirdness. For instance, when given the object that had apparently gone through a solid wall, they would tend to bang it into other objects such as tables; when given an object that had floated in empty space, the babies would tend to drop it. It is as if they are asking: Did that really happen? Can I make it happen again? This focused curiosity and experimentation make the scientist analogy an apt one.

One of the charms of the child-as-scientist view is that it brings us full circle, taking us back to Piaget's project of learning about the origins of knowledge by studying babies and children. If developmental change in children is akin to theory change in adult scientists, then it turns out that there really is a deep connection between developmental psychology and the intellectual growth of our species. Piaget was right—so much of human nature can be revealed through the study of babies and young children. This is not small potatoes.

6

The Ape That Speaks

Language is where the action is. Any successful theory of the mind must meet the challenge of explaining how we learn and use language. This was true for the theories of philosophers such as Locke, Hume, and Leibniz, as well as for the psychological theories developed in the last century. Indeed, as we've seen, the whole behaviorist program largely collapsed due to Skinner's failure to make good on his promise to explain language—or, as he put it, verbal behavior.

And it's true right now. As I write this, there are psychologists who champion different theories of the mind, such as Bayesian learning and predictive processing. The extent to which we'll be taking such views seriously a decade from now will depend, in large part, on how well these approaches deal with the many mysteries of language.

Before we delve into what these mysteries are, we need to be clear about what we mean by "language." The word gets used in all sorts of ways. People talk about the language of birds and chimpanzees, about body language, and about computer languages such as R and Python. Traffic signs can be described as a language, and so can fashion and music and DNA. The linguist David Crystal points out that there are books called *The Grammar of Cooking* and *The Syntax of Sex*.[1] (He adds: "The first was a collection of recipes—as was the second.")

It's fine if you want to talk about language in this general sense. I'm not the cops; you do you. But in what follows, we're going to focus on something more restrictive, what's often called "natural language." This refers to communication systems such as English and Dutch and Hindi and Turkish and Korean, languages that are picked up by young children and are used by almost everyone as their primary mode of communication. If you are listening to this book, we are communicating with just such a language; if you are reading it, we are using a different system—the written word—but one that is based on a natural language.

When I teach the language part of my Introduction to Psychology course in a large lecture hall, I put this English sentence up on the screen:

The girl thinks that the house is big.

And then I walk through the auditorium of five hundred or so students and ask for volunteers to say this sentence in their own native language. It's a diverse group. The first student might say it in Spanish, and then I'll get Russian, Japanese, German, Tamil, and Swedish. I continue, running up and down the aisles of the auditorium with my microphone like a talk-show host in the 1980s until I go through speakers of about twenty-five different languages. I'll make a point of asking for someone who knows a sign language (there are many different sign languages, but in this crowd, it's usually ASL, American Sign Language), and this person will stand up and sign it to the group. Once we're done, the class applauds the volunteers; there's something moving about the experience.

This simple demonstration illustrates the diversity of language—almost all these languages are incomprehensible to me and to most other students in the class. It's not just that I don't know the words; I don't even hear words, just different types of gabbling. And the gabbling is different; I don't know Vietnamese and I don't know German, but I can tell them apart.

This demonstration also reveals what languages share. There are about six thousand spoken languages in the world and about three hundred sign languages, and nobody says, "Well, I can't say that in my language." The sentence I chose for my demonstration conveys a complex idea (what someone thinks about the property of an object), and yet every language can express it.

Languages do this in different ways. The eight-word English sentence I put up on the screen isn't eight words in many languages; some languages put a lot more information into single words than others. Languages also order words in different ways; the English sentence begins with the person who does the thinking, but many languages do not. And if I used a different example, I might run into the fact that there are notions in some languages that don't have natural expressions in others. Yiddish, say, has the word "nachas" (pride in the accomplishments of someone close to you, typically your children), and this has no one-word translation in English. Speakers of ancient Etruscan had no words to express the notions carried by the English "escrow," and "retweet," because there were no mortgages and no Twitter back then, and so nobody coined the words to talk about these concepts. When modern Hebrew was created, it was based on Biblical Hebrew, and to express certain modern ideas it had to borrow from modern languages. Moses didn't use public transit; the word for "bus" in modern Hebrew is borrowed from the French: אוטובוס, pronounced as "autobus."

Still, all languages can capture the same thoughts. This is what makes translation of even the most abstract ideas possible. Think of the simultaneous interpretation done at the United Nations—though some nuance is perhaps lost, we should marvel at how successful this is.

All natural languages have this power—and only natural languages. It would be silly to expect a UN translator to translate a speech by the president of France into a piano concerto, the computer language Python, or birdsong. These other systems are not up to the job.

––––––

There are societies that don't have complex mathematics, where people don't wear much clothing, where all sorts of technology is absent. But no human group lacks language.

Now, it would be facile to conclude from this universality that language is part of human nature.[2] Right now, it's going to be very hard to find a society that lacks soccer or Coca-Cola—these too are universal or near universal. But that's not because the ideas of soccer and Coca-Cola are built into the brain. It is because they were invented at one point and were exceptionally appealing, and so spread through the world. Maybe language is like this. Alternatively, it could be that language is such a good idea that every culture independently develops it. After all, just about every culture uses some sort of instrument to eat food with, such as a fork, chopsticks, or a leaf. This is not because use of eating utensils is innate; rather it is such a useful trick that cultures discover it over and over again. Maybe language is like that.

There are reasons, though, to believe that the universality of language really is because language is part of our nature. This was Charles Darwin's view: "Man has an instinctive tendency to speak, as we see in the babble of our young children; while no child has an instinctive tendency to brew, bake, or write."[3] After all, just about everybody learns language as a child, with the exceptions due to tragic cases of brain damage or genetic abnormality.

Also, as we've seen earlier in this book, parts of the brain are dedicated to language, such as Broca's area and Wernicke's area. There are also specific genes implicated in cases of language disorder in children.[4] All of this supports the notion that language is part of our nature, in much the same way that the inborn communication systems of other creatures such as birds and bees are part of theirs.[5]

We also see the marks of special evolution for language below the brain. Our larynx has been modified over the course of evolution to express the sounds of speech. In most mammals, the larynx is high in the throat, which means that they can eat and breathe at the same time. Ever wonder how a baby can breastfeed for a long stretch without ever coming up for air? Me neither, but it's a good question and the answer is that human babies begin, like the adults of other primate species, with the larynx high up. It lowers as we age. This leads to an increased risk of choking on food, but it allows the developing child to produce the music of language. This suggests that the benefits of vocal communication outweigh the cost of possible death by choking, which is a good argument that the power of speech evolved by natural selection, through the adaptive advantages that it gave us.[6]

Language is not a monolith; it has different parts, including phonology, morphology, and syntax.[7]

Phonology is the aspect of language that directly connects to its physical realization. We can produce many sounds, but only a small subset is used in any given language. We call these phonemes. For most languages, a phoneme can be defined as a sound that distinguishes one word from another. English, for instance, has the phoneme /p/ and the phoneme /b/ and you know this because "pat" and "bat" are different words. Signed languages have phonemes as well—hand configurations and other signals that serve the same function.

Languages have different numbers of phonemes, from fewer than a dozen, as in Rotokas, spoken in Bougainville, an island in Papua New Guinea, to over a hundred, as in Taa, spoken mostly in Botswana. English has around forty-four (the number will vary depending on where it's spoken).[8] There are only twenty-six letters in the English alphabet, and so phonemes and letters can't match up one to one, which is one of many reasons why it's such a pain to learn to read and write.

Sounds make words, and the aspect of language that concerns words is known as *morphology*.

One of the core features of language shows up here—what the linguist Ferdinand de Saussure called "the arbitrariness of the sign." The idea here is that the form of the symbol typically has nothing to do with its meaning. There is a perfectly arbitrary relationship between the sound "dog" and those creatures with four legs and a tail that bark—the word "dog" doesn't look or sound like a dog. So, with just a few weird exceptions like "groan," which does sound a bit like a groan, there's no way you can tell the meaning of a word from its physical form. This is true as well for sign languages. While some signs can be interpreted by their physical structure (the sign for hammering might be a hammering motion), the vast majority are opaque unless you know the language. If you doubt this, watch a speech with simultaneous sign-language interpretation with the sound off and see how far you get.

This principle of arbitrariness shows up in other domains. There is no logical reason why a green light means go, or why moving the corners of your mouth upward can express happiness. These connections just arose over history (cultural for lights, evolutionary for faces), and now it is what it is.

The question everyone asks about words is: How many do people know? History is full of smart people who thought the answer was small. One intellectual in the nineteenth century said that peasants know fewer than one hundred words; they manage to communicate because "the same word was made to serve a multitude of purposes, and the same coarse expletives recurred with a horrible frequency in the place of every single part of speech." The linguist Max Müller guessed that highly educated people know a few thousand words, and other adults know about three hundred. The writer George Simenon said that his books were written simply because the average Frenchman knows fewer than six hundred words.[9]

These are ridiculously lowball estimates; they show that our intuitions are not to be trusted. A better way to answer the question is to test people, and one way to do so is to take a random sample of English words from as complete a dictionary as you can find, ask people about what they know about the words' meanings, and extrapolate. Suppose you have a sample that is one one-thousandth of the dictionary; then you can take the average number of words people know from this sample, and multiply by a thousand. Admittedly, this can give you at best an approximate answer; a lot depends on what dictionary you use and what counts as "knowing" the meaning of a word—is it sufficient to recognize it or have a vague idea of what it's about, or must the person be able to provide a good definition? But regardless of the choices you make here, the number ends up reasonably large. I've seen a range of estimates—for adult English speakers, sixty thousand words is a good ballpark figure.[10] And this is for monolingual speakers. If you know two or three languages fluently, multiply by two or three.

I'm talking so far about words, but this is imprecise. Linguistics call it morphology, not wordology, and everything I've said above about the arbitrariness of sign and the number of words that people know is really about morphemes. Here's the difference: A morpheme is the most basic thing you must learn. "Dog" and "dogs" are both words, but "dog" is a single morpheme, while "dogs" is composed of two morphemes—"dog" and the plural marker "s." Morphemes are generally defined as the smallest units that have a meaning. Some are whole words on their own, and some are prefixes or suffixes that must be combined with other morphemes to make up more complex words.

When you're learning a language, you first have to learn morphemes; many other words then come for free. The instant you hear "dog," you can understand "dogs"; once you hear "tweet," you can use the rules of word creation (morphology) to use and understand "tweets," "tweeted," and "tweeting." So I can now be more exact about the vocabulary estimate above—it's about sixty thousand *morphemes*, not words.

Morphological rules in English are relatively simple, such as "add -s to make a noun plural" or "add -ed to make a verb past tense." In his *Language Instinct*, Steven Pinker notes: "The creative powers of English morphology are pathetic compared to what we find in other languages. The English noun comes in exactly two forms (*duck* and *ducks*), the verb in four (*quack*, *quacks*, *quacked*, and *quacking*). In modern Italian and Spanish every verb has about fifty forms; in classic Greek, three hundred and five; in Turkish, two million!"[11]

Even for a morphologically pathetic language like English, though, we can generate and understand new words, often in creative ways. I recently read the word

doom-scrolling

and was able to use the two morphemes that compose it to infer what it means. (Roughly, scrolling through social media or news sites soaking in all the bad news.) Knowing English morphology, I can now make new words out of it—I *doom-scrolled* for much of yesterday; She really loves to *doom-scroll*.

In the time of COVID-19, when much of this book was written, many new words were coined or resuscitated from the past, like:[12]

Coronospeck (German): Fat gained by staying at home during quarantine.

Hamsteren (Dutch): Hoarding, inspired by how hamsters use their cheeks to store food.

On-nomi (Japanese): Online socializing while drinking alcohol.[13]

———

Words on their own can only get you so far. Language shows off much of its most generative power in syntax, the system of rules that governs how words and phrases are combined to form new phrases and sentences. This is what makes it possible for us to create and communicate and understand a virtually infinite number of thoughts. Most of the sentences in this book, including the one you're now reading, are unique—if you type them into a Google search with quotation marks, you'd get nothing. And yet you understand the ideas that they convey.

We succeed at this through abstract and unconscious rules. When linguists talk about rules of language, they are not typically thinking of so-called prescriptive rules, such as "Don't say ain't" or "Don't end a sentence with a preposition." Rather they are interested in the rules we naturally use when we communicate. Consider this pair of simple sentences:

The pig is eager to eat.
The pig is easy to eat.

When you read the first sentence, you knew, in a fraction of a second, that the eating here is to be done by the pig. And when you read the second, you knew that, although the sentences look a lot alike (they differ by just a word), the meaning suddenly shifts; the eating is *of* the pig, not by the pig. Or consider these:

I heard that it was raining outside.
I heard that was raining outside.

Why does the second sentence sound worse than the first; why is the "it" needed here? The word doesn't seem to refer to anything; rather it reflects a rule in English (but not in languages such as Italian) that verbs need overt subjects, and this rule applies even for a verb like "raining," which doesn't in any real sense have an agent associated with it; there is nothing that rains. But a rule is a rule, so we stick in the pronoun "it."

Such rules of language concern parts of speech, like nouns and verbs; words are involved just insofar as they belong to the linguistic categories. This means that the same linguistic rule that can produce a sensible sentence like

Large angry dogs bark loudly.

can produce a silly one if you put different words into the categories of adjective, adjective, noun, verb, adverb, as in Noam Chomsky's famous example:

Colorless green ideas sleep furiously.[14]

Although nonsensical, we get that this is a proper English sentence.

The interaction of different rules gives rise to the extraordinary generative powers of language. To illustrate this power, imagine a simple language. It has three nouns, two verbs, and one rule.

Nouns: Canadians, Italians, Germans
Verbs: like, fear
Rule: A sentence is a noun followed by a verb followed by a noun

You can use the rule to produce sentences, such as:

Canadians like Italians.

How many sentences can this rule produce? You have three nouns and two verbs. So, if a sentence is noun-then-verb-then-noun, you get $3 \times 2 \times 3 = 18$ sentences.

You had to learn six things, five words and one rule, but got three times as many sentences from it. Consider what you could do with 100 nouns and 50 verbs and the same rule. That's 151 memorized items, and it gives you half a million sentences ($100 \times 50 \times 100$), each expressing a different thought.

Now, imagine another language.

Nouns: Canadians, Italians, Germans
Verb: know
Rule 1: A sentence is a noun followed by a verb
Rule 2: A sentence is a noun followed by a verb followed by a
   sentence

So, for instance, you could evoke Rule 1:

Canadians know Germans.

Or Rule 2 followed by Rule 1:

Canadians know Germans know Italians.

Or Rule 2 followed by Rule 2 followed by Rule 1:

Canadians know Germans know Italians know Canadians.

And so on; because there is no limit to how often you can use this rule repeatedly, there's an infinite number of sentences. (In this example, extremely boring sentences.) This language has the property of *recursion*;

it has one rule feeding into itself, and you can repeat this forever. Recursion allows us to make sentences as complex as needed in order to express our thoughts.

Language is infinite in principle, in that there is no longest sentence, but it's not infinite in practice; time and memory and attention force us back into the finite. But these rules do give us extraordinary expressive powers. Pinker calculates how many possible twenty-word sentences there are, and he comes up with an estimate of 100 million trillion.[15] (And twenty words isn't that long for a sentence; the previous one, "Pinker calculates. . . ," is twenty words long.) At five seconds per sentence, you can say them all in about 100 trillion years.

We can see this creativity expressed in shorter packages. Returning to the COVID theme, the *New York Times* asked readers to send in six-word memoirs of their time during this pandemic. Just six words, yet each tells a story.[16]

"Tired of hearing, 'Mark, you're muted.'"—Mark, Milwaukee
"New baby. New mother. Absent grandmother."—Lydia, North Carolina
"My dad's last breath, on FaceTime."—Bob K., Atlanta
"Graduated college in my living room."—Emma Garcia

Let's stick with the generative powers of morphology and syntax a bit, because these bear on a long-standing debate about how the mind works.

There is little doubt that brains are, at least in part, association machines. This insight was central to the ideas of the British Empiricists, and later, to the behaviorists. So much of learning can be seen as noticing and recording *associations*—the relationships between different parts of experience. You hear a commercial jingle and think of the product; Pavlov's dog salivates at the sound of the bell; Skinner's rat darts toward the side of the maze associated with a delicious treat. Certain modern computational approaches, sometimes described as "connectionism"[17] or "deep learning,"[18] build on this insight, seeing the brain as a powerful statistical learning machine, adept at discovering patterns in the environment. Some would go so far as to say that, in essence, this is all the brain does.

Other psychologists draw upon another philosophical tradition that has its origins in Aristotle and extends through Alan Turing, which sees thought as the manipulation of symbols.[19] This is essential to the operation of logical thought. If you believe that "all Xs are Ys," then you can infer, given an X, that it will be a Y. (All men are mortal, Socrates is a man, so . . .) Rules show up in games: Three strikes and you're out; a pawn, in its first move of a chess game, can go forward either one or two spaces; three of a kind beat two pair, and so on. Rules show up in law: there are strict criteria as to what counts as a married couple, a motorized vehicle, next of kin, and so on.

Often rules and associations sit side by side in our brains.[20] We all have a sense of a typical grandparent, house, and motor vehicle, from having seen a million of them and noticing what they tend to have in common, but we know it's the rules that really matter for what falls into these categories. When the forty-four-year-old Pierce Brosnan portrayed James Bond in *The World Is Not Enough*, he sure didn't look like a typical grandfather, but he was the parent of a parent and so he was. A kind old woman with a twinkle in her eye who bakes cookies may seem like a typical grandmother, but if she doesn't satisfy the rule, she isn't one, not even a bit.

Now, a lot of language, as we've seen, doesn't involve rules. There is memory—you need to remember that the English word for dog is "dog" or that the past tense form of "see" is "saw." And there is association. I say "doctor," you say "nurse"; I say "kick the," you say "bucket." Indeed, we'll see that one of the most important aspects of language learning, the parsing of sounds into distinct words, likely requires recalling the associations between the different sounds that make up words.

But rules are what give language its power to communicate. In English, there is a rule stating that to make the past tense of a verb you add "-ed" to the ending, as when "walk" corresponds to "walked," and another rule that adjectives precede nouns in simple phrases—you say "a round table" not "a table round." (French has a different rule: "une table ronde," not "une ronde table.") We can apply these rules generatively: If I own a flibget and it's very smorky, then I have a smorky flibget; if I like to gorp and do some gorping every day, then yesterday I gorped.

How important are these rules for understanding language? Sometimes you can get far without them. If you hear

Dogs bite men.

you probably guess the meaning just based on the individual words and how they normally go together. (Someone given the words in alphabetical order—BITE, DOGS, MEN—could figure it out.)

But this only goes so far. English speakers can also understand the following (improbable) sentence, and to do so we need to pay attention to the order of words and how they fit with the rules of English grammar.

Men bite dogs.

And it's not just word order, because we can also understand the sentence below, which looks a lot like "Dogs bite men," but is actually the passive form of "Men bite dogs":

Dogs are bitten by men.

As I write this, some AI systems don't use rules to understand sentences and so it's easy to make them stumble.

There's a popular AI system called Delphi that does a good job at answering moral questions. Give it "Cheating on an exam," and it outputs "It's wrong"; give it "Cheating on an exam to save someone's life," and it outputs "It's okay."

But Delphi doesn't fully use syntax to understand the questions. The cognitive scientist Tomer Ullman pointed out on Twitter that you can trip it up by using words that are associated with cruelty or kindness even though they don't influence the meaning of the dilemma in any relevant way. For instance, give Delphi the sentence

Gently and sweetly pressing a pillow over the face of a sleeping baby and it responds with "It's okay"! Getting the syntax right here is plainly a matter of life and death.

———

What about the communication systems of other creatures? Can we call these languages? Sure. Are they languages in the same sense as English, Thai, Russian, ASL, and so on? No. Other creatures have rich communication systems, deserving of much scientific attention, but they are different from human language. Some animals have a finite list of calls, like vervet monkeys who have one call for a snake and another for an eagle. There are all manners of threat displays and mating calls. But what you don't find is phonology, morphology, syntax, arbitrary names, recursive syntax, and so on.

There have been many attempts to teach language to nonhumans, particularly to nonhuman primates like chimpanzees, and there was a lot of enthusiasm about this many years ago. The consensus by now is that these attempts have failed. These animals can develop a limited vocabulary, but they learn words slowly and with difficulty. There is little evidence for anything resembling the syntax of a language like English.[21]

Oddly enough, the nearest thing we have to a success story comes not from our closest evolutionary relatives, but from *dogs*. In one well-known case, a border collie named Rico acquired a vocabulary of more than two hundred words.[22] If you put ten items in a room and asked Rico to fetch one, he would run in and usually bring back the right one. And he could learn words in an interesting way. If you put a new object together with seven familiar objects and asked him to fetch using a name he has never heard before, he would usually retrieve the new object, appreciating, as young children do, that new words tend to refer to new objects. (If I put a cookie and a sock and a weird tool in front of a three-year-old and say, "Give me the flerp," the child will likely hand over the tool, inferring that if I wanted the cookie or the sock, I would have asked for it using the commonly used words.[23])

Rico's abilities suggest that some of the mental apparatus involved in learning language is not unique to humans. But it's informative to look at how different this collie was from a child. A two-year-old has words not just for objects, but also people, properties, actions, and relationships; Rico only responded to words for specific objects. Children can learn words in all sorts of ways, including from overheard speech; Rico only learned through a specific procedure.

Perhaps most of all, when a two-year-old learns a word like "sock," the child can use it to point out a sock ("there sock"), request a sock ("want sock"), comment on the absence of a sock ("no sock!"), and so on. Children know, implicitly at least, that "sock" isn't a command or a response to some stimuli; it's a symbol that refers to something.[24] As best we know, Rico had no grasp of this abstract relationship of "reference" or "about-ness"; what Rico did was associate the word "sock" with the action of fetching a sock—"sock" means get a sock. As the linguist and developmental psychologist Lila Gleitman once remarked in reference to ape language learning, if any child learned words that way, the parents would run screaming to a neurologist.

———

Language learning starts in the womb. Newborns suck a pacifier more to hearing their mother's voice than the voice of a stranger because it was their mother's voice they were most accustomed to hearing before birth.[25] They also prefer the language they heard before they were born. Newborns born into French-speaking families prefer to listen to French rather than to listen to Russian,[26] and babies born into Spanish-speaking families prefer to listen to Spanish rather than to listen to English.[27]

Babies start off with the capacity to distinguish the sounds of all languages. A child born in America to English-speaking parents can tell apart phonemes that are used in Hindi, a capacity that English-speaking adults lack.[28] Similarly, a Japanese baby can hear the difference between "la" and "ra" (distinct phonemes in English, which is why "lamp" and "ramp" are different words)—but a Japanese adult cannot. This is one of the few ways in which babies have greater mental powers than adults.

Babies make mouth sounds like cooing from the get-go, but the sounds of language come in seriously after about a half year, and we call this *babbling*. One of the most fascinating findings in language acquisition is that the same phonological milestone happens in deaf children learning sign language—they babble with their hands, producing the sign language equivalent of ba-ba-ba-pa-pa-pa.[29]

A few months after the onset of babbling, children begin to learn the meanings of words. Place a baby so that it is facing two computer screens, one on the right and the other on the left. Say a word and display images on the screens, only one of which matches the word. For instance, ask "Where's the dog?" and have a picture of a shoe on one screen and a picture of a dog on the other. By six months, babies will tend to look more at the dog, suggesting they have some grasp of the meaning of this word.[30]

At around the child's first birthday—and there's quite a lot of variability here, parents shouldn't panic if it takes longer—children start to produce their first words.[31] There are many records of these first words; developmental psychologists love to record the early speech of their children—I did this for both my sons. There's been enough research to discover that first words look similar around the world. Here are the ten most frequent first words, from three different languages (translated into English):[32]

> **American English:** Mommy, daddy, ball, bye, hi, no, dog, baby, woof woof, banana
> **Hebrew:** Mommy, yum yum, grandma, vroom, grandpa, daddy, banana, this, bye
> **Kiswahili (also known as Swahili):** Mommy, daddy, car, cat, meow, motorcycle, baby, bug, banana, baa baa

You can see here the marks of culture; the sorts of words children learn reflect the environments in which they were raised. Apparently, in Israel, grandparents are heavily involved in the raising of children. And there must be a lot of motorcycles in the parts of Africa where Kiswahili is spoken. But there are also universals. Words that emerge early include names for people, animals, toys, foods, and social words like hi, bye, this, and no. And Mommy is usually first.

By the time children hit their second birthday, they know, on average, about three hundred words. (Many more than that of the smart dog Rico.)

We've talked about babbling, then first words, and now we arrive at the learning of syntax. In English, if you wanted to say that Bill hit John, you would usually start with the person who does the action, then the action, then the person who the action is done to—"Bill hit John." But other languages order the words differently to express the same idea. They might say that Bill hit John, for instance, by saying the word for "Bill," then the word for "John," then the verb "hit."

Children learning English know some of the rules of syntax soon after their first birthday. This has been explored through the same "preferential looking" methods that we just described for word learning. In one study, thirteen-to-fifteen-month-olds were shown two videos, one of a woman kissing a set of keys while holding a ball and the other of the same woman kissing a ball while holding a set of keys.[33] Then they would hear a sentence like "She's kissing the keys!" You can see the cleverness of this study; to figure out which sentence this corresponds to, it's not enough to know what the words mean—both videos contain the woman, keys, and kissing. You need to have some appreciation of the syntax. The infants looked longer at the matching event, where she was kissing the keys, as opposed to the one where she was kissing the ball and there were keys, but where the scene and the sentence didn't match.

Older infants can use syntax even when they don't know what all the words mean.[34] When they hear "Look! The bunny is gorping the duck," with the made-up verb "gorping," they know, like adults, that the rules of English syntax mean that this refers to a video where a bunny is doing something to a duck and not one where a duck is doing something to a bunny.

At about the age of a year and a half, the child starts to increase the rate of word learning and produces two-word sentences. Here are some examples from English:[35]

All dry	All messy	All wet
I sit	I shut	No bed
No pee	See baby	See pretty
More cereal	More hot	Hi Calico
Mail come	Airplane all gone	Bye-bye car

Then smaller words, like "in," "of," and "a," and morphological units like the plural "-s," and the past-tense marker "-ed" gradually begin to appear. "All dry" becomes "It's all dry"; "No bad" goes to "It's not bad," and so on.

At around the same time, children develop a facility with morphological rules, allowing them to produce and understand words they have never heard before. Many years ago, the linguist Jean Berko Gleason created the "Wug test."[36] This involves showing someone a picture of a strange creature, giving it a name—"This is a wug"—and then saying:

Now there is another one.
There are two of them.
There are two _____.

Young children say "wugs," creating a new word by applying rules of morphology.

This use of rules sometimes leads to children making mistakes, so-called overgeneralizations, as when they create words such as "mans," "goed," and "finded," not knowing or not remembering that these words have irregular forms that should be used instead (such as "men," "went," and "found").[37] These mistakes provide a further demonstration that children aren't just parroting back what they hear; they are instead applying generative rules.

There is a certain window of time where one is best at learning language, sometimes called a "critical period" or "sensitive period."[38] One of the most common questions I get from students is how they can better learn second languages. My unhelpful answer is that they should go back in time and learn them when they were children. This is particularly the case for phonology; learn a language too late and you'll always speak with an accent. There is a longer window for syntax, with some recent data suggesting that the period of successful learning stretches all the way to age seventeen.[39] Vocabulary might be exempt from the critical period; we pick up new words throughout our entire lifetimes.

———

What is in the brain of a child that allows it to take in input from the world and come out with knowledge of phonology, morphology, and syntax?

Skinner's solution was operant conditioning—reinforcement and punishment. But this is a nonstarter, for three main reasons.

First, if language is learned through the same mental machinery with which we learn everything else, then there shouldn't be isolated impairments of language. But there are. Just as unfortunate events can happen to you as an adult, like trauma or stroke, that can damage your capacity for language, there are genetic disorders where children have serious problems with language, while certain other sorts of learning are more or less intact.[40] This is difficult to explain with any theory that says that language learning is done through general principles of conditioning—or for that matter, by some sort of intelligence or general desire to communicate.

Second, reinforcement and punishment aren't necessary. Yes, there are parents who believe that you must teach children to talk. But there are cross-cultural differences in how parents relate to their children. In the Tsimané, a society in the Bolivian Amazon, there is a relative indifference to children; mothers carry around their babies but don't interact with them much; they don't even get names until their first birthday. (This might be because of a high mortality rate; perhaps mothers don't want to get too attached to their children.) Adults speak to these children about one tenth the amount that American mothers speak to their children, and they certainly don't train them to speak through reinforcement and punishment. And yet these children, like children everywhere, do acquire language.[41]

And third, as we've just seen, children don't learn behaviors, they learn rules. The two-year-old who hears the word "wug" for a single weird animal and then calls two of them "wugs" isn't generating a behavior shaped by a history of reinforcement. (She never said the word before, so she couldn't have been reinforced for it.) Rather, she is applying a rule of the form: "add -s to turn a singular noun into a plural noun." The three-year-old who hears and understands a sentence that has never been uttered before—like "It's way too late to play with your Pokémon cards"—isn't responding to a familiar stimulus. Rather he is using the rules of syntax to transform a series of words into a meaningful statement. (One that he might respond to in many ways, including by pretending he didn't hear it.)

So much for Skinner, then. Now consider the other extreme. We've seen that language is complex and highly structured, partially separate from other mental capacities. So maybe we should stop seeing the emergence of language as *learning*. Maybe it's growth. Here's the linguist Noam Chomsky presenting a nativist perspective:

> It is a curious fact about the intellectual history of the past few centuries that physical and mental development have been approached in quite different ways. No one would take seriously the proposal that the human organism learns through experience to have arms rather than wings, or that the basic structure of particular organs results from accidental experience. Rather, it is taken for granted that the physical structure of the organism is genetically determined, though of course variation along such dimensions as size, rate of development, and so forth will depend in part on external factors. . . . Human cognitive systems, when seriously investigated, prove to be no less marvelous and intricate than the physical structures that develop in the life of the organism. Why, then, should we not study the acquisition of a cognitive structure such as language more or less as we study some complex bodily organ?[42]

I love that quote. It shows an appreciation of the richness of language; it's a rebuke to simplistic learning accounts. It pays the mind the proper respect.

But it can't be right. Few of us create our own language; we pick language up from the environment. After all, languages differ. They have different phonemes. They have different words. They order words and phrases according to different syntactic rules. And we come to know the phonemes, words, and syntax of our language through attending to the world around us—which isn't at all analogous to the growth of physical structures like arms and hearts. If you're looking for a word to refer to the process of coming to know things through exposure to the right sort of information in the environment, I'd recommend "learning."

Keep in mind, though, that the necessity of learning is not a rebuke to a nativist approach. It just suggests that in some cases what is innate isn't knowledge, but rather the capacity to acquire knowledge. And this capacity is rich indeed. Let's look at two examples of research into the mechanisms of language acquisition.

One problem faced by children, before learning word meanings and syntax, is to figure out what the words are. Now, it might not be immediately clear why this is a problem. Isn't it obvious what the words are? After all, when someone talks to you, you perceive the words as separate. If you are listening to someone read this book aloud, you would hear the final few words of the prior sentence as:

You. Perceive. The. Words. As. Separate.

But this is an illusion caused by the fact that you already know the words. For a nonnative speaker it would be:

youperceivethewordsasseparate

This is easy enough to observe when you listen to an unfamiliar language. If you don't know French and German, these sentences are just mush; it's impossible to tell where one word ends and the next begins.

Je ne sais pas. (French)
Sprichst du Deutsch? (German)

I sometimes find myself in a synagogue listening to Hebrew prayers, and I couldn't tell you what the distinct words were if my life depended on it.

So how does the child, innocent of the language, solve this problem? One promising suggestion is that children succeed by doing statistics with the sounds of language. As a demonstration of this, consider an experiment in which the researchers created a novel language that included four made-up three-syllable words, such as "bidaku."[43] They presented these words to infants in random order, putting them together with no pauses, no rhythm, and no stress differences. And so the babies would hear this:

bidakupadotigolabubidaku . . .

Infants were able to figure out what the words were, and they did so through attending to statistical regularities. If syllables frequently go together, this implies that they are parts of a single word. So, in this example, whenever the child heard "bi," the syllable "da" would follow, and then the syllable "ku," and so the babies could infer that "bidaku" is likely to be a word. In contrast, they did hear "kupa" some of the time, but most of the time "ku" was *not* followed by "pa"—and so they didn't think that "kupa" was a word.

Presumably, then, this is how they break down words in English. The child might hear, say:

You're a good baby! (Youreagoodbaby)
Mommy loves you. (Mommylovesyou)
Uh, the baby just spit up on Mommy. (Uhthebabyjustspitupon-Mommy)

They will notice the frequent connection between "Mo" and "mee" (the two syllables that make up "mommy"), and "bay" and "bee" (the two syllables that make up "baby"), and infer that "mommy" and "baby" are words. Other sounds might also go together (such as "ood" and "bay," in

"You're a good baby!"), but the connection isn't as reliable, and so babies don't learn them as words.

———————

Once children start hearing language as a series of words, how do they come to know what the words mean?

I think this is an interesting question, just as you would expect from someone whose first book was called *How Children Learn the Meanings of Words*.[44] Let's put aside the question of how children learn the function words (such as "a" and "of") or abstract words (such as, well, "abstract"), and take a seemingly simple case—the learning of names for things.

Learning a word like "rabbit" might seem easy. There is a rabbit running around that the child can see. An adult says, "Hey, check out the rabbit," and so the child puts the word and the world together and figures out that "rabbit" refers to rabbits. The philosopher John Locke thought it really was this simple, writing, in 1690,

> If we will observe how children learn languages, we will find that, to make them understand what the names of simple ideas, or substances, stand for, people ordinarily show them the thing whereof they would have them have the idea; and then repeat to them the name that stands for it, as "white," "sweet," "milk," "sugar," "cat," "dog."[45]

But it has to be more complicated than this.[46] There are countless objects in the environment, after all. Locke envisioned the parent carefully placing the object in front of the child as the word is being uttered, but this isn't always what occurs. A simple habit of associating a word with what one is looking at would lead to disaster, if, say, when the adult says something like, "Hey, check out the rabbit," the child is staring at her foot.

A better theory is that children know enough to appreciate that the word refers to whatever the speaker is looking at when using it, and so they attend to the speaker's gaze. An adult says, "Hey, check out the rabbit," the child pays attention to where the adult is looking when uttering the sentence—presumably at a rabbit—and so the child comes to connect "rabbit" with rabbits, through the sort of social understanding or "theory of mind" that we talked about in the last chapter.

There is considerable evidence for this theory, but it too has its limits. For one thing, blind children learn the meanings of words, and they can't follow anyone's gaze.

Also, children have to figure out the exact meaning of the word, and this is harder than it might seem. The problem here is nicely illustrated in an example by the philosopher Willard V. O. Quine.[47] Imagine a linguist in a strange culture; a rabbit runs by and the native says, "Gavagai." The linguist is an adult and possesses full cognitive powers and knows how languages work, and so she interprets the scene as you probably would and writes down "Rabbit" as a tentative translation.

But can the linguist be sure? Well, no. Even if the linguist is right that the utterance is prompted by the rabbit, it might mean something like "Let's go hunting," or depending on local superstition, "There will be a storm tonight." Or maybe it doesn't mean rabbit, but rather refers to the specific rabbit—some words are proper names, after all, like "Flopsy." It could be "mammal" or "animal" or "food," or even "object." It could refer to the color ("white") or the motion ("hopping"). Then there are crazy possibilities—perhaps Gavagai refers to the top half of the rabbit, or its visible surface, or as Quine playfully suggests, "undetached rabbit parts."

How then do children so often get it right? One consideration is that some word meanings are more plausible than others. As one example, when we hear a new word referring to a category, we tend to think of a middle level of abstraction—not too general and not too specific. Not animal, not collie, but dog. Not a piece of furniture, and not a recliner, but a chair. Categories such as dog and chair are said to belong to this *basic level* of categorization.[48] It is natural, then, for the child to assume that the native would use a word like "Gavagai" to refer to the category of rabbits, not the category of animals and not to a subtype of rabbits.

Other logically possible meanings are not psychologically plausible at all. No child is ever going to guess that a new word refers to undetached rabbit parts or just the top half of the rabbit, because our minds don't carve up the world in this way.

Another cue that can guide the word learner is syntax. We rarely name things in isolation. Instead, the native is most likely to say something like (in his or her own language), "Check out the Gavagai," or, if he meant to refer to a property, he would say something like, "That's a very Gavagai thing," or, if he meant to refer to the action, "Wow, look at how it Gavagais." If the linguist knows enough of the language's syntax, she could use this information to make educated guesses as to its meaning. For instance:

If it's a noun, it's likely to refer to the object (rabbit).
If it's an adjective, it's likely to refer to a property of the object (white).
If it's a verb, it's likely to refer to an action that the object is involved in (hopping).

Young children turn out to be sensitive to exactly these cues when learning words. In a pioneering study done in 1957, three-to-five-year-olds were shown pictures of scenes that depicted substances, objects, and actions. When asked, "Do you see any sib?" (mass noun syntax, like "water"), they tended to point to the substance; when asked, "Do you see a sib?" (count noun syntax, like "dog"), they tended to point to an object; when asked, "What is sibbing?" (verb syntax, like "hitting"), they tended to point to the action.[49] They used the syntax to figure out the general meanings of the word.

Much younger children can use syntax to learn whether a word is a common noun or a proper name. In another study, one-year-olds were shown a new doll and either told "This is a zav" or "This is zav." In the first condition, they tended to take "zav" as a common noun, referring to the kind of thing, like "doll"; in the second condition, they tended to take it as a proper name, akin to "Mary."[50]

The psychologist Lila Gleitman has argued that syntactic cues can solve a profound problem in language development.[51] How do children learn words that refer to entities and actions that are invisible? How, for instance, do children learn a verb like "thinking"? You can point to someone who is thinking and say "Thinking!" but unless you're pointing to a Rodin statue, there's no outward appearance or action that really looks like "thinking."

Syntax can come to the rescue. A child might hear "thinking" in contexts like "He's thinking that it's time for supper" or "She's thinking about you," and these unique contexts can suggest to the child that this is a verb describing a certain type of mental state.

The argument here, backed up by considerable experimental evidence, is that once children learn the syntax of their language, they can use it to help acquire word meanings that would be impossible to learn otherwise.

———

Talk of word learning brings us to the question of the relationship between language and thought. This is a fascinating and quite contested area, and I want to get to some of the debates. But before doing so, let's address some questions about the relationship that have easy answers.

First, is there thought without language? Yes. Other animals are capable of thought—they can track prey, navigate through mazes, learn to beg for treats, and so on. A dog chases a squirrel up a tree and barks frantically at the tree, and it's clear that the dog thinks the squirrel is up the tree. We've seen as well that babies, long before they have learned to speak, can reason about the social and physical world.

And adults do all sorts of thinking without language, as when planning to throw a horseshoe, moving a sofa up a flight of stairs, and a million other tasks involving spatial understanding. We often find it impossible to describe what we are doing in words.

Also, we know that language cannot be the medium of thought because language is ambiguous, and thoughts are not. The word "bat" can refer to a flying animal or a baseball bat, but when we think about bats, we think about one or the other. The headline "Patrick Stewart Surprises Fan with a Life-Threatening Illness" has two meanings, one quite sweet, the other darker (someone posted this on Twitter with the remark that "Patrick Stewart gives the worst presents ever"). But it's not that the writer of the headline was confused about what Stewart was doing. The thought was clear; it's the sentence that's ambiguous.

Finally, if there wasn't thought without language, how could we learn language in the first place? How could someone learn the word "dog" if they couldn't think of dogs; how could they learn the word "kissing" without the ability to think about kissing?

More controversially, even abstract thought seems to exist without language. The data here come from a somewhat unusual source—autobiographical accounts of people who lived without language. The most famous case is that of Helen Keller, who became deaf and blind at eighteen months of age, and learned no language until the age of six, when she was taught a tactile language by a remarkable teacher, Anne Sullivan. She learned to read and write and went on to attend college and write several books, including an autobiography, *The Story of My Life*.[52] This biography is sometimes cited as showing the limitations of thought without language, as Keller described herself "at sea in a dense fog." But then again, before learning the tactile language, she was able to develop her own makeshift communication system—for instance, shivering to ask for ice cream. She knew she was different from other people, and that they (somehow) communicated with their mouths. Her own failure to communicate frustrated her. She was able to carry out practical jokes, such as locking her mother in a pantry and laughing as she pounded to get out. Nobody could seriously see her as a mindless automaton.

———

Here is another easy question: Does language *reflect* thought? Do the words and structures of languages reveal how the mind works? Also yes, in countless ways. All languages have ways to refer to objects like dogs and trees, to properties like being large and being green, and to actions like walking and talking. And this is surely because we naturally think about such objects, properties, and actions. There is no word that refers to the top half of a rabbit, and this doesn't have anything particular to do with how language works. It is because of how we think. We see Flopsy hop by, and we think about rabbits, not rabbit halves.

The connection between language and thought means that the study of language can tell us interesting things about how we think. As one example, languages often use the same words to talk about time and about space, as in the English prepositional system—the word "in" can be used for "in a minute" and "in a box," "on" can be used for "on Tuesday" and "on a mat," and so on. This linguistic conflation between time and space likely reflects a deep connection between the two in our minds.[53]

Or take taboos. Forbidden words illustrate certain universals. In every language, any list of forbidden words will include those that refer to sexual acts and parts of the body involved in sex. This tells us something interesting about the special status that people everywhere give to sex. But taboo language can also inform us about cultural change over time. The linguist John McWhorter talks about how, in English, words such as "God," "Jesus," and "damn" once had great emotional weight—you would not find them printed out in a family-friendly book like this one.[54] To deal with these taboo expressions, the English language has evolved less potent substitutes, such as "gosh," "gee," "golly," "jeez," "gee willikers," and "darn." Times have changed; if you were to use these substitutes now, you would seem absurdly prudish. Here language tells us something quite important about our shifting attitudes toward religious matters.

As a final example, linguists have chronicled the metaphorical patterns one finds in many languages, such as the notion that time is akin to money, as in these sentences:

> You're wasting my time.
> This gadget will save you hours.
> I don't have the time to give you.
> How do you spend your time these days?
> That flat tire cost me an hour.[55]

This is not universal—some societies have no money—so here again the patterns of language can be seen as expressing certain culturally specific patterns of thought.

Noam Chomsky was right, then, when he called language "a mirror of the mind."[56]Anyone interested in mental life can make a lot of progress by looking closely into that mirror.

Third, does language *influence* thought? Also yes. It can do so by communicating information. That's what I'm trying to do now—throwing some language at you to make you have certain thoughts.

Now, language isn't necessary to convey thought. You can teach your child to hold a tennis racket in silence; you can glare at a driver who cuts you off in traffic, expressing your attitude without saying a word. But language is a magnificent tool for communication. Indeed, the very design of language—phonology, morphology, and syntax—reflects its role as a complicated device devoted to the task of taking an idea that someone wishes to express (the girl thinks the house is big), turning it into speech or sign ("the girl thinks the house is big" in English, "Nwa agbọghọ ahụ chere na ụlọ ahụ buru ibu" in Igbo, something different in countless other languages), and then transforming the words back into thoughts in the mind of the recipient.

When we talk, our language is crafted for other people, configured to make them think certain thoughts, often in ways that transcend the literal meanings of words and sentences. This brings us to the study of *pragmatics*, which explores the part of language that allows us to say things without literally saying them. Someone tells you

He went to Harvard but he's a real nice person.

And, though it's never explicit, the "but" conveys how they think of most graduates of this fine Ivy League university. Consider two other examples—a couple in the kitchen:

We're out of garbage bags.
I've been busy.

and two professors at a conference:

What do you think of his research?

Well, people say he's got a great personality.

In both cases, if you take the sentence pairs literally, the second has nothing to do with the first. But we understand them, because there is more to language than the direct meanings of sentences. "I've been busy" responds to the speaker's implication that the listener should have picked up garbage bags; "Well, people say he's got a great personality" expresses—by dint of what's left unsaid—that the listener doesn't think highly of the person's research.[57]

It's in pragmatics that language deals with the niceties of human behavior. If I want you to pass me the salt, why do I say "Can you pass me the salt?" instead of just "Pass me the salt"? Asking if you are capable of the act (a question that, taken literally, is silly; it's obvious that you can) is a way of getting around the rudeness of telling you to do something. Such a phrase is the linguistic equivalent of giving someone a gift card as a present (respectable) as opposed to money (tacky).[58]

———

If we all agree that there is thought without language, that language can express thought, and that language can influence how we think, what is there to argue about?

There is one long-standing claim about language that really is controversial. This is the Sapir-Whorf hypothesis. Based on the ideas of the linguists Edward Sapir and Benjamin Lee Whorf, this is the view that language doesn't just change minds by transferring thoughts from one head to another; it configures how people make sense of the world, including about space, time, and causality. As Whorf put it in 1956, "Language is not merely a reproducing instrument for voicing ideas but rather it is itself the shaper of ideas, the program and guide for the individual's mental activity."[59]

This is a powerful idea, and there are several specific hypotheses of what this shaping might be. Perhaps the French contrast between two versions of "you"—"tu" (informal) and "vous" (formal)—guides speakers of

French to think in terms of this social distinction in a different way than speakers of a language like English, which lacks this contrast. Perhaps the binary singular/plural distinction of English—it's either one dog or many dogs—makes us think in terms of one versus many, while speakers of a language that offers more alternatives, such as a special plural marker for exactly two things, might have a more nuanced appreciation. In English, many verbs of motion incorporate the sort of movement, as in walk, jog, hop, amble, creep, spring, run, and so on. Other languages don't work that way; they just have a verb and add more words to specify the manner of motion, using phrases that are the equivalent of "moving in a hopping manner." Perhaps this means that English speakers think more about manner of motion. Perhaps, more generally, the different ways in which languages talk about color, time, causality, and hypotheticals have implications for how their speakers make sense of these fundamental notions.

These are all empirical hypotheses. But testing them is harder than it might seem. To show an effect of language on thought, it's not enough to find that, say, speakers of Korean and speakers of Hebrew have, on average, different worldviews. They plainly do, in all sorts of ways. But this might be because they tend to live in different societies and have different cultures.

At their worst, defenders of the Whorfian hypothesis fail to pull this apart; they describe linguistic differences and simply assume that these reflect differences in thought. The psychologist Gregory Murphy has a biting parody of this sort of argument:

> Whorfian: Eskimos are greatly influenced by their language in their perception of snow. For example, they have N words for snow . . . whereas English only has one, snow. Having all these different words makes them think of snow very differently than, say, Americans do.
>
> Skeptic: How do you know they think of snow so differently?
>
> Whorfian: Look at all the words they have for it! N of them![60]

When careful studies are done of speakers of various languages, they sometimes do find some subtle Whorfian effects—differences in the thought patterns of speakers of different languages that seem to genuinely result from the language itself.[61] Some of the best demonstrations have to do with color. Russian has different words for lighter blues ("goluboy") and darker blues ("siniy"); English mostly uses just one—"blue." One effect of this is that colors might be remembered more accurately in Russian, since we often remember scenes in verbal descriptions, and the descriptions in Russian are more fine-grained.[62] By the same token, compare a wine expert, who remembers a glass as "Merlot with black raisin and entrancing floral notes" and me, a wine moron, who remembers it as "red, from a pretty bottle"; the expert's description will enable them to later recall the flavor in a way that I can't.

It's not just memory, though. If two colors are described by different words in Russian but one word in English, Russian speakers are a fraction of a second faster at telling them apart.[63]

Such findings are of considerable theoretical interest. But they are subtle—the sort of effects you can only find in a lab. As far as I know, there is no support for a strong version of the Whorfian hypothesis, in which profoundly different modes of thought arise through exposure to different languages. As John McWhorter sums up, the effects that we find of linguistic influence are but "a passing flicker, that only painstaking experiment can reveal, in no way creating a different way of seeing the world."[64]

———

Other variants of the claim that language has radical effects on thought are more promising, however. These don't look at differences between languages, but rather explore the effect of possessing a language in the first place.

Think about the difference between two and three. Not so difficult, is it? You can appreciate it in an abstract sense, feeling the moreness of three versus two, or you can easily imagine two cookies versus three cookies. But now think of the difference between ninety and ninety-one. There is not an intuitive contrast: My image of ninety cookies and my image of ninety-one cookies are identical—a lot of cookies. Nonetheless, we appreciate that they are different, that ninety-one is larger than ninety—one more, to be precise.

And this appreciation is possibly because we have learned a language that can encode these number concepts. The cognitive neuroscientist Stanislas Dehaene puts it like this:

> Linguistic symbols parse the world into discrete categories. Hence, they allow us to refer to precise numbers and to separate them categorically from their closest neighbors. Without symbols, we might not discriminate 8 from 9. But with the help of our elaborate numerical notations, we can express thoughts as precise as "The speed of light is 299,792,458 meters per second."[65]

This proposal about the importance of language for understanding number meshes with what we know about nonlinguistic creatures like babies. We saw in the last chapter that even babies can add and subtract: Put one ball behind the screen, add another, then they expect two, not one or three.[66] But their abilities stop around there. Babies don't know, for instance, that 8 plus 8 equals 16. It might be that such understanding must be rooted in a symbolic system, one that is not innate—a symbolic system like natural language.[67]

Or consider the debate from the last chapter over why young children have such problems fully appreciating the minds of other people—so-called theory of mind. Again, some say that the missing ingredient is proficiency in language. There turns out to be a strong correlation between language development and various theory-of-mind skills—the better children are at language, the better they are at reasoning about other minds.[68] And deaf children who have not acquired a sign language have a considerable delay in understanding false beliefs.[69]

Why would language make us better at thinking about other minds? The obvious theory is that engaging in linguistic communication conveys more and more information about the minds of other people. But a more radical theory is that it's the *structure* of language that does the trick.[70] The syntax of language allows for a superior understanding of how the world is seen by others.

Perhaps, then, language is essential for encoding certain complex thoughts about other minds. Just as a creature without language couldn't make sense of:

The speed of light is 299,792,458 meters per second.

perhaps they also couldn't understand dialogue like this (from the sitcom *Friends*):

Oh man, they think they are so slick messing with us! But see, they don't know that we know that they know!

There is a lot more to be said here; there is ongoing debate over the extent to which languages make possible our uniquely human powers of numerical and social thought. What *isn't* a matter of debate, though, is how much language has transformed human life. Without it we would have no culture, religion, science, government, and much else. The philosopher Daniel Dennett once wrote: "Perhaps the kind of mind you get when you add language to it is so different from the kind of mind you can have without language that calling them both minds is a mistake."[71] I don't think this is literally true—surely chimps and dogs have minds—but it's a nice expression of the centrality of this astonishing human capacity.

7

The World in Your Head

The Big Picture

The world makes its way into your head. Sometimes it stays.

Look out the window at the cars on the street, listen to the hum of the air conditioner and the sound of construction, taste the acidity of hot black coffee on your tongue, feel the chair pressing against your behind. These are my own experiences right now. Yours will differ depending on where you are as you read or listen to this—waiting in line at the post office, resting in the woods after a long hike, sitting on your cot in prison, wherever. You will have a sense of the world around you, what philosophers call "a mental representation."

Now close your eyes and cover your ears and try to recover the experience. It's still there—sort of. Not as vivid as when you are attending to it; you know you are remembering and not perceiving. But, still, some of the world remains. Some of what you recall will last for just a few heartbeats; in other circumstances, your memories will remain for the rest of your days.

One explanation for how all this happens goes like this:

There is a real world out there. We experience it because it contacts our sensory organs in the form of light, sound waves, and pressure on our skin. This contact makes neurons fire, and the experience that arises from this firing is known as *sensation*—experiences of light and sound and touch. Your brain chugs away and processes this information, combining sensation with expectations about how the world works, and out of this arises a rich experience of the world around us, and we call this process *perception*.

The world changes—clouds block the sun, the air conditioner clicks off, the cars drive away—and your brain keeps up, changing the representation accordingly. Other changes occur because of your own physical actions—you move your head, squint your eyes, swallow the coffee, walk across the room. You can shift these perceptions through sheer force of will, something we discussed in an earlier chapter when we talked about the science of consciousness. In the privacy of your head, you can draw different aspects of the world into focus and sideline the rest. *Now* I'll count the cars; *now* I'll savor the taste of my Nespresso. We call this process *attention*. William James described attention as "the taking possession by the mind, in clear and vivid form, of one out what seem several simultaneously possible objects or trains of thought."[1] The usual metaphor here is that it is a spotlight.

Finally, when you shut your eyes or cover your ears, some of the information remains. It is encoded in your physical brain. And we call this *memory*. Later, by being reminded or through conscious effort, you can recover some of these memories, and the world of the past can be resurrected.

And so: There is a world. You perceive it. You attend to it. You are conscious of it. You remember it.

This might all seem obvious, but to be fair, I should note that not everyone would accept this story. For one thing, it refers to information in the head, and so you've lost the behaviorist. Perception, attention, and memory, the behaviorist will tell you, are made-up constructs that have no place in our mature science of psychology. We should stick to stimulus and response.

Also, we've turned mental life into an unfolding physical process, with nothing else added, and so we've also lost the dualist. Under our account, light bounces off things in the world, hits your retina, makes neurons fire, the brain chugs away, and soon you're thinking, whoa, those cars are moving fast. But what about the soul? Where's the transition when these experiences exit the physical brain and go to some spiritual realm? There is none, and so Descartes would say that you're missing something.

Finally, our theory assumes that there's a physical world out there, something we make contact with and come to know about, and so you've lost certain skeptics. If we all live in the Matrix, then we don't experience the external world at all. (There is no spoon, the boy tells Neo.) Long before Keanu Reeves, the philosopher George Berkeley claimed that all that exists is the mind of humans and of God: *Esse est percipi*, he wrote—things only exist when they are perceived.[2] And there are more than a few scholars right now who carry on the skeptical tradition of doubting that there is a world out there independent of our perception of it.[3]

I think we can confidently respond to some of this criticism. We shouldn't be behaviorists (they are wrong—our brains contain a representation of the world) and we shouldn't be dualists either (they are wrong—our minds are our brains). A proper science of the mind is nonbehaviorist and nondualist; it will include processes of perception, attention, and memory, and will understand them as the workings of our very physical brains.

What about skepticism? Well, we can dismiss the radical version. There is an external world, and we apprehend it. There is the story in Boswell's *Life of Samuel Johnson*, published in 1791, where Boswell and Johnson discuss Berkeley's claim that there is no such thing as physical matter, and then Johnson responds by kicking a stone, and saying, "I refute him thus."[4] Even skeptics accept, at some level at least, the existence of a world outside the head. Thomas Reid in 1785 wrote, "I never heard that skeptic run his head against a post . . . because he did not believe his eyes."[5] What is reality? I like the answer of the novelist

Philip K. Dick: It is whatever, when you stop believing in it, doesn't go away.[6]

Indeed, the reason we have sense organs at all is that natural selection favors creatures who get things right. Other things being equal, animals that know what's real outlive and outreproduce those that don't.

And yet, a milder form of skepticism is true. We should reject the position, sometimes called *naive realism*, that our senses inevitably capture the world as it is. In the picture below—known as the café wall illusion—the horizontal lines are straight, but they don't look that way. This illusion and countless others show that we can get things wrong in systematic ways.

Similarly, you might think that a pile of dog poop is disgusting, your child is adorable, and the top of the mountain you are hiking toward is depressingly far away. But such interpretations are products of your mind. Color itself is a fascinating case; our experience is not a direct transcription of actual wavelengths in the world. Recall the internet meme of a few years ago, where some people saw a dress as black and blue and some saw it as white and gold, and everyone was astonished that others perceived it differently than they did (for a discussion, and theories of what happened, see the endnotes[7]). Radical skepticism is wrong, but so is naive realism. There is some truth to the aphorism of Anaïs Nin: "We don't see the world as it is, we see it as we are."

Sensation and Perception

Let's start our journey from world to mind by considering the contrast between sensation and perception. Sensation is the first stage; when receptors in the skin, the retina, the ears, and elsewhere take the world and transform it into experiences of pressure, color, and so on.

We usually don't have conscious access to our sensations. My experience when I sip what's in my mug is that of a certain brand of dark, strong coffee. The faces of the people I love are beautiful to me; I can't see my sons without seeing them *as my sons*. If you tell me "It's raining outside," I don't hear the actual sound of your words; my knowledge of English infuses them with meaning and a certain rhythm; I can't help, for instance, hearing your sentence as composed of a series of three distinct words—It's. Raining. Outside.—while for someone who didn't know English, it would be a garble. As the philosopher Jerry Fodor once put it, "One has practically no access to the acoustics of utterances in languages that one speaks.[8] (You all know what Swedish and Chinese sound like; what does *English* sound like?)"

Or take vision. Stanislas Dehaene writes,

> We never see the world as our retina sees it. In fact, it would be a pretty horrible sight: a highly distorted set of light and dark pixels, blown up toward the center of the retina, masked by blood vessels, with a massive hole at the location of the "blind spot" where cables leave for the brain; the image would constantly blur and change as our gaze moved around.[9]

No matter how much you try, you can't experience this. Rather you get:

> a three-dimensional scene, corrected for retinal defects, mended at the blind spot, stabilized for our eye and head movements, and massively reinterpreted based on our previous experience of similar visual scenes.

The world comes to us preprocessed.

Some argue that it was different when we were babies. Remember Piaget's claim that babies don't see the world in terms of persisting objects, or William James' view that for the baby, all is a "blooming, buzzing confusion." As we've seen, though, this view is probably wrong. A baby can't help seeing objects as objects, not fleeting retinal images. Babies *perceive*.

Still, there are cases where even adults have sensation without perception. To return to Fodor's example, my ignorance of Swedish and Chinese means that I just hear the sounds—not much perception here. Psychologists can shine a light in your eye or put a weight in your hand, and your experience is again mostly pure sensation.

There is a science of how sensations relate to the stimuli that produce them, called *psychophysics*. One focus of this science has to do with the limits of your sensory powers. If you hold two weights, one in each hand, what's the difference you can notice? How much sugar can I put in your tea before you can tell that I'm adding it?

The best answer to these questions is known as Weber's Law, thought up by Ernst Weber in the 1830s.[10] This law says that our experience of change in the stimuli is determined in proportions, not absolutes.

To get a sense of this, put a weight in your left hand. How different does a weight in your right hand have to be so that you'd be able to tell there is a difference? If you're looking for an answer in grams or ounces, Weber's Law tells you that you're looking in the wrong place. The amount you need to add is not a constant—it's a proportion of the first weight. If you hold a quarter in your right hand and two quarters in your left, you are likely to notice the weight difference—it's a 1:2 ratio. But put a brick in one hand, and a brick with a quarter on it in the other, and they feel the same, even though the absolute difference in weight is identical to the first case.

Or consider a noise that gets louder in absolute increments, going from 1 to 10. We experience the difference between 1 and 2 as greater than the difference between 2 and 3. This is because the first is a 100 percent change, and the second is just a 50 percent change. (The difference between 1 and 2 is, psychologically speaking, the same as between 2 and 4.)

The relevance of proportions extends to more abstract assessments. A fancy coffee at my neighborhood café costs $4. If they raised the price to $104, I would be shocked and would never return. But if I were buying a house, and saw a house advertised for $426,000 and then checked the next day and saw that the price was raised to $426,100, I'd laugh. Who cares! It's only $100 more. This suggests that I'm not freaking out in the café about $100 (if I was, I'd be equally upset at the rising price of the house); I'm freaking out over a coffee that costs twenty-five times more than it normally does.

(In general, context matters when we think about money. In *Double Down*, a memoir of gambling addiction, Frederick and Steven Barthelme write, "At the table, losing our money, we were all smiles, as if it were nothing. In fact, it felt like nothing. Money isn't money in a casino. At home you might drive across town to save a buck on a box of Tide, but at the table you tip a cocktail waitress five dollars for bringing a few Cokes. You do both these things on the same day."[11])

Let's end the discussion of sensation with a puzzle, one that harkens back to the mysteries of consciousness we discussed before. There is nothing more different than experiences from separate senses—between, say, the smell of a rose versus the feeling of a kiss, between looking at a rainbow versus listening to a toddler giggle. But all these experiences come into the brain in the same way, as neurons firing. As a team of psychologists point out, "The signals don't have some secret handshake to say they belong to the vision or hearing club."[12]

And so the phenomenological difference has to arise through distinct sorts of neural computations. The occipital lobe takes the fire . . . fire . . . fire and turns it into the experience of color and light; the parietal lobe takes fire . . . fire . . . fire and turns it into taste and smell, and so on.

And now we get to a puzzle. What would happen if you put the optic nerve from the eye into the parietal lobe and the neurons from the ears into the occipital lobe? Would you hear light and taste sound? What if you switched other neurons; could you feel the aroma of chicken vindaloo or taste a rainbow? This is not entirely fantasy; some people experience synesthesia, where letters and numbers have sensory experiences associated with them. In an extreme version, one famous neurological patient known as S. had such powerful synesthesia that he didn't read the newspaper while eating breakfast because the flavors he perceived while reading the words would ruin the taste of his meal.[13]

––––––––––

In 1966, a prominent AI researcher at MIT was working on an intelligent system, and he wanted the machine to interface with the world. And so he gave an undergraduate student a summer project to connect a computer to a camera so that it could "describe what it saw."[14]

Psychologists love this story, because it nicely expresses how simple the problem of perception looks and how hard it really is. (Of course, the undergraduate never finished the project, though I'll add that he—Gerald Sussman—is now a leading scholar in artificial intelligence.) In fact, we've been working on the project of object recognition by computers for over half a century and progress has been slow and difficult.

I used to tell students that this research program has been a massive failure—but I don't say this anymore. There has been progress. There are algorithms that can name pictures with some degree of accuracy; there are self-driving cars that have some success at making their way through traffic. Perhaps in the next year or so, there will be algorithms from Facebook or Google or some other corporation that really can recognize objects as well as a person can.

Or not. We're certainly not there yet. Just as no computer can speak or understand language at the level of a three-year-old, no machine can perceive the world as well as a toddler. As I'm writing this, websites screen out bots by asking us to identify letters written at funny angles, a task that is trivial for a person but beyond the capacity of most algorithms.

The problem of perception feels so easy because we solve it so naturally. This is an example of something that the evolutionary psychologists Leda Cosmides and John Tooby called "instinct blindness": "Intuitively, we are all naïve realists," they write, "experiencing the world as already parsed into objects, relationships, goals, foods, dangers, humans, words, sentences, social groups, motives, artifacts, animals, smiles, glares, relevances and saliences, the known and the obvious."[15] The processes that give rise to these experiences are automatic and unconscious and so it takes a lot of work to appreciate how complicated they are. It's only when we try to implement these processes into a computer, to build something that finds the words in the sound stream or picks out the smiles in a sea of faces, that we are forced to confront how difficult these problems really are.

———

Light hits your eyes, often because it bounces off something. The light passes through your corneas, goes through the lenses (which have muscles allowing you to shift the focus to close or faraway objects), and then hits the back of the eyes, which are filled with photoreceptors. There are a lot of them: each eye has about 120 million rods on the outer ring (which are highly sensitive and handle vision at low light) and about 6 million cones (mostly in the center, which deal with color vision).

From these rods and cones, you get neural firing that goes to the back of the head (which is sort of strange, because you would think the processing would happen closer to the eyes). This leads to experience. In the movie *The Matrix*, trained operatives look at computer screens with seas of moving numbers and in this they see a world. If we think of patterns of firing neurons as expressing a numerical code, this is precisely what our brain does, all the time. It infers a three-dimensional world from a moving two-dimensional array.

And this is a staggeringly difficult problem. Among other things, a good perceiver has to recover information about fundamentally static properties. But our perception is in constant flux. Imagine watching a dog as it runs past you down the street. The image of the dog gets smaller and smaller on your retina as it retreats, but you shouldn't conclude that the dog is shrinking. As the dog turns a corner, you move from an image of its tail and back legs to the full dog in profile—a radical change in shape—but you shouldn't infer that the dog itself is changing shape. When the dog darts under the shade of a tree, there is less light reflecting from its body, but you shouldn't see the dog as changing color. Somehow, the brain takes shifting information and calculates that there is a singular unchanging being that is moving through space and time. (To make this problem even harder, we need to be able to perceive objects in the world that actually *do* change, as when we watch someone blow up a balloon, twist clay into a statue, or paint a canvas.)

If we knew how the mind did all this, we could program computers to do the same, but we don't and so we can't. But we do have a general insight here, which is that successful perception involves the coordination of two sorts of information.

1. The input to the visual system; the neuronal firing based on light hitting the retina (sometimes called "bottom-up" information).
2. Your assumption about the world. Some of these might be part of the visual system itself, wired into the visual cortex; some of it might come from your memories and expectations ("top-down" information).

Some simple examples: You are sitting in a room, and it's filled with light and then it goes totally dark, then it lights up again, then it goes dark. This is your bottom-up input. What's happening here? Well, maybe something is going on with the world; maybe the lights are going on and off. But if you know that you are squeezing your eyes open and shut, then you would infer the world is unchanging. You have a better way of making sense of your experience. Or imagine that you're in a dark room and a white circle in front of you is expanding. You would see it as approaching—unless you were walking toward it, in which case you would see it as stable.

Here's a less fanciful example. We see a lump of coal as black and a snowball as bright white. The simple theory of how we do this is that the snowball reflects more light than the lump, and our eyes register the difference. But this is just part of the story; there's more to perceiving the brightness of an object than the amount of light that bounces off it. We are sensitive to the context the object is in. At dusk, a snowball might be darker (reflect less light) than a lump of coal on a bright day.

As a dramatic illustration of how context influences perceptual experience, consider this example from cognitive scientist Edward Adelson. Look at the tiles marked A and B.

A looks darker than B. Is this because A is sending less light to your eye than B? Nope, they are producing the same amount of light; they are the same color on the screen or on the page. (If you don't believe me, cover up the image so that only the two tiles are showing.)

So why do they look different? Because B is in shadow. Since shadows make surfaces darker, your brain compensates and you perceive it as brighter. To put it in the framework we just introduced:

1. **Bottom-up:** The squares emit the same amount of light.
2. **Assumptions:** But one is in shadow. We know that shadow makes something look darker than it is, and so we infer that it must really be lighter.

Think now about the broader project of segmenting the visual world into distinct objects and appreciating the relationships between them. You might look at a table and see a coffee cup, a book, and an apple, all lined up in a row. On your retina, though, all that exists is an undifferentiated smear of light. How do you get from this smear to the perception of these three distinct objects?

One general approach to this problem was developed in the early 1900s by a community of scholars known as the Gestalt psychologists.[16] They proposed perceptual grouping rules that guide us to make sense of the world, to organize our perceptions in an adaptive way. As a simple illustration, check out the Os below.

O O O O   O O O O

From a logical perspective, there are many ways to group these eight items. Perhaps they fall into three groups. One group might be the first two Os ("O O"), another might be the next four Os ("O O O O"), and a third group might be the final two Os ("O O"). But nobody would ever see it this way. We are prone to segment items based on factors such as *proximity*, and so we naturally see this as:

Group 1:   O O O O   Group 2:   O O O O

Or consider this, with the letters equally close to one another.

B B B B T T T T T W W

Again, there are many ways to break this up, but there is a natural tendency to see this as three groups, one of Bs, the other of Ts, the third of Ws, clustering them on the basis of *similarity*.

These Gestalt principles capture regularities in how objects are normally distributed in the world. To go back to our example, the color and texture of the coffee cup will be different from that of the book, which will be different from that of the apple. Furthermore, the three objects will be likely separated in space. And so, the same principles of similarity and proximity will parse your perception into three separate things: Cup. Book. Apple.

Or consider the scene below. It's perfectly possible that you are looking at a complex figure with three pieces: a block, and two separate moving bars, one on top and one on the bottom.

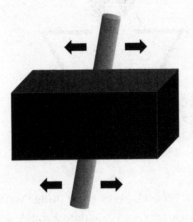

But this is unnatural. It would be quite the coincidence for two different bars to line up so nicely and move together in such perfect synchrony— and the mind assumes that there are no coincidences. So, guided by the Gestalt principle of *good continuation*, we see this as a single bar with its middle hidden behind the block. (Babies, when shown this display, also see it as one bar, not two.[17])

A clever perceptual psychologist can set up scenes where the Gestalt principles motivate the perception of a form that isn't there. On the following page is the Kanizsa Triangle illusion.[18] This is consistent with a triangle lying on top of three circles, and it makes so much sense that there is a real triangle there that the mind puts one in, creating an object even though no such thing actually exists.

In addition to segmenting the world into objects, we also place the objects in a three-dimensional world—we see them in depth. How does this work?

One solution exploits the fact that we have two eyes that are separated in space. Put your hand close to your face and look at your thumb, closing one eye and then another. Then look at something far away (maybe very far, like the moon). When the object is close, the eyes get different im-

ages; when it is far, they see much the same thing. Your visual system uses this cue—binocular disparity—to calculate a rough sense of how distant objects are in space.

You can see the world in depth with even one eye, though. One way we do so involves the typical size of things. If a woman approaches you and her image on your retina takes more space than that of a house, it's a safe bet that the woman is closer than the house, because houses are normally bigger than women. Another cue is interposition. If the outline of the woman is complete and blocks part of the house, she is likely to be in front of the house. A third cue has to do with the speed of motion. Why do birds seem to move quicker in the sky than airplanes? Because airplanes are farther away—objects that are distant create images that move slower on the retina than those that are closer. We can use this to infer distance. (This has a neat name—*motion parallax*.)

———

We've talked about how we assess brightness, parse the world into objects, and locate them in space. The specifics are different, but the general account is identical: there is bottom-up input from the world that is messy and incomplete, and then the brain works to make sense of it.

This interplay between bottom-up information and beliefs and expectations is not special to visual perception. It also happens in language processing. Consider what's called the *phonemic restoration effect*.[19] You take a recording of a sentence such as:

It was found that the wheel was on the axle.

and you remove a phoneme and replace it with the sound of a cough, so that the actual sentence played is:

It was found that the [cough] eel was on the axle.

But this is not what people hear. For an English speaker, it is clear from the context what the missing sound is, so your brain fills it in and you actually hear the "wh"; the perception of the sentence includes the word "wheel" and a simultaneous cough.

It was found that the wheel was on the axle.
[cough]

This mental filling in of gaps is what makes proofreading difficult. Scott Alexander gives the example below:

I

LOVE

PARIS IN THE

THE SPRINGTIME

and then writes,

This demonstrates how the top-down stream's efforts to shape the bottom-up stream and make it more coherent can sometimes "cook the books" and alter sensation entirely. The real picture says "PARIS IN THE THE SPRINGTIME" (note the duplicated word "the"!). The top-down stream predicts this should be a meaningful sentence that obeys English grammar, and so replaces the the bottom-up stream with what it thinks that it *should* have said. This is a very powerful process—how many times have I repeated the the word "the" in this paragraph alone without you noticing? [20]

Sometimes it is difficult to notice where your mind did a fix. During a bitter online debate about vaccination during the spring of 2021, someone posted this on Twitter as something that made them laugh:

When the time comes, I 100% support mandatory vacations for everyone. If anyone refuses they should be FORCED.

At first I simply didn't get the joke, and I read it over and over (was there a repeated "the" somewhere?). Finally, someone had to tell me to look closely at the word after "mandatory" and then I realized that it was "vacations"—not the word I expected and therefore saw: "vaccinations."

It makes sense that perception works that way. Much of our experience of the world is ambiguous, and a well-working perception system favors the plausible over the implausible. It's more plausible that the person said "the wheel was on the axle" and we missed the "wh" because of the cough—as opposed to saying "the eel was on the axle"—so it makes sense to hear the word as "wheel." It will trip us up occasionally—it's THE THE, not THE; it's "vacations," not "vaccinations"—but the efficiency gains of favoring the likely interpretation outweigh the costs of occasionally getting it wrong.

Also, importantly, if there is a clash, bottom-up eventually wins.[21] When I was a child, I read a book in which aliens came to Earth, and the idea was that only children noticed, because adults don't believe such things as aliens are possible and so literally couldn't see them. I remember being entranced by the idea—very flattering to me as a child—but it's silly. If our perceptual system could only tell us about what we expect, then we could never be surprised, and plainly we can. I really don't expect to turn around and see a gorilla standing behind me, ready to plunge a knife into my neck, but if I turned, and there he was, I'd notice it. This sort of thing—telling us what we don't already know—is what perception is for.

Memory

I sometimes do a class demonstration where I ask a volunteer to pretend that they have totally lost their memory. The phrase "lose your memory" comes with a certain connotation, as in movies like *Vertigo* or *Dead Again* or the excellent Jason Bourne series with an amnesic superspy played by Matt Damon. When I ask, "So what's your name?" the volunteer will often ham it up, make a confused face, and say, "I don't know!"

But there's a trick here, because memory underlies *everything* we know, not just our autobiography. Bourne speaks English, drives a car, fights skillfully, makes love, walks, puts on his clothes, and uses the toilet. All of this involves him drawing on experiences from his past that are somehow stored in his brain—memories. If my volunteers really lost all of their memories, they wouldn't be able to answer my question. They would be lying on the floor, crying and pooing.

Some philosophers, like John Locke, argue that our memories are in large part what we are—if you woke up with my memories and I woke up with yours, I would be you and you would be me.[22] I'm skeptical of this theory myself, but there's something right here—a complete loss of all of one's memories would be something close to an obliteration of the self.[23]

There are different types of memories. There is *autobiographical memory*, the memory of personal experiences. This is what Bourne is said to have lost, and what we typically talk about when we talk about "losing our memory." But Bourne retains what's called *semantic memory*—he is aware that Paris is the capital of France, that dogs usually have tails, and so on. He retains *procedural memory*—memory about how to do things. He walks, reads maps, drives a car. Woken on a bench by two policemen, Bourne quickly knocks them unconscious, but he is startled that he could do this; he knows how to fight (he has procedural memory) but doesn't know he knows (he lacks autobiographical memory). And Bourne can certainly acquire new memories; a glance at a map and he's good to go.

He has had a devastating memory loss, then, but certain types of memory are perfectly intact.

One of the great discoveries of psychology, with much help from neuroscience, is the taming of memory—our understanding of its many different parts, their limits, and their powers. Let's continue our trip from the outer world into the mind, and ask: How do we remember what we perceive?

———

Here is the standard model of memory.[24] This model is simplified, but it is a useful place to begin.

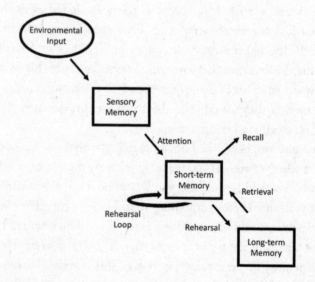

We begin with *sensory memory*. Some of this is visual. If there's a flash of lightning, and you close your eyes and see an afterimage, that's a form of sensory memory. To experience sensory memory with your eyes open, play with sparklers and write your name in the air. The letters don't remain in the world—if your brain was wired differently this trick wouldn't work. Your perception arises because certain experiences endure in visual sensory memory.

Then there's auditory sensory memory, which lasts a few seconds longer. If someone is telling you a story, and you are bored and glance at your phone, they might demand, "What did I just say?" Although you weren't really listening, the words remained in your mind, as if by magic, and you can repeat them.

You are always forming sensory memories, but unless attended to, they are quickly lost. When you do attend, the information moves to the next stage, short-term memory—also known as *working memory*. Many psychologists identify working memory with consciousness; we can point to that box and write next to it "You are here." It's the *self*. Working memory is the interface between world and mind; it can receive input through attention (a car drives by and you notice it) or from long-term memory (you are reminded of what you had for breakfast and you now think about it).

You can hold information in working memory, but its capacity is limited. If possible, have someone read to you, at normal speaking pace, this list:

14, 59, 11, 109, 43, 58, 98, 487, 25, 389, 54, 16

Now stop. What numbers do you remember?

You'll remember some of it—if you are taking this task seriously, you will repeat the numbers over and over in your head, something known as *maintenance rehearsal*. Some of what you rehearse will drift into long-term memory, at least for a while, and these are often the earlier items, because you have more time to rehearse them—the *primacy effect*. And you'll often remember the last few items, because they'll still be in short-term memory when the list stops—the *recency effect*. So when you plot the success of memory against the position of the items in the list, you get a U-shaped curve—a bump for primacy effect, a dip in the middle, then a bump again for the recency effect.[25]

How many items could you remember? How much can short-term memory store? There is a classic 1956 paper by the psychologist George Miller, and it answers this question in the title: "The Magical Number Seven, Plus or Minus Two: Some Limits on Our Capacity for Processing

Information."[26] So, between five and nine. Later estimates suggest that the truth lies on the low side; the storage of memory is about four items.[27]

Four what? Four meaningful units, sometimes described as "chunks." The role of understanding here means that the powers of one's short-term memory—how much information can be held in consciousness—will differ depending on knowledge. Suppose you get this string of letters:

L   A   M   A   I   S   O   N

That's eight units; too much to hold if you treat them as eight distinct items. But since you are an English speaker, you might notice something; it's four words, within your memory capacity.

LA   MA   IS   ON

And if you know French, you know this means *house*, and that's easy peasy, a single unit.

LA MAISON

Expertise influences the power of your memory. Can you remember the configuration of three complete chess boards? If you don't know chess, hell no. Thirty-two pieces on an 8 × 8 grid, three times over, absolutely impossible. But if you do know chess and the pieces fall into a logical, sensible configuration (such as, say, a position likely to appear after a few moves of a popular opening), you can remember it easily. A master player can hold multiple games in memory. Expertise enhances your storage capacity; it expands the scope of one's consciousness.[28]

———

Now we arrive at *long-term memory*. This is similar to what Freud called the preconscious; it's always there, but we are not typically aware of it. All of what we've talked about—sensory memory, short-term memory—

counts as memory in the technical sense that they involve storage of information, but long-term memory is *memory* as we usually use the term.

The most obvious way in which long-term memory and short-term memory differ is in capacity. We've seen that short-term memory is sharply limited. Long-term memory, however, has virtually unlimited storage. Every word you know is in long-term memory, every face you recognize, every story, every joke, every victory, and every humiliation—they are all in long-term memory.

Now, long-term memory has to have some limit. It is encoded in the physical brain, and brains are finite things. It would be neat to compare its capacity to that of computers, but estimates should not be trusted, because we know so little about how the brain encodes information. Still, some brave souls have taken a shot at it, and one estimate is that long-term memory is 2.5 petabytes, about a million gigabytes. If you had a computer with that much memory, you could hold three million hours of video.[29]

———

How do we get information into long-term memory? (To put this in more casual terms: How do we learn stuff?) Often learning is unconscious and automatic. There are all sorts of experiences in my head that got there without effort. Yesterday I left a coffee shop and stepped into the street without looking and a car slammed to a halt and the driver glared at me. I remember that! And I remember that it took a long time to get the coffee and the barista was an older woman. But I don't remember who else was there or what she was wearing. If I didn't write this down—and relive it as I edit the book—I would forget this all in a few days. Other experiences stick for longer. I remember when I got on one knee to propose to my wife, and unless something bad happens to my brain, this will remain with me until the day that I die.

But what if you want to get something into memory, something that wouldn't get there otherwise? What tricks are there to remember new information?

One is to exploit what's called *depth of processing*: the deeper you think about something, the more sense you try to add to it, the easier it is to remember. There is a classic study that illustrates this.[30] You show people a series of words one by one and have them answer questions about them. (They are not told to remember the words.) One group is asked whether the words are written in capital letters. A second group is asked whether the words rhyme with "weight." And the third is asked whether they would fit into the sentence "He met a ____ in the street." Then they get a surprise memory test. It turns out that what they were asked to do with the words influenced their subsequent memory of them. The subjects who remembered them best were the ones who were asked to think about whether it could appropriately appear in a sentence. More generally, the deeper you think of something—focusing on meaning rather than superficial aspects, like capital letters—the better you will be at remembering.

Another technique is to make your experience *vivid*, make it stand out, make it interesting. If you want to know the function of the hippocampus, you might try to remember:

The hippocampus is involved in the memory of spatial environments.

But you would do a lot better with this:

The hippocampus helps you find your way around campus.

Not the best rhyme, but now, perhaps for the rest of your life, you'll know that the hippocampus helps you remember where things are.

This is nothing compared to what memory pros can do. They can use an insight that's been known at least since the ancient Greeks—the power of vivid images.

Can you remember the order of a shuffled deck of fifty-two playing cards, looking at each one for a moment? Probably not, but you can learn to. In *Moonwalking with Einstein*, Joshua Foer describes how he came to win the U.S. Memory Championship.[31] The trick is to first learn to associate each card with a vivid image. For him, the five of clubs is associated with the actor Dom DeLuise, a karate kick is associated with the nine of clubs, and so on for all fifty-two cards. Then, when shown a series of cards, the job for Foer is to create a story of wild and memorable interactions corresponding to the characters and actions associated with the specific cards. Show Foer the cards . . .

> Five of clubs, nine of clubs, three of diamonds, five of spades, six of diamonds, king of hearts, two of clubs, king of clubs, six of spades . . .

and he remembers them like this:

> Dom DeLuise, celebrity fat man (and five of clubs), has been implicated in the following unseemly acts in my mind's eye: He has hocked a fat globule of spittle (nine of clubs) on Albert Einstein's thick white mane (three of diamonds) and delivered a devastating karate kick (five of spades) to the groin of Pope Benedict XVI (six of diamonds). Michael Jackson (king of hearts) has engaged in behavior bizarre even for him. He has defecated (two of clubs) on a salmon burger (king of clubs) and captured his flatulence (queen of clubs) in a balloon (six of spades) . . .

Gross and profane—but that's the point. This is a lot easier to remember, later on, than five of clubs, nine of clubs, three of diamonds, five of spades, six of diamonds, king of hearts, two of clubs, king of clubs, six of spades.

These are powerful methods to store seemingly arbitrary information in your head. I'll add, by the way, that these techniques make you better at intentional memory, not everyday memory. Years ago, Foer was kind enough to come lecture to my Introduction to Psychology class about his memory techniques. A few hours after he left, he sheepishly emailed me to say that he left his phone in the class.

If depth of processing and rhyming and visual imagery seem like a lot of work (and they are) a gentler bit of advice is to get some sleep. After a memory is formed, processes called *consolidation* embed it into the brain.[32] Sleep, and particularly dream states, seems to assist in the consolidation of memory.

———

Some memories are readily available, as if written on a page in front of you—all you have to do is look. The capital of France? Check. My wife's birthday? Check. Who directed *Pulp Fiction*? Check. And so on, for an incredible—though ultimately finite—number of facts.

Often memory is elicited by experience, called *retrieval cues*. If I have a dentist appointment this afternoon and I forgot about it, brushing my teeth this morning could remind me.

The most famous description of activated memory is from *Swann's Way*, the first book of Marcel Proust's *Remembrance of Things Past*, where the narrator talks about the flood of memory after biting into a madeleine dipped into tea:

> No sooner had the warm liquid mixed with the crumbs touched my palate than a shudder ran through me and I stopped, intent upon the extraordinary thing that was happening to me. An exquisite pleasure had invaded my senses, something isolated, detached, with no suggestion of its origin. And at once the vicissitudes of life had become indifferent to me, its disasters innocuous, its brevity illusory—this new sensation having had on me the effect which love has of filling me with a precious essence; or rather this essence was not in me it was me. . . . Whence did it come? What did it mean? How could I seize and apprehend it? . . . And suddenly the memory revealed itself. The taste was that of the little piece of madeleine which on Sunday mornings at Combray (because on those mornings I did not go out before mass), when I went to say good morning to her in her bedroom, my aunt Léonie used to give me, dipping it first in her own cup of tea or tisane. The sight of the little madeleine had recalled nothing to my mind before I tasted it. And all from my cup of tea.[33]

Retrieval hews to the compatibility principle: Memories come back easier in the same context in which they were acquired. The compatibility principle works not just for sameness of the physical environment, but sameness of the psychological state. If you are mildly buzzed when you study, you will remember the information better if you are mildly buzzed while tested[34]—though, I'll be quick to add, being drunk while you study causes other problems. If you are sad, you do tend to be better at remembering experiences you had when you were sad; if you are happy, happy experiences are quicker to come to mind.[35]

Sometimes you will never be able to retrieve an experience.

You might forget because you never stored the information in the first place. Michael Connelly—who writes wonderful crime fiction—has a scene in one of his novels that assumes exactly the wrong theory of memory.[36] A witness is being hypnotized and is told to imagine holding a remote control that operates a TV showing his past. He is put into a trance and asked to go to January 22, and then, after some preliminary questions, the detective asks him this:

> "Okay, James, okay," McCaleb said, interrupting him for the first time. "That's good. Now what I want you to do is take your special remote and back up the picture on the TV until the point that you first see the car coming out of the bank's parking lot. Can you do that?"
>
> "Yes."
>
> "Okay, are you there?"
>
> "Yes."
>
> "Okay, now start it again, only this time run it in slow motion. Very slow, so you can see everything. Are you running it?"
>
> "Yes."
>
> "Okay, I want you to freeze it when you get the best view of the car coming at you."
>
> McCaleb waited.
>
> "Okay, I got it."

"Okay, good. Can you tell us what kind of car it is?"
"Yes. Black Cherokee. It's pretty dusty."
"Can you tell what year?"

Great novel, but I cannot emphasize enough that this is not how memory works. It is not a recording that you can go back to whenever you are properly motivated.

Sometimes what you experienced, including what was once right in front of your face, is never attended to and so is lost forever. Sometimes the memory is stored but then forgotten. Memories are encoded in brains and brains are physical things, and so there is degradation over time. Sometimes the memory persists, but it's hard to find. In such cases, we might benefit from searching strategies. We've all had the experience of struggling to remember something, everything from the name of an actor on a television show to where we put our car keys, trying to approach the memory from different directions (What other shows was she in? What did I first do when I returned to the house last night?), and then the answer pops into our heads. It's as if memories leave tendrils and if you grab hold of one of the tendrils, you can trace your way back to the source. But often you'll never get there.

———

You can also lose memories through brain damage. The most frequent case is *retrograde amnesia*. Someone is hit on the head, in a car accident perhaps, and all the memories a few minutes prior to the experience are gone forever. The consolidation process has been interrupted; the memories are never embedded into the brain and are gone forever. With more serious brain damage, memories of a few days or even weeks prior to the accident might be lost, though some of them can eventually return.

Then there is *anterograde amnesia*, where one loses the ability to create new memories. Psychologists have learned a lot about anterograde amnesia from the study of one of the most famous patients ever, perhaps more so even than Phineas Gage, whom we met earlier. He was known as H. M., but after his death, his real name was revealed to the world—Henry Molaison.[37]

In 1953, when he was twenty-seven, Molaison went through an operation intended to treat his severe epilepsy. Doctors removed large parts of his brain, including his medial temporal lobes, the hippocampus, and most of the amygdala. His epilepsy was cured, but the effects were devastating. He lost his memory of a few years prior to surgery (retrograde amnesia), but worse, suffered from severe anterograde amnesia. From then on, he lived in a perpetual present. He met regularly with different doctors, including Suzanne Corkin, who first saw him in 1962 when she was doing her doctoral dissertation. Each time Molaison saw Corkin, for years and years and years, he would politely introduce himself to her, because for him, they were strangers.

Molaison could not remember personal experiences or factual information, but some other kinds of memories stuck. At one point, he was introduced to the difficult task of drawing a figure while looking at a reflection of it in a mirror. Each time, when he was asked to give it a try, he had no memory of doing it before, but he got gradually better, showing that his procedural memory was at least partially intact.

The discovery that someone with anterograde amnesia can nonetheless form certain sorts of memories predates Molaison.[38] In the late 1800s, a Swiss neurologist named Édouard Claparède did a cruel, albeit clever, little experiment with an amnesic patient of his. He hid a pin in his palm as he shook her hand; when he met her again, later that same day, she had no conscious memory of the pinprick, or even of meeting the doctor, but when he extended his hand, she pulled her own hand back, though she could not explain why.

———

As a last example of forgetting, we suffer from what Freud called "a peculiar amnesia"; we forget our experiences as a baby and young child.[39] Something seems to veil these early memories. Freud thought this extended to age six, which isn't even close—on average, people's earliest episodic—or autobiographical—memories go back to about age two or three.[40]

Nobody quite knows why earlier memories are lost. It may be that they are not properly stored; perhaps because the relevant brain areas are not fully developed. But this can't be entirely right, since children themselves can remember their pasts; you can talk to a two-year-old about something that happened to them a little while ago. More likely they are stored but not later accessible to us as older children or adults.

Whatever is going on here, language and culture appear to play some role. North Americans have earlier childhood memories than children in China, for instance, possibly because of differences in how adults in these cultures talk to children about the past.[41]

Now, you might think that you do have an early memory from when you were a baby. Perhaps you are an exception. But it's more likely that you are illustrating one of the most important facts about memory, which is that our memories can be distorted and shaped by factors we're unaware of. Many of our memories, even those we are confident about, are false.

Jean Piaget provides a nice illustration of this:

One of my first memories would date, if it were true, from my second year. I can still see, most clearly, the following scene, in which I believed until I was about fifteen. I was sitting in my pram . . . when a man tried to kidnap me. I was held in by the strap fastened round me while my nurse bravely tried to stand between me and the thief. She received various scratches, and I can still vaguely see those on her face. . . . When I was about fifteen, my parents received a letter from my former nurse saying that she had been converted to the Salvation Army. She wanted to confess her past faults, and in particular to return the watch she had been given as a reward on this occasion. She had made up the whole story, faking the scratches. I therefore must have heard, as a child, this story, which my parents believed, and projected it into the past in the form of a visual memory. . . . Many real memories are doubtless of the same order.[42]

This sort of thing happens all the time, and it gets to the more general problem with Connelly's story of hypnotic reconstruction, where the memory sits in the mind like a book in a library or a video on a computer hard drive. Memory is not a veridical recording of the world, stored intact, ready to be recovered through introspection or dreams or hypnosis. Rather, our experience of the past is molded by all sorts of other processes.

When we discussed perception, we saw that our experience of the world reflects a balance between what impinges on our senses and what we expect to perceive. The same sort of thing happens for memory. So, just as psychologists can create visual illusions, they can also create situations crafted to generate false memories. In one study, subjects were asked to remember a string of words, presented at a rate of about one word per second.[43]

> bed, rest, awake, tired, dream, wake, snooze, blanket, doze, slumber, snore, nap, peace, yawn, drowsy

When later asked to recall the words, people will often remember the word "sleep," even though it was never spoken. All the words are sleep-related, and so it's such a reasonable word to include on the list. Memory, like perception, is sensitive to plausibility.

There are studies in which psychologists tell people stories about somebody who has a meal in a restaurant, and later ask their subjects what they remember about the story. People add plausible details. Someone might remember that they've been told the diner paid the bill even if it wasn't in the story, just because this is what people do in restaurants.[44] Just as with the similar phenomenon in perception, this sort of "filling in" is a rational way for the mind to work.

Another example of this is from a classic paper published in 1989, called "Becoming Famous Overnight."[45] The first paragraph summarizes the findings, and it's so well written (at a level that I have very rarely seen in a scientific journal) that I'm going to quote it here:

> Is Sebastian Weisdorf famous? To our knowledge he is not, but we have found a way to make him famous. Subjects read a list of names, including *Sebastian Weisdorf*, that they were told were non-famous. Immediately after reading that list, people could respond

with certainty that Sebastian Weisdorf was not famous because they could easily recollect that his name was among those they had read. However, when there was a 24-hr delay between reading the list of nonfamous names and making fame judgments, the name *Sebastian Weisdorf* and other nonfamous names from the list were more likely to be mistakenly judged as famous than they would have been had they not been read earlier. The names became famous overnight.

We see this sort of misattribution all the time in the real world. I once told a story to friends about a funny but somewhat stressful experience I had a few years ago, and later on, my wife gently reminded me that, while the details were correct, it happened to her, not to me.

––––––

The fuzziness and malleability of memory has serious consequences, particularly as it connects to law. The most influential work on this topic comes from the lab of the psychologist Elizabeth Loftus.

Some of her research explores how leading questions can influence memory. In one study, undergraduates watched a movie where a car hit a pedestrian. Some of them were asked, "How fast was the car traveling when it passed the yield sign?" Later on, these subjects were more likely to remember a yield sign in the scene, even though it was really a stop sign. Given that the question assumed the existence of the yield sign, subjects obligingly updated their memory.[46] People are also more likely to later remember a broken headlight when they had been previously asked, "Did you see the broken headlight?" (which presumes that there was one) than when asked, "Did you see a broken headlight?" Similarly, asking, "Did you see the children get into the school bus?" makes subjects more likely to remember seeing a school bus.[47]

In other research, Loftus and her colleagues met with college students' family members and got information about events from their childhood. The students were reminded about these events and interviewed about their memories. The twist is that for each student, one of the events had never happened; it was made up by the researchers. Such events included being lost in a shopping mall, nearly drowning, spilling punch on a bride's parents during a wedding, and being attacked by a vicious animal.[48] In all these cases, some of the subjects—though not a majority of them—came to remember these false events as actually occurring.

This research has led to a revolution in the law. We now appreciate that police interrogations that are intended to tap memories can instead shape and create them. There are numerous cases where people have been put in prison and later exonerated through DNA evidence due to false eyewitness testimony shaped by such interrogations.[49] More strikingly, there are cases where people have come to believe, falsely, that they themselves have committed terrible crimes even if there is clear evidence that they are innocent.[50]

It's hard to study these sorts of false memories in a lab. For obvious ethical reasons, no psychologist can do a study where they try to trick someone into believing that they had been raped by a family member or that they had murdered someone years ago. But the real-world cases suggest that such extreme memory distortions occur, and by looking at more banal instances of induced false memories—remembering words you haven't previously seen, being lost in a shopping mall, and the like—we can get some insight into what's going on. This might guide us to modify the justice system so that memory distortion is less likely to happen in situations where it really matters.

———

This has been quite the journey, from sensation to perception to memory. There is much we've learned. We don't simply absorb sensory data from the world; we don't simply store what we perceived. Rather, our perception and memory are shaped and informed by intelligent guesses of how the world should be—what patterns of color and light typically correspond to a single object, what people typically do in a restaurant. Because our interpretations are based on what's probable, we sometimes get it wrong, and our mistakes reveal themselves when we encounter visual illusions and tell false stories about our pasts. But in a world of limited and fragmentary information, there might be no better way for the system to work.

These processes are, for the most part, hidden from consciousness. As psychologists, we can study phenomena such as how shadows influence our perception of brightness and how the workings of short-term memory limit how much we can hold in our minds. But in our everyday lives, we see what we see and remember what we remember. We carry the world in our heads, effortlessly and unconsciously.

8

The Rational Animal

Aristotle defined our species as the rational animal, but he had never heard of the Third Pounder.[1] In the 1980s, the A&W restaurant chain came up with this sandwich as an in-your-face response to McDonald's popular Quarter Pounder. The Third Pounder did better in taste tests, but its big selling point was that it had more meat—a third of a pound, not merely a quarter. Just check out the name. And as the story goes, that's why it failed. Customers felt it must have *less* meat, because the three in "Third" is smaller than the four in "Quarter."[2]

People are often irrational. We believe weird things. We screw up regularly and consistently, and we are vulnerable to those who know about our biases and blind spots and use them to exploit us.

The study of our irrationality is where psychology connects to fields such as probability theory and behavioral economics; it's where psychologists win Nobel Prizes and get invited to advise governments and multinational corporations, to hobnob with billionaires and despots at places like Davos. It's heady stuff.

I am a huge fan of this work and can't wait to talk about it. One of the charms of this part of psychology is that the findings are easy to appreciate. Much of the research I've talked about so far is removed from everyday life. Most of us haven't been in an fMRI machine to get our brains scanned; we haven't watched a baby participate in a looking-time study; the only psychoanalytic therapy we've experienced is on television and in the movies. But I'm going to demonstrate the failures of rationality on *you*, right now. Also, many of the other main findings of psychology are subtle; they might be real, but you need to test a hundred or a thousand subjects to see them. But the demonstrations of irrationality are more like visual illusions; they just jump out at you.

What does it all mean? I once spoke to a prominent psychologist about her own Introduction to Psychology course, and she told me that the main message, the single point she tries to get across to her students, is: *You aren't as smart as you think you are.* Contemporary psychologists are often embarrassed about Freud, but they would agree with him about the fundamental irrationality of our mental systems, and they see the work summarized in this chapter as an illustration of this central point: People are dumb.

My own position is the opposite. One of the main messages of this book is the extent of our marvelous cognitive powers. If I was forced to give this book a cheesy self-help title, it would be: *You Are Even Smarter Than You Think You Are.* I don't doubt the glitches and failures, but I think that these findings have been overblown and misunderstood. As I discuss this research and its implications for everyday life, I'll suggest that, with suitable caveats and qualifications, Aristotle was right after all.

———

If you go to Wikipedia, you'll find an entry called "List of Cognitive Biases," and it goes on for a while, spanning from "agent detection" to "the Zeigarnik effect."[3] Some of these supposed biases are quite specific. There is, for instance, the "rhyme as reason" bias, which states that we are more likely to believe rhyming statements than nonrhyming equivalents (one such study finds that people are more persuaded by a sentence like "What sobriety conceals, alcohol reveals" than one like "What sobriety conceals, alcohol unmasks").[4] This bias is said to explain in part the acquittal of accused murderer O. J. Simpson, where the defense made the closing argument: "If the gloves don't fit, you must acquit." It rhymes, so it's extra convincing.

But I'm not going to go through a hundred or so biases, because I don't think there really are that many. I'll reduce the list to four, and then, later in the chapter, will introduce a fifth; I think most of the rest follow from these. This list of biases might resemble the list of Freudian defense mechanisms from an earlier chapter, and this field has its own Freud—two of them, actually: the friends and collaborators Amos Tversky and Daniel Kahneman. (Kahneman, in 2002, won the Nobel Prize for this work; Tversky would have received it too, but he passed away from cancer at age fifty-seven in 1996.)

———

Let's start with what's called the *availability bias*. Like many of the biases, it has to do with how we figure out the probability of an event. What are the odds that a swimmer in Cape Cod will be eaten by a shark; who is most likely to win the next election; how much can I count on my investments going up in value? We implicitly make probability judgments all the time. When I step into the car to drive to the airport to pick up a friend, I am tacitly making assumptions about the likelihood of getting killed in a car crash on the way and the likelihood of my friend arriving on the scheduled plane. (I assume that the odds of the first are low and the odds of the second are high; if I didn't, I wouldn't go.)

Probability theory tells us how to properly think about odds. For instance, there is the multiplication rule: the probability of two unrelated events = the probability of one event × the probability of the other event. The odds of getting a five when you throw a die are 1 in 6, the odds of getting a five in another throw are 1 in 6, so the odds of two fives in a row are $1/6 \times 1/6 = 1/36$ or 1 in 36.

This might seem straightforward, but we often get probability wrong. Try this:

Is a word more likely to begin with the letter K or to have K as the third letter?[5]

If you thought the first option was more likely, then you're in the majority—about 70 percent give this answer. But it is wrong; there are about twice as many words where K is the third letter than the first. Given the way the mental lexicon (the storage of words in our head) is structured, it's easier to come up with words based on the first letter. It's easier to think of KITE than RAKE. Given the availability bias, if something comes to mind quickly, we infer that it is more likely.

You might object that getting the above wrong isn't a fallacy of reasoning; it's just a quirk of how words are recovered from memory. But now consider this:

Is it more likely that a seven-letter word has "n" as its sixth letter or ends with "ing"?[6]

People often say "ends with 'ing'" because, again, many more words of this sort come to mind here. But if you think about it, this must be wrong. Every seven-letter word that ends with "ing" has "n" as its sixth letter, and so "ing" words can't be more frequent. (They are actually less frequent, because of words like COMMENT.)

Here is a similar case:

What is the chance that a massive flood would occur somewhere in North America?

What is the chance that a massive flood would occur due to an earthquake in California?[7]

This is the same trick as the "ing" case. People often say that the California scenario is more likely. But this can't be right. Every event in California is also an event in North America, so the former can't be more likely.

Perhaps the most significant real-world instance of the availability bias is our tendency to overestimate the likelihood of events that have emotional force, such as plane crashes, shark attacks, and violent crime. People keep their children home because they worry about school shootings but have no qualms about long cross-country drives, which are, statistically, far more likely to end their offspring's lives. But when families die in car crashes, it doesn't end up on the front page of national newspapers. It is not as available and so we mistakenly think it's not as frequent.

———

If you are estimating the likelihood of a single event, you should base your estimate on the overall probability of such types of events happening—their base rates. Suppose there are a dozen bagels in a bag and you reach in to grab one, and you think it just feels more like poppy seed than sesame seed. If you had no other information, you'd bet on poppy. But it should matter what the distribution of bagels in the bag is—is it half poppy and half sesame, or one poppy and eleven sesame—or twelve sesame and no poppy? Base rates matter.

But we often don't focus enough on them; we often suffer from what's called *base-rate neglect*. Here is a rather vivid example.

> You are tested for a terrible disease. The test never misses the disease; if you have it, you will test positive. But 5 percent of the time, the test will say that you have it when you don't. (In other words, there are 5 percent false positives.)
>
> You test positive.
>
> How much should you worry?[8]

Many people will say A LOT: I mean, holy cow, there is a 95 percent chance of having the disease.

But no, there isn't. You have no idea of the odds of having the disease because I didn't tell you about the base rates.

Suppose 1 out of 1000 people tested have the disease. Now, what are the odds that you have it?

Let's work out the math. Imagine that 200,000 people are tested.

- On average, 200 out of the 200,000 (1 in 1,000) will have the disease. They will test positive.
- 199,800 of them will not have the disease. Since the test is wrong 5 percent of the time, then one twentieth (9,990) of these people will also test positive.
- We see that 10,190 people tested positive (200 + 9,990), and you're one of them. But only 200 of them have the disease. And so your odds of having the disease are really 200/10,190, about 2 percent, which is a lot less worrying.

It's simple math when you work it out, but it doesn't seem natural. (Some of us would still freak out when we got this test result.) When Kahneman and Tversky asked people about this scenario, the average answer about the probability of having the disease was 56 percent and half the people said 95 percent, ignoring the base rates entirely.

Here's a less mathematical example.[9] Suppose you hear about an American named Rob who is an opera buff who likes to tour art museums and was totally obsessed with classical music as a child. Which is more likely, that he plays trumpet in a major symphony orchestra . . . or that he is a farmer?

He sure seems like a stereotypical musician. But now look at the base rates. In the United States, there are (this will be approximate, but you get the point) about three hundred trumpeters in major symphony orchestras and about two million farmers. So regardless of what we know about him, Rob is almost certainly a farmer. Similarly, regardless of how much your next-door neighbor seems like a serial killer, he probably isn't, because there just aren't that many serial killers.

———

A third phenomenon is that we are sensitive to how information is framed. Tversky and Kahneman asked participants to decide between two treatments for six hundred people who contracted a fatal disease.[10]

Treatment A would result in 400 deaths.

Treatment B had a 33 percent chance that no one would die but a 66 percent chance that everyone would die.

Which treatment is better? Treatment A ends up with 400 deaths for sure. If you repeat Treatment B over and over again, you will get on average (2/3 × 600) = 400 deaths. One's preference, then, rides on whether you would rather have an awful outcome or a small chance of a much better outcome combined with a large chance of a somewhat worse outcome. This is a tough call; neither choice is, in itself, wiser than the other. The math doesn't give you the right answer.

In general, the probabilities and the payoffs—the expected values—never tell you, by themselves, the best decision to make. Everything depends on your tolerance for risk and on what you think of the costs and benefits. Suppose someone offers to flip a coin with me: If tails, I lose my life savings. Everything I own is gone. But if heads, I end up with three times everything I have. If a hundred people took the bet, and each started with $100,000, then about half would go broke, and about half would get $300,000, and, on average, people would end up with $150,000, which is $50,000 more than they started with. Just looking at the numbers, then, this seems like an excellent bet. But personally, this would be a *terrible* bet. The pain I would get from losing all my savings will make me far more unhappy than the windfall will make me happy. Other people might be less risk averse. The point is that the decisions that one makes depend on all sorts of considerations, including how you would be affected by gains and losses.[11]

Still, one can make dumb choices here, and one way to do so is to think about the problem differently depending on how the options are framed. Return to the example above and compare these two framings, keeping in mind that six hundred people have caught this disease:

Positive framing: "Saves two hundred lives" (Treatment A) versus "a 33 percent chance of saving all six hundred people, 66 percent probability of saving no lives" (Treatment B)
Negative framing: "Four hundred people will die" (Treatment A) versus "a 33 percent chance that no people will die, 66 percent probability that all six hundred will die" (Treatment B)

These are the same dilemmas just described in different ways, and so it would be rational to make the same choice for both. But Treatment A was chosen by 72 percent of the people who got the positive framing and by only 22 percent of the people who got the negative framing. Basically, saving two hundred lives sounds good and consigning four hundred people to die sounds bad—even though they are describing the very same event.[12]

Framing effects show up all over the place. It's better to describe condoms as 95 percent effective than as failing 5 percent of the time. If a conference is going to charge people differently depending on when they sign up, it's nicer to describe this as a discount for early registration than as a penalty for late registration. If you wanted to get people to use some service you were offering, and it takes forty-nine minutes on average, best to say "We get it done in less than fifty minutes!"

As a final example, imagine that you had to rule on a custody case, where only one of the parents can get custody of the children.[13] Here is the information about the parents:

- Parent A is average in every way—income, health, working hours—and has a reasonably good rapport with the child and a stable social life.
- Parent B has an above-average income, is very close to the child, has an extremely active social life, travels a lot for work, and has minor health problems.

Who should get the children? Well, I don't know, but I do know that the specific framing of the question shouldn't matter—and it does. If you ask who should be awarded custody, people are more likely to say B . . . and if you ask who should be denied custody, people are also more likely to say B! The explanation for this is that when asked about awarding custody, you notice special factors in B's favor (income, closeness to child), while there's nothing especially good about Parent A. But when asked about denying custody, you look for negative considerations, and you also find them in Parent B (social life, travel, health problems), while there's nothing especially bad about Parent A. Plainly, something has gone wrong here—B cannot be both the best choice to award custody and the best choice to deny custody—and framing effects are to blame.

———

Finally, we often seek out information that fits our biases—a *confirmation bias*. We look for information that supports our beliefs and values. When people go to a doctor and get good news—"You don't have cancer"—they are unlikely to demand a second opinion. We normally read books and go to websites that support our prejudices, not ones that challenge them.

In the laboratory, this shows up in several ways. The psychologist Peter Wason developed what's come to be known as the Wason Rule Discovery Task, where you must find a rule that captures a series of numbers.[14] To start you off, I'll tell you that the following sequence satisfies a certain rule:

2   4   6

Your job is to figure out what the rule is. To do so, you should think up other strings of numbers and I'll tell you whether these also satisfy the rule.

A typical guess is that the rule is a sequence of even numbers, and so people often propose something like:

4   8   12

And yes, that fits, and so the guesser thinks: Got it! But the trouble here is people rarely try to *falsify* their hypotheses. If you think the rule is ascending even numbers, why didn't you try ascending odd numbers such as 5 7 9 or descending numbers such as 12 8 4? If the answer is no for these series, then you have evidence for your hypothesis. Indeed, the actual rule that Wason thought up is simply that the three numbers are ascending, and many of his participants couldn't work out the rule, because they only tested sequences that confirmed their hypothesis, not ones that could disconfirm it.

Confirmation bias is not something that happens only in logic puzzles. We saw earlier that psychologists such as Skinner and Freud went awry in part because of confirmation bias—they sought to prove their theories true, not to test whether they might be false.

As another example of the confirmation bias, consider the Wason Selection Task.[15] Here we are moving away from judgments of probability and onto deductive logic, where one reasons about what necessarily follows from the premises. But we stumble with this as well.

Here are four cards. Each has a number on one side and a letter on the other side. You only see one side; the other is hidden.

Now consider this rule:

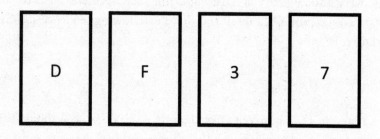

If a card has a D on one side, it has to have a 3 on the other side.

Which of the cards above do you have to turn over to see if the rule is satisfied?

Most people answer D, and it is right—if you turn the D and there's not a 3 on the other side, the rule isn't satisfied. But what else did you choose? If you chose 3, you're mistaken. (The rule doesn't say anything about cards with 3s on one side; the other side could be a D or not a D, it doesn't matter.) The other right card to turn over is the 7. If the other side of the 7 is a D, then the rule is false. But only about 10 percent of people get this right.

I can't resist sharing a joke that the writer Jon Ronson made on social media: "Once I heard about the confirmation bias, I can't help but see it everywhere."

A while ago, the psychologist (and my undergraduate mentor) John Macnamara noted that these failures of reasoning reveal two things about our
minds.[16] Most obviously, they illustrate irrationality, how we mess up.
This is the conclusion that psychologists usually draw. But what's sometimes missed is that they also show how intelligent we are. After all, we
know that they are mistakes. Upon reflection, we appreciate the relevance
of base rates, we acknowledge that there cannot be more floods in California than floods in North America, and we agree that asking about
who gets custody and who doesn't get custody are really different ways
of asking about the same thing. When we hear the story about the Third
Pounder, we get it; we shake our heads at how dumb people can be.

Every demonstration of our irrationality, then, is also a demonstration of how smart we are, because without our smarts we wouldn't be able
to appreciate that it's a demonstration of irrationality in the first place.
(When a psychologist says, "People are so dumb," the assumption here is
that at least one person is smart—the psychologist.) This is one of the remarkable capacities of the human mind, that we can recognize and reflect
on our mistakes and avoid making them in the future.

Why are we so vulnerable to such mistakes in the first place? One
approach is to think of them as akin to our susceptibility to illusions, as in
the Kanizsa triangle, discussed earlier in the context of vision. Our visual
system builds in expectations and biases that usually allow us to see things
as they are, but clever artists and psychologists can arrange scenes where
these expectations go awry. Similarly, our reasoning has evolved to do
well with real-world problems (more precisely, real-world problems that
arose in the environment in which we evolved), but unusual or hypothetical situations trip us up—reasoning illusions instead of visual ones.

Some reasoning illusions are created by clever psychologists, as with
the Wason Selection Task. But most arise in the real world. Someone
trying to get people to believe that crime is increasing will tell stories of
shocking criminal acts, appreciating, at least at an intuitive level, how
the availability bias distorts our judgments of frequency. Someone trying to discourage people from taking a drug that is 99 percent safe will

emphasize the 1 percent risk factor, framing the statistics in the way that best makes their point.

Other times, there is nobody to blame. There is no conspiracy afoot to make people believe that shark attacks are relatively frequent; this is just a product of how the media works. There is no hidden agenda to get people to ignore base rates in medical diagnoses; this just reflects the difficulty we have with problems that are framed in terms of percentages.

We do better when faced with simpler cases. Consider the base rates in the following scenarios: *You wake up with the sniffles and a headache.* These are the symptoms of a cold or flu. They are also the symptoms of bubonic plague. *There is a thick letter in the mail, with your name and address in fancy script.* An offer from the Marriott hotel, or have you just been given a knighthood from the king of England? *Your friend is late for a meeting.* A bit of traffic, or has she been eaten by bears? In such cases, the situation—the sniffles, the letter, the late friend—is compatible with both options, but when you make your likelihood estimate, you are smart enough to attend to the base rates of different options. The classic reminder to medical students—if you hear hoofbeats, think of horses, not zebras—captures this logic.

Often our performance improves when the same problems are framed in more natural ways—natural in the sense of matching the conditions in which our minds have evolved, and natural in the sense of meshing with everyday experience. Consider again the Wason Selection Task, and remember how hard it was to confirm the rule "If a card has a D on one side, it has to have a 3 on the other side." But now phrase the same problem in a way that has more real-world context.[17] You are a bouncer at a bar and the rule is,

If someone is drinking beer, they have to be over twenty-one.

These cards tell us someone's age on one side and what they're drinking on the other. Which of the cards on the following page do you have to turn over?

Most people get this right—you have to check the person drinking beer *and* the person who is fifteen. Translated into a real-world context, this becomes much easier.

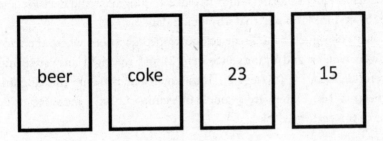

One conclusion here is that we get things right in natural contexts and wrong in unnatural ones. But that would still be understating how smart we are. As Macnamara pointed out, we can, upon reflection, get the unnatural ones right. We can think like logicians.

There is a very old theory that posits the mind contains two opposing parts, one that is emotional and instinctual, the other that is capable of careful deliberation. The best-known modern version of this theory was developed by Daniel Kahneman and his colleagues, and it is nicely summarized in the title of Kahneman's bestselling book, *Thinking, Fast and Slow*.[18]

For Kahneman, there are two systems, broken up as follows:

System 1	System 2
Fast	Slow
Parallel	Serial
Automatic	Controlled
Effortless	Effortful
Associative	Rule-governed
Slow-learning	Flexible
Emotional	Neutral

"System 1" versus "System 2" is now very much in the language of the working psychologist. A friend of mine remarked on the awkward dancing of older scholars at a conference on judgment and decision making, by saying, "You see, that's System 2 dancing."

My colleague Shane Frederick developed questions where the wrong answer is quick and intuitive (System 1) and the right answer requires careful deliberation (System 2). These questions make up the Cognitive Reflection Test.[19] Here are a couple of examples from a more recent version of this test.[20]

> If you're running a race and pass the person in second place, what place are you in?
> Emily's father has three daughters. The first two are named Monday and Tuesday. What is the third daughter's name?

When faced with the first question, many people respond by saying "first place," but if you think about it, it's second place. The intuitive answer for the next question is "Wednesday," but nope, it's . . . Emily. If you got the wrong answer, you're not dumb, you're just not reflecting deeply. You're going with your gut. You're in thrall to System 1. The extent to which you rely on System 1 for such questions connects to other things of importance, such as how likely you are to believe conspiracy theories on social media[21] and how likely you are to believe in God.[22]

How can we encourage better reasoning? One focus has been on so-called fake news—stories that spread like crazy but aren't true. How can you make people think harder about the stories they encounter, to use their System 2 and not their System 1?

Looking at false claims about COVID in 2020, one team of scholars found that people get more focused on truth if you simply give them some sort of reminder, direct or indirect, to focus on accuracy.[23] For instance, if you get people to rate the accuracy of a neutral headline (having to do with the discovery of a new star, for instance), it seems to bring out System 2—afterward, people are about three times more discerning when deciding what sort of information to share on social media.

The conclusion here is threefold, then. First, we can be irrational in important ways. Second, this irrationality is more likely in unnatural conditions, of the sort that our mind hasn't evolved to deal with. And third, even in such cases, we have the potential to do better.

———

If we're so smart, why do we often seem so dumb? Take conspiracy theories. There are those who deny the existence of the Holocaust, who believe that the September 11 attacks were an inside job, who think that the COVID pandemic is a fraud. Many Americans currently believe in QAnon, a conspiracy that rests on claims supposedly made by an anonymous individual (or a group of them) known as Q. Believers think that Donald Trump was recruited by top military generals to bring to justice a cabal of Satanist cannibalistic pedophiles that includes Barack Obama, Joseph Biden, Hillary Clinton, Pope Francis, the Dalai Lama, and, most implausibly, beloved actor Tom Hanks.

Many conspiracy theories are motivated by politics, and politics is an area where we seem the most vulnerable to irrational thought. I'm writing this book during a pandemic, so I'll lead with this example: In the United States, beliefs about the coronavirus became highly politicized. To put it crudely, as I write this, Democrats worry about this disease a lot and are in favor of vaccines, masks, and lockdowns, and Republicans don't worry about it as much and are relatively less in favor of such interventions. At this point, this looks like a simple disagreement, the sort that reasonable individuals with different priorities might have. But it led to serious errors on both sides. In a poll published in March 2021, about one third of Republican voters said that people without COVID symptoms could not spread the virus (they actually could), and that COVID was killing fewer people than the seasonal flu (it was killing about fifteen times more people).[24] On the other hand, many Democrats believed that a high proportion of COVID patients had to be hospitalized (the correct proportion was between 1 and 5 percent), and that a considerable share of COVID deaths occurred in children (actually, people under eighteen accounted for less than one thousandth of COVID deaths).

Perhaps Republicans and Democrats are just listening to different experts. But there is abundant evidence that, even when given the very same information, our beliefs and preferences are distorted by our political affiliations.

In one demonstration of this, subjects were told about a proposed welfare program, which was described as being endorsed by either Republicans or Democrats, and were then asked whether they approved of it.[25] Some subjects were told about an extremely generous program, others about an extremely stingy program, but this made little difference. What mattered was the political affiliation of the person being asked: Democrats approved of the Democratic program, Republicans approved of the Republican program. When asked to justify their decision, though, participants insisted that party considerations were irrelevant; they felt they were responding to the program's objective merits.

A recent meta-analysis reviewed fifty-one studies of this sort and the evidence is strong: Liberals and conservatives respond to a scientific conclusion or a proposed policy more based on whether it supports their political beliefs and affiliations than on its actual merits.[26] Our political minds seem to support the aphorism from Anaïs Nin we quoted in an earlier chapter: "We don't see the world as it is, we see it as we are."

This all illustrates what can be seen as the mother of all biases—a variant of the confirmation bias that psychologists call the *myside bias*.[27] The writer and prominent rationalist Julia Galef has a different term, though.[28] She calls this style of thought the *soldier mindset*. She notes that this military metaphor for our beliefs and attitudes pervades our thought and our language:

> It's as if we're soldiers, defending our beliefs against threatening evidence. . . . We talk about our beliefs as if they're military positions, or even fortresses, built to resist attack. Beliefs can be *deep-rooted, well-grounded, built on fact*, and *backed up by arguments*. They *rest on solid foundations*. We might hold a *firm* conviction or a *strong* opinion, be *secure* in our beliefs or have *unshakeable* faith in something.
>
> Arguments are either forms of attack or forms of defense. If we're not careful, someone might *poke holes in* our logic or *shoot*

down our ideas. We might encounter a *knock-down* argument against something we believe. Our positions might get *challenged*, *destroyed*, *undermined*, or *weakened*. So we look for evidence to *support*, *bolster*, or *buttress* our position. Over time, our views become *reinforced*, *fortified*, and *cemented*. And we become *entrenched* in our beliefs, like soldiers holed up in a trench, safe from the enemy's volleys.

And if we do change our minds? That's surrender. If a fact is *inescapable*, we might *admit*, *grant*, or *allow* it, as if we're letting it inside our walls. If we realize our position is *indefensible*, we might *abandon* it, *give it up*, or *concede* a point, as if we're ceding ground in a battle.

We see the soldier mindset at its extreme in politics. We tend to affiliate more with members of our political group, interact with them more on social media, and read and transmit their views to our friends.[29] This leads to a powerful form of groupthink. We are soldiers, and our political communities are the armies we belong to.

But Galef summarizes research suggesting that this mindset extends more broadly. Do you identify as a "gamer"? Then you will be more skeptical about studies that show violent video games are bad for you.[30] Do you feel solidarity with the Catholic faith? Then you are more skeptical when you hear that a Catholic priest is accused of sexual abuse.[31]

Consider intuitions about a good legal system. Here are two questions about the same policy, framed in different ways.

1. If you get sued and you win the case, should the person who sued you pay your legal costs?
2. If you sue someone and you lose the case, should you pay his costs?

Eighty-five percent say yes to the first, 44 percent say yes to the second.[32]

Many believe that these findings and observations show that we are irrational when it counts. But I think there are reasons to be cautious about this conclusion.

First, it is not irrational to hold *some* conspiracy theories. Some of them are correct. *The Washington Post* did a quiz on "know your conspiracy," and here are some of their items: Which of these do you believe really happened?[33]

A. The FBI kept tabs on civil rights leaders such as the Rev. Martin Luther King Jr., attempting to find compromising information and damage his reputation.

B. During Ronald Reagan's presidency, government officials secretly and illegally sold weapons to Iran and used the money to fund Nicaraguan revolutionaries.

C. Hillary Clinton conspired to provide Russia with nuclear materials.

D. The U.S. government secretly dosed Americans with LSD in an attempt to develop mind-control technology.

E. The U.S. government knew hundreds of Black men in Alabama had syphilis, but told them they had "bad blood" and withheld treatment as part of a medical experiment.

F. The dangers of genetically modified foods are being hidden from the public.

G. Fossil-fuel companies like Exxon knew about climate change for decades, but spread misinformation about the issue to deflect blame and influence environmental policies.

According to the *Post*, C and F are false; all the rest are real. But it's not obvious from the descriptions that it's stupid to believe in C and F and reasonable to believe in the others.

In some cases, the jury is still out. Consider these two:

Jeffrey Epstein, the billionaire accused of running an elite sex trafficking ring, was murdered in prison to cover up the crimes of powerful politicians.

Donald Trump colluded with Russians to steal the presidency in 2016.

The *Post* says these are both false and I agree, but I know smart people who believe in each of them, and nothing about their beliefs seems particularly unreasonable. When George W. Bush was president, about half of Democrats thought he let the September 11 attacks happen to give him an excuse to start a war in the Middle East, and as I write this, most Republicans believe that Donald Trump won the 2020 election. These theories are again probably both false, but they don't indicate minds that are working badly—after all, politicians *will* scheme to go to war; elections *are* stolen. It's a poor psychology that defines "irrational" as "believing in different conspiracy theories than the ones I believe in."

But what about QAnon, or the idea that the moon landing was a fraud, or the notion that school shootings such as the one in Sandy Hook were faked by the government to drum up support for stricter gun laws? What about the view that the world is run by shape-shifting lizard people from another planet? Some conspiracy theories really do border on delusional in their disconnection with the facts.

There is a sense in which believing in these conspiracies is irrational. But philosophers and psychologists have pointed out that the weirder theories that people endorse have a funny psychological status.[34] Steven Pinker notes that some conspiracy theories involve

the world beyond immediate experience: the distant past, the unknowable future, faraway peoples and places, remote corridors of power, the microscopic, the cosmic, the counterfactual, the metaphysical. People may entertain notions about what happens in these zones, but they have no way of finding out, and anyway it makes no discernible difference to their lives. Beliefs in these zones are narratives, which may be entertaining or inspiring or morally edifying.[35]

You can see these wilder theories as similar to salacious celebrity gossip, judged more as entertaining stories than as facts about the world.

There is little tangible cost to holding these theories. If I have the wrong idea about how to fill my car with gas, or what clothes to wear for a formal wedding, or how to change a diaper, I will pay for my mistakes. But if I believe in QAnon, it might not matter that much in my personal life. So long as I'm surrounded by people who share these conspiratorial views and don't ostracize me for holding them, I'll get along just fine.

This conclusion might seem perverse. How can it be rational to believe stupid things or rely on biased sources of information?

A lot depends on what is meant by "rational." The notion of rationality I have been assuming here involves properly using knowledge and logic to achieve your goals.[36] If you're stepping outside and it's raining and you don't want to get wet, it's rational to bring an umbrella. But critically, the rationality of the choice is grounded in the goal of not getting wet. If this goal changes—if you decide you want to joyously dance naked in the downpour—then bringing an umbrella isn't rational at all.

Rationality defined in this manner is separate from goodness. Kidnapping a rich person's child might be a rational way to achieve the goal of getting a lot of money quickly, so long as you don't have other goals, such as obeying the law and not being a horrible person.

Rationality is also separate from seeking the truth about the world. I was at a dinner once when Donald Trump was president, and we were all complaining bitterly about him. Someone mentioned the latest ridiculous thing he did, and we were all laughing, and then a young man, no fan of Trump, politely pointed out that this event didn't really happen the way we thought it did. It was a misreporting by a partisan source; Trump was blameless. People pushed back, but the man knew his stuff, and gradually most of the room became convinced. There was an awkward silence, and then someone said, "Well, it's just the sort of thing that Trump *would* do," and we all nodded, and the conversation moved on.

Was the young man's contribution a rational one? It depends on his goals. What was he most hoping to accomplish—to know and speak the truth, or to be liked? If your goal is truth, then having a soldier mindset, being biased to defend the positions of your group because of loyalty and affiliation, is plainly irrational. Truth-seeking individuals should ignore political affiliation when learning about the world. When forming opinions on gun control, evolution, vaccination rates, and so on, they should seek out the most accurate sources possible. And when it comes to conspiracy theories like QAnon, they should laugh them away—such theories are ridiculous, and one shouldn't believe ridiculous things.

But we are social animals. While one of the goals that our brains have evolved to strive for is truth—to see things as they are, to remember them as they really happened, to make the most reasonable inferences based on the limited information we have—it's not the only one. We also want to be liked and accepted, and one way to do this is by sharing others' prejudices and animosities.

———

One can step back here and ask what goals one *should* have. I would be mortified if one of my sons thought that our ancestors rode dinosaurs, even though I can't think of a view that matters less for everyday life. I believe that we *should* aspire to the truth. It's an important goal in its own right. Science, history, and similar pursuits are of value even if the truths they yield might not improve our lives in any tangible sense.

And truth seeking typically does have a payoff. We'll talk about some possible exceptions later on—perhaps it's adaptive to have an overly positive impression of one's own abilities, or an unrealistic worry about certain dangers—but for the most part, accurate perception is better than hallucination, grounded belief is better than delusion, the proper application of the laws of probability and logic is better than systematic errors in reasoning.

Finally, the rejection of truth in political thought leads to a collective action problem. In 1833, the economist William Forster Lloyd introduced one version of such a problem—what's come to be known as "the tragedy of the commons."[37] Lloyd imagined a pasture open to all. It's in the interest of each herdsman to keep his cattle grazing there—to hold back would be to suffer a cost—but the consequence of overgrazing is that the resource will be depleted and eventually destroyed, and then everyone will suffer. Such problems are everywhere. It might be best if most people vote, for instance, but voting is in nobody's individual interest; it's virtually impossible for any single vote to make a difference, so better to sit home and not waste the time and energy.

Our style of political thought can be seen as a cognitive tragedy of the commons. If everyone around you has silly views, then it might be more useful for you, as an individual, to go along with the crowd. With some important exceptions—such as refusing to get vaccinated because of the conspiratorial views of your community—you're better off conforming. But there will ultimately be tragic consequences if most people are indifferent to the facts, particularly when public opinion plays a large role in determining how nations deal with issues such as climate change, economic policy, and war.

But I don't want to end this chapter on a sour note. We have here the same combination of bad news and good news we saw in our earlier discussion of cognitive biases. We are not angels, our cognitive powers are limited, and we make all sorts of mistakes. But we are also smart enough to notice that they are mistakes and to take steps to address them. In the case of these collective action problems, societies can work to solve them, through encouraging individual restraint and creating regulations, laws, and customs. Yes, we are the animal that rejects the Third Pounder, believes in QAnon, and selfishly destroys the commons. But we are also the animal that is capable of acting intelligently for the long-term good, that votes and recycles and works out a schedule for grazing, and that, sometimes, decides to abandon the soldier mindset and tries to see things as they really are.

APPETITES

9

Hearts and Minds

O ne of my favorite articles in all of psychology speaks to the question of human motivation. It was published in 1937 by Edward Thorndike, a leading behaviorist whose work on the Law of Effect we discussed earlier.[1] But the article, entitled "Valuations of Certain Pains, Deprivations, and Frustrations," doesn't read as if it was written by a behaviorist or anyone with a strong theoretical ax to grind. It seems motivated by sheer curiosity.

Thorndike was interested in "pains, discomforts, deprivations, degradations, frustrations, restrictions, and other undesired conditions." To explore this, he did a survey. He made up a long list of unpleasant activities and experiences, and asked people how much money they would accept to do them or go through them.

Below are some samples from his chronicle of bad tidings, ordered from least bad to worst. The numbers that follow are the median responses by students and teachers in psychology. Keep in mind that the money corresponds to U.S. dollars in the 1930s; to translate it to current dollars, multiply by about sixteen. "No sum" means that the median response is to refuse to accept any amount of money at all.

Spit on a crucifix. [$300]
Suffer for an hour pain as severe as the worst headache or
toothache you have ever had. [$500]

Have one upper front tooth pulled out. [$5,000]

Lose all hope of life after death. [$6,500]

Choke a stray cat to death. [$10,000]

Have the little toe of one foot cut off [$10,000]

Spit on a picture of your mother. [$10,000]

Eat a live beetle one inch long. [$25,000]

Attend Sunday morning service in St. Patrick's Cathedral, and
in the middle of the service run down the aisle to the altar,
yelling, "The time has come, the time has come," as loud as you
can until you are dragged out. [$100,000]

Become unable to smell. [$300,000]

Become entirely bald. [$750,000]

Have to live all the rest of your life on a farm in Kansas, ten miles
from any town. [$1,000,000]

Eat a quarter of a pound of cooked human flesh (supposing that
nobody but the person who pays you to do so will ever know
it). [$1,000,000]

Eat a quarter of a pound of cooked human flesh (supposing that
the fact that you do so will appear next day on the front page
of all the New York papers). [No sum]

Have one ear cut off. [No sum]

Become unable to chew, so that you can eat only liquid food. [No
sum]

Have to live all the rest of your life shut up in an apartment in
New York City. You can have friends come to see you there,
but you cannot go out of the apartment. [No sum]

Some of these responses make sense. People want to avoid pain and
mutilation. We want to keep our senses in working order and retain our
freedom to travel. We value our reputations. I, too, wouldn't want to be on
the front page of the *New York Times* with the story of how I ate a quarter
pound of human flesh.

But ten thousand dollars—well over a hundred thousand dollars
today—to spit on a picture of one's mother? This seems like a lot—the
same as the sum required to have a little toe chopped off! Or consider the

high price required to choke a stray cat to death, twice as much as what subjects wanted for the agonizing experience of having a front tooth pulled out. If someone were to tell you that people only care about avoiding pain and getting pleasure, these findings make the start of an excellent rebuttal.

By looking at what we would pay to avoid, we can learn about our priorities. Of course, one cannot simply invert these items—just because getting a toe chopped off is aversive doesn't mean that people would pay to get an extra toe. But Thorndike's list shows the value we give to our dignity and our reputations, to being morally good, to having a diversity of experiences, to showing respect to those we love. It illustrates how motivation is complex and diverse; we want many things.

The rest of this chapter explores the mysteries of motivation. It starts with general theories of goals, instincts, and emotions—What are they for? How did they evolve?—and then turns to two of the most interesting aspects of human nature: sexuality and morality.

———

Some scholars believe that, at root, humans just want one thing. Indeed, many believe that the ultimate question of human motivation can be answered in just a few words.

One early proposal is that the goal of the human brain is homeostasis, or:

to maintain a consistent fluid bath around the cells of our body. [2]

This sounds odd, but there is a logic to it. To survive, animals must keep their internal environment within certain parameters—the right amount of glucose to provide energy, the right amount of salt, the proper temperature and so on. Scholars in the nineteenth century proposed that more specific drives such as hunger and thirst serve this more general goal of balance or homeostasis.

A more recent proposal is that the brain seeks to:

minimize prediction error.

This comes from a theory known as predictive processing (or PP).[3] The idea here is that we aspire to make increasingly correct predictions about the world. We've touched on this before when talking about perception. Any visual experience is consistent with all sorts of hypotheses as to what one is seeing, and the goal of the visual system is to make the right guess, to resolve this uncertainty.

The PP theory gets radical when it comes to explaining action. The commonsense view of why, say, someone reaches for a glass of water is (a) they want to drink water (they have a desire) and (b) they think that bringing the glass to their mouth will satisfy this desire (they have a belief). In general, we act because we have desires and we believe that our actions will satisfy these desires. What could be more obvious?

The PP theory has a different account. Under that theory, what goes on in this particular action is that the mind generates the currently false prediction that the person is now drinking a glass of water, and then, to make the prediction come true, the person reaches for the glass. Don't talk about someone wanting to own a donut shop; talk instead about someone predicting that they own a donut shop and struggling to fix up the mismatch between this prediction and their current state by becoming the owner of a donut shop.

To close it out, here's a third theory that needs no elaboration and rings true for many people:

Avoid pain / Seek pleasure.

––––––

These are plausible enough goals. We do need to attend to our fluid bath; the brain does make predictions about the world; we do avoid pain and seek out pleasure.

The problem is that none of these theories provide a comprehensive account of motivation. The homeostasis theory has the virtue of being specific—which means that we can easily see how it's incomplete. It works well for hunger and thirst, but doesn't explain motivations such as curiosity, sexual desire, concerns about status, and much else. None of these have anything to do with one's fluid bath.

The other theories suffer from a different sort of problem. Consider one of the Thorndike findings—his subjects didn't want to eat a live beetle. What would a PP theorist say about this? Well, an easy response is "Our brains predict that we are not eating live beetles and so we work to make this prediction come true." What would an advocate of the pain/pleasure theory say? Also easy: "Eating live beetles is painful and not eating them is pleasurable."

I hope you agree that these responses aren't satisfying. They aren't explanations; they are redescriptions. Suppose that Thorndike had discovered instead that his subjects love eating live beetles. One could just as well say that the brain predicts beetle eating and wants to make the prediction come true, or that this activity gives pleasure, not pain. When you can easily account for any conceivable course of events, you're not explaining things at all.

This sort of worry should be familiar from the behaviorism chapter, where we talked about Chomsky's critique of Skinner. In a critical discussion of the PP theory, the psychologists Zekun Sun and Chaz Firestone make this connection to Chomsky concrete, and it's worth quoting them at length:

> Children and adults, Chomsky noted, do things like talk to themselves when nobody is around, make music in private, or imitate the sounds of cars and airplanes—none of which is typically a "rewarded" behavior. So why, on behaviorism, do we do such things? Skinner's answer was "self-reinforcement": we talk to ourselves because it feels rewarding to do so, such that we are the reinforcers of our own behavior. Chomsky replied, rightly, that appeals to self-reinforcement actually undermine behaviorist explanations, because they are either (i) false (is talking to oneself really "rewarding"?), or (ii) trivially true—a panacea that could explain any behavior imaginable. And mechanisms that can explain anything ultimately explain nothing, because they become empty or

unfalsifiable: "When we read that a person plays what music he likes, says what he likes, thinks what he likes, reads what books he likes, etc., because he finds it reinforcing . . . the term 'reinforcement' has no explanatory force."

We worry that "self-prediction" shares this property with self-reinforcement, and so risks a similar dilemma for PP accounts. Either the self-prediction account is: (i) false . . . or (ii) trivially true, accommodating any possible behavior. Why do we dance? Because we predict we won't stay still. Why do we donate to charity? Because we predict we will do good deeds. Why do we seek others? Because "the brain has a prior [prediction] which says 'brains don't like to be alone.'" In some moods, these answers will land as deep and profound truths about the mind; but in others they will simply be nonexplanations that lead one to repeat one's question.[4]

The same critique applies to the theory that we just want to get pleasure or to avoid pain. This is either false—there's no literal sense in which poking at a sore tooth or weeping at a tragedy is pleasurable—or trivially true, because the notions are used so broadly that anything one willingly does counts as pleasure seeking and pain avoiding. Either way, it's not a satisfying account.

———

Let's move away from one-sentence theories, then, and consider the idea that we possess many motivations. Some are shared with other creatures; some are uniquely human. Some are for the benefit of the individual; others are directed toward others.

One such approach was outlined by William James in his great book *The Principles of Psychology*, published in 1890.[5] It's clear, James notes, that other creatures possess *instincts*, defined "as the faculty of acting in such a way to produce certain ends, without foresight of the ends, and without previous education in the performance." Spiders spin webs; birds build nests; all sorts of creatures have their ways of mating, hunting prey, avoiding predators, taking care of their young, and so on, and all this is largely unlearned, the product of biological evolution.

Some of James' contemporaries, and quite a few of my own contemporaries, think that, through some quirk of evolution, these innate capacities have vanished in the millions of years since our species split off from other primates. Humans are special in that we are blank slates. We have no instincts.

James treats this claim with the proper scorn. He doesn't deny that we have capacities for learning and for the formation of habits—in fact, as we saw in the chapter on consciousness, he has a lot to say about these topics. But he rejects the notion that humans lack what every other creature possesses—"On the contrary," he writes, "man possesses all the impulses that [other animals] have, a great many more besides."[6]

———

The best theory of the origin of these instincts is natural selection. Their existence was shaped by the same evolutionary process that led to the development of mental powers such as memory, perception, language, learning, and reason. Instincts exist because rudimentary versions of these traits and capacities were associated with greater reproductive success.

As a quick summary of how natural selection works for a creature like us, the general logic goes like this: There is variation in the bodies of animals, and this is due in part to mutations in our genome and in part to the shuffling of genes that occurs during mating. This variation influences the animals' fates—some reproduce more than others. And this causes the genes that are related to an increase in reproductive success—*fitness*—to become more common in the next generation. And so the population will evolve to contain animals that are better adapted to their environment.[7]

That's it. It might seem simple, and in a sense, some of it is old news. Animal breeders long appreciated that by allowing only some animals to breed, they could create future generations with certain sorts of traits, as when pigeon fanciers "evolve" different types of pigeons, some specialized for flying, others for beauty. Part of the brilliance of Darwin's discovery (as well as the brilliance of the lesser-known Alfred Wallace, who had the insight at the same time) is that the same process of selection can be driven by the environment, without any human intervention—you can have natural selection along with artificial selection. This process isn't limited to the modification of existing species but can create new ones. And then there is the further insight that, over the course of many generations, this process can lead to what Darwin described as "organs of extreme perfection and complication," including, most relevantly for us, the brain.

Natural selection is not a random process. While randomness does play a role, as with the genetic mutations that lead to variation, natural selection is responsive to the environment in a decidedly nonrandom way. As a textbook example, the peppered moth in Great Britain used to be both light and dark, but during the Industrial Revolution, the trees became covered in black soot, making the light moths stand out, easy prey for the predators. Darker moths survived longer, so they had more offspring. This led to the genes for dark colors becoming more populous, and after a few generations, just about all peppered moths were dark. When pollution was reduced in the mid-1900s, the soot went away, and now very dark moths—which stand out—are rare. This example illustrates the gradual character of natural selection. It's not just that, poof, moths went from one color to another. Rather, the populations of moths gradually evolved to converge (in a mindless way) on the optimal solution of dealing with predation in a radically new environment.

Natural selection is the best explanation we have for the existence of complex biological systems, including psychological ones, and we'll be drawing heavily on it in the discussion that follows. But it's critical to realize that not all beneficial traits are adaptations.[8] Flying fish have the very useful feature of returning to the water once they leap out of it, but this is not because the fish that flew off into space reproduced less! There is a species of wading bird that uses its wings to block reflection from the water while looking for fish; a useful trick, but not likely what the wings were adapted to do.[9] And some other traits confer no fitness advantages, such as the redness of blood, or the fact that there are a prime number of digits on our hands and feet. Many people sneeze when they are in bright light, and this has nothing to do with reproductive success.

Similarly, many psychological motivations, including some that influence our lives in significant ways, are not the direct products of natural selection. Many people spend a lot of time with pornography, but it's not because our ancestors who enjoyed porn reproduced more than those who didn't. Some of us eat a lot of chocolate, but it's not because of the reproductive edge that chocolate eating provides. In these cases, the traits are by-products of more general traits—sexual desire, hunger for sweet food—that are themselves adaptations.

In other cases, the psychological phenomena of interest are neither adaptations nor the by-products of adaptations. They arise due to cultural forces. For whatever reason, some societies—not all—associate boys with blue and girls with pink.[10] We wouldn't be able to learn these associations without brains capable of forming cultural associations, but still, these are just arbitrary facts, and evolutionary theory has nothing much to say about them.

Those are easy cases; here are some harder ones: Do religious beliefs and practices show up in every human culture because they led to advantages in our ancestors or are they by-products of aspects of mental life that have evolved for other purposes? What about music? What about the pleasure we get from visual art? Humor? There is vigorous argument about the role of evolution in the origin and nature of all these human universals. The business of figuring out what evolved through natural selection, what is a biological accident, and what is the product of culture is an essential part of coming to understand human nature.

———

Thinking in terms of evolution helps us be clear about the sorts of motivations one would expect humans to possess. Return, for a moment, to homeostasis theories, which assume that the purpose of the brain is to keep the body alive. Some versions of the pleasure/pain account make the same assumption. The comparative psychologist George Romanes wrote this in 1884: "Pleasure and pain must have been evolved as the subjective accompaniment of processes which are respectively beneficial or injurious to the organism, and so evolved for the purpose or to the end that the organism should seek the one and shun the other."[11]

I've seen this cited approvingly by contemporary psychologists, but it's mistaken in an important way. Think about the pleasure we get from sex, think about the exquisite suite of adaptations that drive us to care for our offspring (physical for women, who feed babies from their bodies, but psychological for both men and women). None of this has to do with what's "beneficial or injurious to the organism"; our bodies would hum along just fine if we didn't feel lust or weren't motivated to care for our children.

These pleasures and pains illustrate how the force shaping our minds isn't survival, it's *reproduction*. We are drawn to have sex and care for our offspring because creatures in the past who possessed such motivations had more reproductive success than those who didn't. Survival is critically important, of course—but only insofar as you must be alive to have offspring and then care for them.

Thinking in terms of evolution is an essential part of the psychologist's tool kit; it helps us think clearly about all sorts of things. But there are misconceptions that arise when we bring together evolution and psychology, and it's worth quickly discussing a couple of common ones.

The first is to confuse the evolutionary motivation with the psychological motivation. From the standpoint of evolution, we eat to sustain our bodies and have sex to produce offspring. But this doesn't mean that these are our personal psychological motivations. They plainly are not. James again: "Not one man in a billion, when taking his dinner, ever thinks of utility. He eats because the food tastes good and makes him want more. If you ask him *why* he should want to eat more of what tastes like that, instead of revering you as a philosopher he will probably laugh at you for a fool."[12]

Similarly, from the standpoint of evolution, we love our children because of the amoral forces of natural selection, for reasons that are, in some metaphorical sense, selfish. But this surely doesn't mean that when parents take care of their offspring, they are driven by selfish motivations. To paraphrase James, not one person in a billion, when rocking their baby to sleep, ever thinks of utility. A mother or father is usually motivated by love.

The second misconception is to assume that evolutionary goals are valuable goals, what we should be doing. This also doesn't follow. To make the jump from *This is how things are* to *This is how things should be* is so confused that it has its own name: *the naturalistic fallacy*.

It's not hard to see how it's a fallacy. The belief that the most important thing for a person to do is spread their genes implies that the doctor who secretly used his own sperm to impregnate women who were seeking to have children with their husbands (true story) was living the very best life. And that a woman who gives birth to many children and treats them terribly (but they do survive and reproduce) is living a much better life than if she adopted children and treated them with love and respect. The relationship between evolution and morality is intricate, but to simply assume that "survival and reproduction" is the ultimate moral goal is bad philosophy worthy of a cartoon supervillain. We are intelligent creatures and can make our own goals, including moral goals.

———

Any discussion of human motivation will soon bring us to the emotions.

There are many ways to distinguish the emotions. They are related to distinct sorts of conscious experience. There is a certain feeling of what it is to be afraid, which is different from what it is to be angry or sad. Some emotions come with distinctive physiological responses. Embarrassment makes your face turn red; fear can make you tremble; grief can make you weep. Lust can lead to physical reactions that are entirely different from those of, say, disgust. And many emotions, including the six so-called basic emotions of anger, disgust, fear, happiness, sadness, and surprise have their own facial expressions.

Much of this is innate. These same facial expressions, with some variations, show up everywhere, even in small-scale societies isolated from the rest of the world.[13] Studies of congenitally blind people find that their expressions are typically indistinguishable from sighted people, suggesting that they are not learned through experience.[14]

On the other hand, it's a mistake to assume, as some do, that there is a direct and universal connection between an emotion and a certain feeling and a certain response in the body and the face. It's more complicated than that.

For one thing, some of the stronger claims about universals have been challenged.[15] Comparisons between Japanese people and Canadians, for instance, find that the Japanese people are less likely to express emotions of anger, contempt, and disgust and more likely to express happiness and surprise. In some countries, people grin and laugh a lot; in others, they don't. In part, these reflect cultural norms—"display rules"—concerning what emotions one should show to the world.[16]

Also, there is often a paradoxical flavor to certain emotional responses, when we respond with the seeming opposite response to what one would expect.[17] We laugh at what we find funny, but we also laugh when anxious or embarrassed. We grin when happy, but sometimes we grin when angry. Smiling is associated with joy, but when researchers asked people to watch a sad movie scene—the part of *Steel Magnolias* where Sally Field's character is speaking at the funeral of her daughter—about half of the subjects smiled.[18] Another study finds that people sometimes find it difficult to distinguish the face of someone in agony from the face of someone having an orgasm.[19]

A particularly strange reaction that's always fascinated me concerns our response to adorable babies. Some people get an odd feeling. They want to pinch and squeeze. They sometimes nibble on babies and say they are going to eat them. Your friend shows you his one-year-old baby, and you grab the baby's toes, gnaw on them, and growl, "I want to gobble you up!"—and nobody thinks you're crazy, not even the baby. One survey found that most people agreed to statements like these:

If I am holding an extremely cute baby, I have the urge to squeeze his or her little fat legs.

If I look at an extremely cute baby, I want to pinch those cheeks.

When I see something I think is so cute, I clench my hands into fists.

I am the type of person that will tell a cute child "I could just eat you up!" through gritted teeth.[20]

One theory of these strange reactions is that they arise when your feelings become overwhelming. You need to calm the system down, and so, to compensate, you generate expressions and actions that counteract them, that go in the opposite direction. These odd emotional reactions are like putting cold water on a fire that might get out of control.

There is an idea in popular culture that emotions are unnecessary or, worse, an impediment to successful thought and action.[21] This idea shows up in the science fiction series *Star Trek*, of which I am a huge fan. Two of the most beloved characters—the half-human, half-Vulcan Mr. Spock (from the original series) and the android Data (introduced in *Star Trek: The Next Generation*)—are described as excellent Star Fleet officers. But they are said to be, to different degrees, without emotion. Spock rarely experiences emotions because he was raised in a society that decided to give them up, and Data lacks them entirely because he doesn't have an "emotion chip."

But under any reasonable theory of the emotions, this can't be right. They had to have emotions. Here is how the psychologist Steven Pinker makes this point:

> Something must have kept Spock from spending his days calculating pi to a quadrillion digits or memorizing the Manhattan telephone directory. Something must have impelled him to explore strange new worlds, to seek out new civilizations, and to boldly go where no man had gone before. Presumably it was intellectual curiosity, a drive to set and solve problems, and solidarity with allies—emotions all. And what would Spock have done when faced with a predator or an invading Klingon? Do a headstand? Prove the four-color map theorem? Presumably a part of his brain quickly mobilized his faculties to scope out how to flee and to take steps to avoid the vulnerable predicament in the future. That is, he had fear. Spock may not have been impulsive or demonstrative, but he must have had drives that impelled him to deploy his intellect in pursuit of certain goals rather than others.[22]

Elsewhere, the philosopher Richard Hanley notes that Data, over the course of just a few episodes of the television series, displays regret, trust, gratitude, envy, disappointment, relief, bemusement, wistfulness, pride, curiosity, and stubbornness.[23]

When people say that Spock and Data lack emotions, they are referring in part to their expressionless faces and modulated voices. They don't use contractions, so they'll say "this is not" rather than "this isn't," which makes them sound formal and controlled. They don't laugh, cry, pout, or grin. And they tend to make good decisions, ones that are aligned with their goals. But as Pinker's discussion makes clear, any functioning individual must have emotions, because without them, they wouldn't pursue any goals at all.

One way to explore the role of emotions would be to look at people who entirely lack them. No such people exist, but as a rough approximation, we can consider the fates of some unfortunate individuals who suffer damage to the prefrontal cortex, which leads to a blunting of certain emotional responses. We've already talked about one such individual— Greg F., who was studied by Oliver Sacks—and saw that he was helpless, almost infantile; his father described him as "scooped out, hollow inside."[24] Another famous case study, explored by the neuroscientist Antonio Damasio, is of a man named Elliot, who had a brain tumor in the frontal lobes. The tumor was removed, but the damage had been done. Elliot remained an intelligent man, but, echoing the famous phrase used to describe Phineas Gage, Damasio writes: "Elliot was no longer Elliot. . . . He needed prompting to get started in the morning and go to work. Once at work he was unable to manage his time properly; he could not be trusted with a schedule." Damasio blames these failings on Elliot's relative loss of emotions: "The cold-bloodedness of Elliot's reasoning prevented him from assigning different values to different options, and made his decision-making landscape hopelessly flat."[25]

When considering specific emotions, fear is a good place to start. William James observed that fear "has bodily expressions of an extremely energetic kind, and stands, beside lust and anger, as one of the three most exciting emotions of which our nature is susceptible."[26] (We'll get to lust and anger later.)

Like any other emotion, fear involves multiple brain areas, but it is most related to a specific structure known as the amygdala.[27] It has a characteristic facial expression and bodily response—the release of adrenaline, an increase in heart rate, more blood flow to the muscles, slowing or shutting down of the digestion. Goose bumps—piloerection—might occur, something which is useless for contemporary humans but helped our hairier ancestors, as it made them look larger—a useful trick when one is under threat. Fear leads to arousal and a focusing of attention. It is usually unpleasant, but it's rarely boring.

And it has a certain function. I had some critical things to say about *Star Trek*; now I'll complain about a different science fiction classic, Frank Herbert's *Dune*. The hero of this book, Paul Atreides, has a litany he recites to himself during times of crisis.

> I must not fear.
> Fear is the mind-killer.
> Fear is the little death that brings total obliteration.[28]

Too much fear really can be a serious problem. But fear is not the mind-killer. It is an adaptation for dealing with threat. This function explains the associated physical responses, which reflect the sympathetic nervous system preparing for danger, activating the body for fight or flight. Psychologically speaking, the experience of fear motivates us to take these threats seriously. Fear is important and useful.

We see the stamp of evolution in what elicits our fear. A recent review of the prevalence of "specific phobias" gives the following list:[29]

Animals
Heights
Storms
Water
Flying

Crowds
Closed spaces
Blood
Dentists

("Animals" is too vague a category. I never met anyone who was afraid of robins, goldfish, or pandas. More typical is fear of snakes and spiders.)

If you think the mind is a blank slate, this list is impossible to make sense of. How can we explain the fear of spiders and snakes? How many people do you know who have been seriously injured by a spider?

A better account here is that we are predisposed by evolution to fear (or quickly come to fear) those creatures, things, and situations that were dangerous to us in the ancestral environment. This nativist theory is supported by studies in which six-month-olds were presented with pictures of different creatures. Without any experience with spiders and snakes, these pictures caused the babies to get excited, more so than if they were shown pictures of flowers and fish.[30]

But dentists? Presumably there were no dentists wandering around the African savanna waving their drills and their little mouth mirrors. Here we see the flexibility of fear; evolution bequeathed us not just with a list of what to be afraid of, but also gave us the capacity to learn to fear new things.

Finally, there is an intelligence to fear. Our expectations, inferences, and beliefs can make us afraid of what would normally be harmless. You don't usually become afraid if you hear someone whistling, but you might if it was the middle of the night and you thought you were alone in the house.[31] Not wearing a parachute isn't normally a fear elicitor—I'm not wearing one now and I'm right as rain—but if I had jumped out of a plane and discovered that my parachute was gone, it would be the most terrifying discovery of my life (and the last). What frightens us is more than a list of things and experiences, then; our fear can be influenced by our conscious appreciation of what puts us in danger.

———

Every psychologist has a favorite emotion, and for me it's not even close. I've long been fascinated by disgust.

Let me remind you what it feels like.[32] Imagine opening a food container and discovering immediately from the smell that it is hamburger gone bad. Most people would get a certain unpleasant feeling at this point, one that might include a bit of nausea—and you might get it from simply imagining this event. This feeling is accompanied by a special facial expression (a "yuck face"—nose scrunched, mouth shut, tongue pushed forward) and a distinctive motivation: *Get that away from me.*

What disgusts us?[33] The psychologist Paul Rozin is the preeminent researcher on the topic and, along with his colleagues, he created a disgust scale.[34] Check out some of their items and monitor your own reactions.

> Your friend's pet cat died, and you have to pick up the dead body
>   with your hands.
> You see a bowel movement left unflushed in a public toilet.
> You see a man with his intestines exposed after an accident.
> While you are walking through a tunnel under a railroad track,
>   you smell urine.

Your mileage may vary. When I read these aloud in classes and presentations, some people wonder what the fuss is about; others gag; and once a student ran out of the lecture hall.

Some of what triggers disgust is universal; people everywhere are repelled by blood, gore, vomit, feces, urine, and rotten flesh—these evoke what Rozin dubs "core disgust." Unfortunately for us, these substances are also the stuff of life. As the title of a well-known children's book says, *Everybody Poops*. All manner of substances squirt, drip, and ooze from our bodies and from the bodies of those we love.

But we don't start off disgust sensitive. As Freud put it in *Civilization and Its Discontents*, "The excreta arouse no disgust in children. They seem valuable to them as being a part of their own body which has come away from it."[35] If left unattended, young children will touch and even eat all manner of disgusting things. In one of the coolest studies in all of developmental psychology, Rozin and his colleagues did an experiment in which they offered children under the age of two something that was described as dog feces ("realistically crafted from peanut butter and odorous cheese"). Most of them ate it.[36]

Then, sometime in early childhood, a switch is thrown, and children become like adults, disgusted by much of the world. Psychologists have often wondered what motivates this change, and many follow Freudian theory and blame the trauma of toilet training. But this is a nonstarter. Other societies have very different practices when it comes to urination and defecation (and some don't even have toilets)—yet disgust is universal. Blood and vomit and rotten meat are disgusting but have nothing to do with toilet training.

A more plausible theory is that core disgust serves an adaptive purpose. According to this theory, disgust isn't learned but rather emerges naturally once babies have reached a certain point in development. But if disgust is an adaptation, what is it an adaptation for?

The most popular explanation is that it evolved to ward us away from eating bad foods. Indeed, the English word itself derives from the Latin, meaning "bad taste." This theory has much to support it. First, as Darwin observed, the facial expression of disgust corresponds to the acts of trying not to smell something, blocking access to the mouth, and using the tongue to expel anything already within. Actually, the "yuck face" is the same expression one gets when actually retching, and this may be its origin. Second, the feeling of nausea associated with disgust serves to discourage eating and sometimes leads us to expel foods we've already eaten, through vomiting. Third, even controlling for an overall increase in the rate of nausea during pregnancy, pregnant women are exceptionally disgust sensitive during the same period that the fetus is most sensitive to poison.[37] Fourth, the anterior insular cortex, which is implicated in smell and taste, becomes active when people are shown disgusting pictures.[38]

There may be other functions of disgust, though. The anthropologist Valerie Curtis and her colleagues surveyed more than forty thousand people from 165 countries to find out what images disgusted them.[39] They found that pictures indicating potential disease were rated as particularly gross. People were also somewhat disgusted by someone made up to look feverish and spotty-faced. This disease-avoidance theory also captures why the smell of an unwashed stranger can be so repulsive—being unclean is a sign of illness.

Like fear, then, disgust has universal and adaptive roots. But also like fear, there is learning involved. People vary considerably in what disgusts them. The idea of eating rats or dogs makes me gag, but people raised in some societies find these foods perfectly yummy. Humans face what Rozin has dubbed the *omnivores' dilemma*—we eat a huge range of foods, but some of them can kill us—so we need to learn what we can and cannot eat in our local environment. In the course of this learning, food, and particularly meat, is guilty until proven innocent. Nobody ever told me that it is gross to eat fried rat; I find it gross because, during the critical period of childhood, people around me never ate it.

————

A potentially more pleasant emotion to look at is lust. From the standpoint of evolution, this is a no-brainer. Lust motivates sexual behavior. The engine that drives natural selection is reproductive success—and you don't need a doctorate in evolutionary biology to appreciate that reproduction has a lot to do with having sex.

But how complicated does this evolved adaptation have to be? Perhaps we've just evolved some vague instinct for mating, one that sparks up in adolescence and gently fades in the older years—and that's it.

After all, there are certain ways in which lust is different from the other motivations we've discussed. Take away food from someone, and soon enough, they will have a powerful and all-consuming need to eat, and if this need isn't satisfied, they will die. But remove opportunities for sex, and, well, people differ. Some might feel the same sort of powerful and all-consuming need—though, despite what some might say, nobody has died from abstinence. Others are troubled by celibacy but cope with it, and still others are fine with going without sex for their entire life. I know people who have given up on sex; I never met anyone who has given up on eating.

Also, hunger is directed toward behaviors that are nicely aligned with a certain evolutionary goal—getting food into the body. But there is a lot of sex that people seek out that just isn't going to make babies, and there exists a sizable minority of people whose primary sexual desires are non-procreative, who have no interest in acting in ways that facilitate contact between sperm and egg. Consider as well that human females, unlike other primates, can be sexually receptive all the time, not merely when they are ovulating and capable of getting pregnant. Put all of this together, and it might seem like human sexuality is unmoored from human evolution.

———

I don't think this conclusion is correct, however. There is an intricate psychology of sexual desire, one informed by evolutionary theory.[40] And there is a lot about our sexual psychologies that can only be properly appreciated through a Darwinian lens.

Consider the curious limits of desire. When I taught Introduction to Psychology at Yale, a famous social psychologist would regularly give a guest lecture on the psychology of love. (I would schedule his lecture as close as possible to Valentine's Day.)[41] He would tell the class about the Big Three—the major factors that determine who you are most likely to be romantic partners with. These are:

Proximity
Familiarity
Similarity

He was right. We tend to fall in love with people who are physically close to us and personally familiar, and we tend to be drawn to people who are similar to ourselves in every conceivable way.[42] (The worst bit of folk wisdom ever is "opposites attract"; it's true only in rom-coms.)

But there is an intriguing exception to the Big Three. Think about it: For much of your life, who were you physically closest with, most familiar with, and most similar to?

The answer for many people: your siblings. The Big Three predicts that your brothers and sisters should be, for you, the hottest people in the universe. Accordingly, they are the ones you should most want to have sex with and to spend your life with. And I don't need to tell you that this isn't the case. Many of us have siblings that we might grudgingly acknowledge are attractive in some objective sense, but it's rare to be personally aroused by them. There is no movement to change the law to allow siblings to have sex and get married, and this is because almost nobody wants to have sex with or marry their brother or sister.

There is an evolutionary explanation for this. It's a bad idea to have children with your close relatives, because they share too many of your genes. This is known as "inbreeding depression"—the risk that recessive genes become more likely to be homozygous, which raises the odds of bad outcomes. Many creatures, including humans, show a strong bias to mate outside of the family.

Here we see another illustration of how evolutionary forces and psychological forces pull apart. Few of us avoid sexual intercourse with our brothers and sisters because we explicitly worry about genetically damaged offspring. Rather, we find ourselves disgusted at the very thought, and this does the trick, in the same way we avoid consuming rotten meat without ever worrying about microorganisms and contamination.

This evolutionary account raises a psychological question, though: How do we know who our kin are? Some creatures duck this problem and just leave the area of their birth, so that they aren't close to their siblings when they become sexually mature. (In some cases, only the males or only the females leave.) Other animals stick around and use certain cues to find their kin—squirrels, for instance, seem to estimate genetic relatedness by smell and restrict their mate selection accordingly.[43]

The anthropologist Edward Westermarck proposed that humans solve this problem in a different way: We tend not to be sexually interested in those we are raised with as children.[44] This disinterest overrides any conscious knowledge. You might know perfectly well that someone isn't biologically related to you, but if you spent your childhood with him or her, it's a libido killer. Conversely you might know perfectly well that someone is close genetic kin—as in the rare cases where someone meets, as an adult, a sibling that was separated from them at birth—but if you have never lived with that person when you were young, the idea of sex with them doesn't necessarily gross you out.

To explore the Westermarck hypothesis, the psychologist Debra Lieberman and her colleagues asked adults a series of questions—about whether they were raised with their siblings, how much they care about them, and how disgusted they are (if at all) at the thought of having sex with them. As predicted, duration of coresidence is the big factor—the longer you live with the sibling, the more sexual aversion.[45]

This theory makes a neat prediction about the real world. There should be false alarms, cases when the "Ugh, sex with you is disgusting" light goes on even though the individuals bear no genetic ties, just because they lived together as children. And such cases exist: the most studied examples of this are Israeli children who are raised together on a traditional kibbutz. In such a case, sexual and romantic relationships don't tend to occur later in life.

———

Let's turn now to what people want, what we find attractive. Anyone who has ever visited an art museum or watched old movies or traveled the world will appreciate that there is variation here. For one thing, cultures differ a lot in the range of body weight that is found most appealing. And of course, there are differences in the tastes of individuals of the same culture: What you find attractive, I might find meh, and vice versa.

But it's not that anything goes. There are universal features that are associated with attractiveness.[46] If you go to different societies and consider what they find good-looking, the same criteria come up. One can find faces and bodies that meet these criteria—or create them with computer programs—and people all over the world will find them appealing.

Focusing on faces, we tend to be drawn to those that have features such as smoothness of complexion and symmetry. These are cues to youth and health. Clear skin for humans has been likened to bright plumage in birds and a shiny coat of fur in other mammals.[47] Since injuries, poor nutrition, and parasites make faces less symmetrical, a face where the right side is a mirror image of the left is a biological success story and looks good to us.[48] Even newborn babies prefer to look at the faces that meet these criteria.[49]

A more mysterious item on the list of attractive features is being average. If you pick out one hundred faces at random and morph them together, the result would be rated as quite good-looking.[50] Perhaps averageness is another cue to health, on the grounds that most deviations from normal are bad. Perhaps it's a cue to genetic diversity, which is another good thing. Perhaps average faces are in a literal sense easy on the eyes; they require less visual processing than nonaverage faces, and we tend to prefer visual images that are easier to process. One qualification here is that average faces look good, but not terrific; these morphs give you a fine face, but not one with movie-star potential.[51] Perhaps, too, it isn't that average faces are positively attractive; it's that nonaverage faces run more of a risk of being unattractive.

———

I've said nothing about sex differences so far. Both men and women veer away from sex with close kin, and both men and women tend to find the same sorts of faces attractive. The logic of natural selection here works the same for everyone.

In other regards, though, the optimal strategies for reproductive success differ for biological males and biological females of many species, including our own, and this leads to certain psychological sex differences.

One issue is fertility. We would expect animals to be most sexually drawn to those whose faces and bodies suggest that they are of an age at which reproduction is possible—those you can make babies with. This is the immediate answer to why few of us are sexually attracted to prepubescent individuals; they aren't candidates for baby making.

Women are fertile during a certain limited period of their lives; men can bear children late in life, though there is a decline in fertility with age. And just as one might predict, our psychologies are sensitive to this fact. When it comes to partner choice, heterosexual men have a strong preference for young women; while heterosexual women are less discriminating about age, often preferring men who are slightly older or the same age as themselves. This existence of age difference in preferences for a long-term partner is one of the most robust findings we have in our field, originally found in a study with thirty-seven cultures[52] and then recently replicated in forty-five countries.[53]

Other studies that just look at judgments of attractiveness show a more extreme difference.[54] Data from OkCupid, a dating site, finds that, on average, heterosexual women's judgment of the age of the most attractive men tracks their own age or, in contrast to the partner-choice data, skews slightly younger—twenty-year-old women are most attracted to twenty-three-year-old men, fifty-year-old women are most attracted to forty-six-year-old men, and so on. Heterosexual men show a different pattern. Regardless of their own age, they tend to be drawn to a narrow age range corresponding to peak fertility. Twenty-year-old men are most attracted to twenty-year-old women; fifty-year-old men are most attracted to twenty-three-year-old women, and so on.

A second issue concerns differences between male and female mammals in the best strategy for reproductive success.[55] To create a child, the minimum that a male must do is have intercourse. Once this is over, the male can, in theory at least, find another mate and create another child. The minimum a female must do is . . . *a lot*. There's intercourse, then the baby is grown inside her body, then, in the environment in which we evolved, the baby is fed from her body. This takes years, and during most of that time, she can't create another baby.

This means that while a male can create many children by mating with many females; a female has no similar advantage in mating with many males. The evolutionary biologist Robert Trivers noted that this discrepancy leads to a male-female difference in optimal reproductive strategy.[56] Females should be prone to invest more in their offspring than males, because they can have fewer, and so each one matters more, and this leads to greater choosiness by females as to whom to reproduce with.

Importantly, under Trivers' theory, it's not being male or being female that leads to this difference in optimal strategy; it's just that males and females tend to have different degrees of investment in any given offspring. This leads to an interesting prediction: In a species where males are more invested than females, the sex differences should flip. And they do. There are species such as jacanas—sometimes known as Jesus birds because they seem to walk on water—where the males take care of the eggs and, later, the chicks.[57] As a result, females are larger than males, fight each other for access to them, and sometimes keep harems of males. (The size and aggression difference exists because the sex that benefits more from multiple partners—here, the females—often fights it out for access to the sex that is pickier—here, the males.)

And if you have a species in which parental investment is equal, either because the male and female work together to protect extremely fragile offspring (penguins) or because they just spray their sperm and eggs into the sea and offspring don't need care after that (some species of fish), you can't tell, just by looking, who is male and who is female, and sex differences in mating strategy disappear.

Humans are not Jesus birds or penguins or fish, though. Our reproductive pattern follows the standard mammalian plan of requiring greater female investment, and this predicts, among other things, a certain sex difference—men should be more interested than women in having multiple sex partners.

Is this prediction true? There are many ways to explore this question in the real world, but the simplest is to just ask men and women how many sexual partners they would like to have. Since we're interested in both universals and cultural variation here, a good study needs to test a diverse population, so here are results from a study with over sixteen thousand people from fifty nations: Just as with age preferences, there are large and consistent effects.[58] Over the next month, men, on average, claimed to want twice as many partners as women (about two versus one). Over the next decade, men want an average of six; women want an average of two. It is being male or female that drives the effect, not sexual orientation—gay men show the same pattern as straight men; gay women show the same pattern as straight women.

———

I am more confident in the findings described above—about incest avoidance, about what we find attractive, about sex differences in the age of the ideal partner, and about sex differences in preference for sexual variety—than I am in most research in psychology. The data here, drawn from hundreds of studies, is strong and robust.

But I said earlier that people often get confused about evolution. This is particularly true for the evolution of sex differences, so I want to clear up a few things there.

First, the evolutionary story of human sexuality is more complicated than it might first seem. One consideration here is that human babies are tiny and powerless, needing extensive protection from adults for years and years. Fathers matter. Correspondingly, human females will prefer males who are the sort to care for their offspring—and this makes such traits good for males to have.

A further twist is that human females can have, and enjoy, sexual intercourse anytime during the menstrual cycle. One theory of why this unusual trait evolved is that it facilitates pair bonding. If human females can mate all the time, and if it is unpredictable when this mating will lead to a child, it behooves males to stick around to ensure that any child they end up raising contains their genes.

More generally, humans bond. We fall in love and team up to raise these so-precious and so-fragile offspring. Because of this, as one evolutionary-minded psychologist put it, "In certain ways, we're more like the average bird than the average mammal."[59]

Then there's culture. Our minds are shaped by cultural history as well as by evolutionary history, and you can see that in some of the research we just discussed, such as the study that asked men and women about the optimal number of sexual partners they wanted. In South America, 35 percent of the men and 6 percent of the women wanted more than one sex partner per month; while in East Asia, this drops to 18 percent and 2.6 percent. We still see a powerful sex difference, but one that sits alongside large cultural effects.

There's also a lot of individual variation within cultures. Differences between human groups, including sex differences, are always an "on average" sort of thing. On average, men are larger than women, more interested in a variety of partners, and more concerned about youth in their sexual partners—but this is perfectly compatible with the obvious reality that there are women who are larger than the average man and men who are smaller than the average woman; that there are women who prefer multiple partners and men who don't; and that there are women who are only sexually attracted to young men and men who are indifferent to their partner's age.

Finally, a lot of the research on the evolution of sexual desires focuses on features such as age and physical attractiveness, in part because these are easy to study. It's easy to forget that looks aren't everything when it comes to either desire or decisions about commitment. The logic of natural selection says that we should be attracted to those who have certain relevant traits—relevant to caring for children and giving them the best shot at the future—and a lot of those traits don't show up on the face or body. Smart and kind people do well in the world, and so do their children. So even from the most cold-blooded Darwinian perspective, one would expect us to look for deeper qualities when we seek out partners. And to add the obvious, we aren't only sexual creatures. We have other goals we hope to satisfy, and so there are other things we look for in our partners.

Given all this, we shouldn't be surprised that our choices of who to be with, particularly for the long term, are influenced by concerns such as how much we enjoy their company and how much we trust them. In study after study, when people are asked about what they want in an ideal partner, they talk about traits like intelligence, sense of humor, and honesty. And the number one trait? In one of the studies mentioned earlier, looking at people in thirty-seven cultures, the most important factor for both men and women is . . . kindness. [60]

The topic of kindness brings us to the last part of this chapter. The emotions we've looked at so far motivate us to satisfy our own priorities—avoiding what's bad for us (as with fear and disgust) or guiding us to get what's good for us and our genes (as with lust). But there is more. We also have moral emotions and moral motivations. We feel compassion, guilt, shame, gratitude, and outrage. We are driven to help others, even at a cost to ourselves; we cooperate; we judge; and we are motivated to condemn, and even sometimes punish, those we see as cruel or unfair.

We'll see that there are aspects of our moral psychology that are genuinely mysterious. But we can start with something that is less mysterious—the love we feel for our children and the time and energy we devote to them.

This is as much of an evolutionary no-brainer as lust was.[61] Suppose (and this is of course a radical simplification) that there are two variants of a gene: Variant A creates an animal that cares for its offspring—finds its babies adorable and wants to feed them and hold them and keep them safe and care for them until adolescence. Variant B creates an animal that only cares for itself and thinks of babies as a potentially valuable source of fat and protein. If you check out the population a generation later, you will find only Variant A. The purely selfish variant is a loser in the Darwinian sweepstakes.

Our kindness to our children is a special case of what's known as "kin selection," and it extends to relatives more generally. The genetic tie one has with cousins and aunts and uncles is weaker than between parent and child and between siblings—all who share, on average, half of their genes—but this is a difference of degree, not of kind. From the perspective of natural selection, then, a gene that guides an animal to help its relatives can spread through the population even if this helping is costly to the animal itself. Evolution favors animals that, to some limited degree, love others as they love themselves.

The story, most likely apocryphal, goes that the biologist J. B. S. Haldane was asked if he would give his life to save his drowning brother, and he responded that he wouldn't, but he would happily do it for two brothers or eight cousins. (He shared, on average, half of his genes with each brother, and one eighth with each cousin, so the math worked.) Of course, Haldane was being playful in his calculations; few of us are consciously motivated by an explicit desire to preserve our genes; we don't take out a calculator if we're asked to donate blood to our niece. But the different degrees of genetic overlap do shape our psychological intuitions, explaining how our love for our children can be so intense and overwhelming, while our ties—our kinship—to the children of our brothers and sisters might be strong but are not usually as strong.

———

Kindness is not limited to kin. We do all sorts of good things for people we are not related to. I'm lucky enough to live in a neighborhood where we help each other out—pick up one another's mail when we go out of town, shovel the sidewalks for the older people who live on the street, that sort of thing. And plainly friends are kind to one another, providing all sorts of support and love, making sacrifices for one another, even without the slightest hint of genetic relatedness. What sort of sad world would it be if this weren't so?

We are also, to a lesser extent, kind to the strangers we encounter. In many places in the world, if you stand on the street befuddled, someone will walk up and ask if you need directions, and if you scream for help, people will come. I've had strangers help push my car out of snow and done my own share of car pushing. If nobody gave to people asking for money on the street, people wouldn't ask for money on the street.

I live in Toronto now, but one study on this topic was done in my previous home of New Haven, Connecticut, by Stanley Milgram. He is better known for a famous experiment he conducted on obedience, which we'll get to soon enough, but in 1965, Milgram was interested in kindness.[62] He scattered stamped and addressed letters all over New Haven, dropping them onto sidewalks and placing them in telephone booths and other public places as if they had been forgotten. He found that most letters got to their destinations, which means that the good people of New Haven had picked them up and put them into mailboxes—simple acts of kindness that could never be reciprocated. The kindness was selective in a way that suggests a moral motivation—letters that had an individual's name on the front, like "Walter Carnap," tended to be mailed, but letters addressed to recipients like "Friends of the Nazi Party" were not.

The extent of such helping varies from society to society, but it's not entirely a cultural practice. [63] Other animals groom one another and help care for one another's offspring. Sometimes they help at a risk to themselves. Blackbirds and thrushes give warning cries when hawks are above, which gives other birds a chance to escape. When a gazelle notices a dog pack, it bounds away in a stiff-legged way, known as strotting. Strotting slows down the strotter and makes it easier prey, but other gazelles who notice it run away and are less likely to be caught.

We also find kindness early in life. I wrote a book about the emergence of morality in babies and children,[64] and there's so much to say here, but the upshot is that even the youngest children care about others, and often try, in their own limited ways, to make things better. Some experiments explore this by getting adults to act as if they are in pain—an experimenter might pretend to get her finger caught in a clipboard—and then seeing how children respond. Often they try to soothe the adults, hoping to make their pain go away. Other studies find that toddlers will help adults who are struggling to pick up an object that is out of reach or struggling to open a door. The toddlers do so without any prompting from the adults, not even eye contact, and will do so at a price, walking away from an enjoyable box of toys to offer assistance.[65]

In addition to kindness, there is cooperation—individuals working together for a common cause. In the modern world, just about everything that matters, from business to art to science, requires cooperation at a massive scale. We are reliant on others. In fact, this has always been true. In many societies in the past, hunting was a communal endeavor; people worked together to bring down a large animal. (This is not uniquely human, by the way—dogs and chimps also team up to hunt.) Childcare was communal as well; one or two adults raising a child by themselves is not the usual way that our species takes care of offspring. Rather, children were looked after by communities, including close kin (especially sisters and grandmothers) as well as distant relatives and people who aren't relatives at all.[66] It really does take a village to raise a child, at least if you're a hunter-gatherer.

———

You might think that the explanation for all this is simple. Things work out better if there is mutual aid. If we each hunt alone, nobody gets to capture large animals; if we coordinate a group hunt, we all get to have meat tonight. If I take care of your baby when you're sick and you do the same when I'm sick, both of our babies do better in the long run. As Adam Smith wrote: "All the members of human society stand in need of each other's assistance, and are likewise exposed to mutual injuries. Where the necessary assistance is reciprocally afforded from love, from gratitude, from friendship, and esteem, the society flourishes and is happy."[67]

But the challenge lies in explaining how this love, gratitude, friendship, and esteem could ever evolve. The obstacle here is that natural selection doesn't care for the good of the group, or for the benefit of society, or the general maximization of happiness, or anything so high-minded. It just cares about the spreading of genes. And—as we saw at the end of the last chapter when we discussed the tragedy of the commons—in a world where everyone else is being kind to one another, a free rider will do best of all.

A free rider is someone who takes the benefits without paying the costs. It is a hunter-gatherer who has others take care of her children but doesn't care for those of others. It is a gazelle who notices the strotting of others but doesn't itself take the risk of strotting. Such animals will benefit relative to the kinder individuals who surround them, and to the extent that their behaviors are influenced by their genes, it is these genes that will prosper and spread through the environment.

To see the problem more generally, consider a Utopia—a population of creatures whose genes guide them to be nice to everybody. And then, as you did for the kinship case above, imagine two variants of a gene: Variant A makes an animal care for itself and its kin and everyone around it. Variant B makes the animal care just for itself and its kin. This time, it's the selfish variant, Variant B, that prospers. Biologists talk about *evolutionarily stable strategies*—strategies that, when adopted by a population, are locked in because no invader with a different strategy can do well in that society. Tragically, a society of indiscriminate altruists is *not* an evolutionarily stable strategy; free riders can bust in and mess up the whole system, so that, in a matter of generations, the population turns selfish.

———

Given the success of selfish variants, how did kindness to non-kin ever evolve? The best answer we have, derived in part from the work of Robert Trivers, whose ideas we discussed earlier in the context of sex differences, is that kindness and cooperation of the sort we've been discussing can only evolve if they are accompanied by something that makes it costly to be a free rider.[68] This might involve a desire to overtly punish the free riders, or it could involve shunning them, refusing to bestow your kindness on them—not sharing your meat with the hunter who didn't share his meat with you; not caring for the child of the mother who refused to care for yours. In a situation where animals need to interact with other animals to survive, this form of shunning is a powerful, sometimes lethal, punishment.

Trivers' theory leads to a psychological hypothesis. Wherever you find a motivation for kindness and cooperation among nonrelated individuals, there should be a corresponding motivation to make life difficult for those who don't play by the rules. This is not trivial; it requires the capacity to recognize the bad apples, remember who they are, hold this memory across time, and be motivated to punish and shun. As the theory predicts, numerous studies find that these capacities are present in both humans and nonhuman animals.[69]

There's a rich irony here. It turns out that the best explanation of our finest instincts—love, gratitude, friendship, and esteem, as Smith listed them—entails that these can only persist if we also have feelings such as anger and resentment. What could be seen as the worst parts of human nature turn out to be essential for the creation of a good, loving, and cooperative society. If the appetite for revenge had been stripped from our ancestors long ago, we would never have arrived at where we are today.

———

There's much about kindness, cooperation, and morality that we won't have the time to touch upon. This field of moral psychology is rich and interdisciplinary, connecting to fields such as game theory, behavioral economics, sociology, anthropology, primatology, law, and theology. And it's exciting, just because there are so many puzzles that remain.

One puzzle concerns the theory we just outlined. We've explored how our outrage toward those who cut in line, slack off on the group hunt, and so on, makes possible a society of kind and cooperative individuals. We punish those who make it worse for all of us. But punishment of this sort is itself an altruistic behavior, often carrying some risk. So why would animals be motivated to do it themselves, instead of holding back and letting others make this effort? (It can't be "because by punishing, everyone benefits"; we've seen that natural selection doesn't work that way.) Put differently, we tried to solve the free-rider problem by proposing that punishment makes it costly to free ride, but now we need to explain how to deal with those who free ride when it comes to punishing free riders. One might propose that we would evolve an instinct to punish those who fail to punish free riders, but there seems to be an infinite regress here. Do we also need a further instinct to punish those who fail to punish those who fail to punish free riders?[70] Something is going wrong here.

A second puzzle concerns the sorts of activities that stir up our moral outrage. Consider sex. The sexual acts of others do no direct harm to me or my kin, and there's no obvious reasons why we should be so troubled by them. Actually, from a Darwinian perspective, men seeking female partners should welcome other men who willingly leave the heterosexual mating market by choosing celibacy, or exclusively homosexual relationships, or whatever. Less competition! But it doesn't work that way. In the book of Leviticus in the Hebrew Bible, sex between men is punishable by death, and this attitude is prevalent in most countries and most religions in the world, and in the minds of most people through history. Societies that see nothing wrong with such acts are the exception. This fact about our moral psychologies admits of no easy explanation.

———

A final puzzle involves self-sacrificial moral acts. Some people give up airplane travel because they are worried about climate change and its effects on future generations. Some stop eating meat, even if they enjoy the taste of animal flesh, because they believe that it is wrong to be complicit in the suffering of animals. There are those who send money to help others in faraway lands and those who fight for the rights of others even when this involves giving up some of their own privileges, as when millionaires argue that their own taxes should be raised to support the poor, or when members of majority groups fight for the rights of minority groups. If asked why they are doing these things, people will say that they are motivated by morality; these are the right things to do.

This has fascinated many scholars. Earlier in the book, we struggled with certain mysteries. We began with the mystery of intelligence—how can a physical thing, a brain, be capable of rational thought?—and then we moved to the mystery of consciousness—where does sentience come from; how is it possible that we can feel? And here, in our discussion of human motivation, we arrive at a mystery that may seem just as challenging. How can a creature evolved through the amoral force of natural selection be capable of real moral insight and true moral action?

Just as with intelligence and consciousness, some have been tempted by nonmaterialist explanations. In 1869, the codiscoverer of natural selection, Alfred Russel Wallace, observed that humanity has transcended evolution in many regards, including in our "higher moral faculties," and concluded that there must be some superior intelligence shaping the development of our species.[71] (Darwin was horrified by what he saw as Wallace's apostasy.) More recently, Francis Collins, the former director of the National Institutes of Health, argued that an appreciation of objective moral truth—a "Moral Law"—cannot be explained in Darwinian terms, and constitutes an argument for the existence of God.[72]

Many others, though, believe that we can explain this without appeal to a deity. The case with morality isn't so unique. There are many other ways in which human brains, working together in societies over thousands of years, have come to transcend our initial capacities. Our senses have evolved to perceive things such as rocks and trees, and yet we have come to appreciate the very small (such as atoms) and the very large (such as galaxies). We are born with the capacity to think about one, two, and three, but everyone reading this can appreciate large numbers, fractions, negative numbers, and to some degree, at least, infinity. We can develop rich systems of logic, mull over metaphysical issues such as free will and consciousness, and create scientific theories of all sorts, from physics to psychology.

At the start of this book, I quoted Immanuel Kant, who wrote: "Two things fill the mind with ever new and increasing admiration and awe, the more often and more steadily we reflect upon them: the starry heavens above me and the moral law within me." Over history, people have come to learn a lot about both of these things, not just about the starry heavens, but also about the moral law.

One critical discovery here has been made over and over in human history. It is about the importance of impartiality. The philosopher Peter Singer points out that explicit statements of this principle show up in virtually every religion and moral philosophy.[73] They are expressed in the various forms of the Golden Rule, as in Christ's command, "As you would that men should do to you, do ye also to them likewise," or Rabbi Hillel's statement, "What is hateful to you do not do to your neighbor; that is the whole Torah; the rest is commentary thereof." When Confucius was asked for a single word that summed up morality, he responded, "Is not reciprocity such a word? What you do not want done to yourself, do not do to others." Kant himself proposed as the core of morality: "Act only on that maxim through which you can at the same time will that it should become a universal law." Adam Smith appealed to the judgment of an impartial spectator as the test of a moral judgment, and Jeremy Bentham argued that, in the moral realm, "each counts for one and none for more than one." John Rawls suggested that when ruminating about a fair and just society, we should imagine that we are behind a veil of ignorance, not knowing which individual we will end up as.

Once we possess this notion, it can combine with our evolved moral capacities to take us to some surprising places. We are angry when we are betrayed and want to punish someone who assaults our child, but now, armed with an appreciation of the principle of impartiality, we can come to appreciate that it's also wrong when *anyone* is betrayed and when *any child* is assaulted. Just as other motivations can go in directions that evolution could never have anticipated—consider a child's curiosity about other planets, or a reader's sadness about the fate of Anna Karenina—moral emotions such as compassion, guilt, and righteous anger can, in concert with our rationality, lead us toward decisions and actions that take us far away from the project of reproductive success.

None of the above is intended as a complete solution to the puzzle of self-sacrificial morality, of course. There is so much more to say, and many counterarguments to address. But it does illustrate how one could approach this issue without relying on divine intervention.

————

I'll end with a competing theory for why we act in such seemingly self-sacrificial ways. This is the idea that we do so to bolster our reputations to others, to look more appealing to sexual partners, allies, and friends. We don't give to the poor or abstain from eating meat because we see these acts as good in themselves; rather the propensity to engage in such acts has evolved as a costly signal (one that is likely unconscious) of our fairness and our kindness. From this perspective, a man who marches for the rights of women, say, is ultimately driven to do so not because of a wish for a better and fairer world, but because of the advantages that come with being seen as the sort of person who wishes for a better and fairer world.[74]

This view has its merits. Among other things, it can help solve the earlier mystery of why we are motivated to make the seemingly costly decision to punish free riders—there is empirical evidence that we do so not out of a self-sacrificial urge to improve the situation for everyone, but rather because it makes us look good to others.[75] People admire those who do moral acts, including moral acts of punishment, and prefer to associate with them.

My own view is that the reputational theory is, in the end, too cynical to be a complete account of self-sacrificial morality. Often people really want to do good for its own sake. But I'll admit that for many examples of a disinterested moral act, there's a good explanation from fans of the reputational account. Richard Alexander, an evolutionary biologist known for his work on the origins of morality, describes an argument he had with his mentor. Alexander was trying to make a case for pure moral motivations, and he described how he went out of his way to avoid stepping on a line of ants. Isn't that truly altruistic? And his mentor responded: "It might have been, until you bragged about it."[76]

RELATIONS

10

A Brief Note on a Crisis

Social psychology is the part of the field that deals with our social natures. It's not literally true, as some social psychologists have boasted, that "all cognition is social cognition."[1] We've already seen a lot that's interesting about the mind that isn't about other people—such as depth perception, fear of heights, and the disgust we feel toward rotten food. But social psychology does include many of the topics that matter the most, such as prejudice, conformity, and persuasion. It's so interesting that I couldn't squeeze it into a single chapter.

There is practical value to this line of research. Social psychologists are involved in real-world interventions, working to make societies kinder, helping companies market their products and structure their organizations, and collaborating with governments in all sorts of ways. During the COVID epidemic, social psychologists came out in force, brimming with advice on how to persuade people to wear masks, get vaccinated, and so on.[2] This is the branch of psychology that is most likely to get in other people's business, and has the potential, when done right, of making the world a better place.

So what's not to like?

Well, the field has been in crisis. Over the last few decades, many social psychology findings, including some of the classic results that you may have heard of, have failed to replicate. If you redo the studies carefully, the results don't hold.[3] One project, looking at one hundred psychology studies published in top journals, found that only about 40 percent of the replications got significant results—and the social psychology studies fared worse than the others.[4]

Now a failure to replicate can happen even if an effect is real. For reasons of random variation, not every effect shows up when you look for it. Aspirin really helps with headaches, but it's not perfect, and so you can do a study in which you bring in two dozen people with headaches, give half of them aspirin and the rest nothing, and then, an hour later, test them and find no difference—the aspirin group isn't feeling any better than the control group. This just means that sometimes, by chance, real effects don't show up in a finite sample.

But there have been many failed replication attempts by now, and it's clear that these failures occur because, at least some of the time, the findings of these previous studies were spurious and do not reflect real psychological phenomena. And this is because psychologists have been doing their studies wrong.[5]

In a viral blog post, the social psychologist Michael Inzlicht summarizes what happened:

> We have abused our inferential tools by massaging our data to make them say what we want as opposed to letting them reveal their truths. . . . I am not talking about fraud or scientific misconduct. We did not think there was anything wrong with these practices. We thought we were merely allowing the truth that was baked in the data to rise. We did not yet understand how badly these practices warped our scientific inferences. If your analyses don't work out as planned, play around a little with your variables and see what happens: What happens if you add things to the mix? What happens if you exclude people with troublesome data? The point was to massage the data until it relaxed enough to reveal what everyone wanted: statistical significance.

These so-called questionable research practices, referred to more poetically as p-hacking in reference to statistical p-values, were commonplace in psychology. P-hacking was not something unscrupulous researchers did after dark when their more principled colleagues left the building. No, respectable and eminent scholars p-hacked out in the open, unashamed. My Ivy League professors explicitly encouraged us to p-hack. And because scientific writing is still an act of persuasion, we were taught to frame all the fruits of our exploration as if they were fruits of confirmation, as if we had predicted these baroque patterns all along.[6]

Here's an example of what Inzlicht worries about. Suppose you read about a study that finds that people who work in pink rooms are more creative than those who work in white rooms. The article reassures you that this is a robust and statistically significant finding. Pretty impressive study with real practical implications, right? Off to paint my study pink.

Now imagine that when the experimenters began their studies, their hypothesis was that green rooms (not pink ones) would make people persevere longer at a task (not be more creative). And they tested it, and it didn't work. Then they changed their hypothesis to "black rooms will make people persevere more," and they tested it, and nothing. Then purple rooms, nothing. Then pink rooms—and again it didn't work, but they were doing a set of other tests at the same time, just messing around, and they noticed that people did somewhat better on a creativity measure when in the pink room. So they took their findings and wrote an article saying "pink rooms make you more creative." Under these circumstances, the reported statistical effect can't be taken seriously. With multiple statistical tests, the odds of getting an effect—any effect—are much higher. The experimenters drew the bull's-eye after they fired the arrow.

This is an exaggeration, of course, but it illustrates the logic of the problem. The more analyses one does, the greater the odds of a spurious finding. And yet doing more analyses is just what many psychologists (including me) believed we were supposed to do—keep plugging away at the data, try to find something, publish studies that "work," ignore those that fail. There was some concern even before the crisis that this was a bit suspect, that we were inflating our odds of getting false positives, but as one set of critics noted, "Everyone knew it was wrong, but they thought it was wrong the way it is wrong to jaywalk." It turns out, they add, that "it was wrong the way it is wrong to rob a bank."[7]

Now we're in trouble. Similar concerns arise in other areas of psychology, as well as in psychiatry, economics, particle physics, and perhaps most troubling, medical research. It's shocking to read a *Nature* paper reporting a failure to replicate significant experiments in cancer research in forty-seven out of fifty-three cases.[8]

There is something very human about this crisis. People respond to incentives, and publishing interesting findings in top journals is how to get a job and then get tenure. Some incentives are less tangible. We are social animals, sensitive to status, and so even people who already have jobs and tenure want to impress others, to make a splash.

Most of the crisis is caused by shoddy research practices, but a notable exception was the work of the psychologist Diederik Stapel, who wrote several publications with fabricated data. When I read his autobiography, written after he was exposed, I was struck by how understandable his motivations were, even for us nonfrauds. As he described it,

> I published all kinds of things, but for whom? My hand-crafted, minor discoveries were too trivial to make it into textbooks, and my theories were too dry for anyone to want a chapter on them in the handbook they were editing. At conferences I was still in the small-to-medium rooms off to one side. I'd never been invited to be the keynote speaker. I was bored. I wanted to come up with a line of research that everyone would want to follow. I didn't want to clean up after other people any more. I wanted to stop picking apart balls of yarn and start winding some of my own. I wanted to get out the spinning wheel and weave my own research. I wanted to do something really important and sensational, to really make a contribution. I wanted to be one of the stars.[9]

A lot of scientists share Stapel's motivation. We want to be stars. This can get us into all sorts of trouble.

————

The replication crisis might not be the worst of it. A related concern is that almost all our psychology has been based on a narrow and unusual population—WEIRD people, where WEIRD is an acronym meaning Western Educated Industrialized Rich Democracies. Even though only about one eighth of the world is WEIRD, just about all the subjects in psychology research are. Indeed, many are from an even narrower population—undergraduates. As Joseph Henrich and his colleagues put it, in an article that introduced this issue to a broad audience of scholars, "a randomly selected American undergraduate is more than 4,000 times more likely to be a research participant than is a randomly selected person from outside of the West."[10]

This situation is perfectly understandable—almost all psychologists are WEIRD, and it's just so much easier to run studies in our own communities or online. But it's a serious problem, and I'm often surprised that my colleagues aren't more upset about it. Suppose that the conclusions of psychological research had been based on studies with 95 percent male subjects. This would plainly be ridiculous. Right now, if you did research that didn't include women—without some special reason for the exclusion—you wouldn't be able to publish it or get funding for it. But with the notable exception of sexual preference, the difference between American undergraduate men and American undergraduate women is much smaller than the difference between American undergraduates and people from the rest of the world.[11]

What kind of differences are we talking about? Henrich, in a book called *The WEIRDest People in the World*, goes on about some of the oddities of people like me and him:

> Unlike much of the world today, and most people who have ever lived, we WEIRD people are highly individualistic, self-obsessed, control-oriented, nonconformist, and analytical. We focus on ourselves—our attributes, accomplishments, and aspirations—over our relationships and social roles. We aim to be "ourselves" across contexts and see inconsistencies in others as hypocrisy rather than flexibility. . . . We see ourselves as unique beings, not as nodes in a social network that stretches out through space and back in time.

When acting, we prefer a sense of control and the feeling of making our own choices. When reasoning, WEIRD people tend to look for universal categories and rules with which to organize the world, and mentally project straight lines to understand patterns and anticipate trends.[12]

The specialness of WEIRD people is an issue that affects the whole field, but as we'll see in a bit, it's particularly relevant to claims made by social psychologists.

————

But enough kvetching; I want to end this interlude on a cheery note. Some people hear about the problems in psychology and give up on the whole field. This is, to put it mildly, a mistake. For one thing, in the last decade or so, social psychologists have been getting their act together, taking the lead on reforming scientific practices. There is more preregistration, where you announce ahead of time what analyses you are going to do, so as to avoid the problem of spurious findings. It is more common now to share your procedures and your data so that other scientists can double-check them. And there is a lot more research with non-WEIRD populations.

For another, despite the failures to replicate and despite the problems with our narrow subject population, psychology has made robust and important discoveries about universal features of the mind.

Examples? Sticking with what we've talked about up to now, we've seen the following: Different parts of the brain are reliably related to specific aspects of thought. We know, for instance, that the hippocampus is related to spatial understanding and that the parietal lobe has to do with hearing. Partial reinforcement makes behaviors hard to extinguish, for both rats in a cage and humans in a casino. Babies are surprisingly smart about the physical world, and children, before they speak, can understand word meanings and even rudimentary syntax. Our expectations influence both how we see the world and how we remember what we've seen. We overestimate the likelihood of infrequent but conspicuous events, and we do poorly at reasoning about problems framed in terms of percentages—

but much better when the tasks are made more realistic. There is a universal list of things and experiences that frighten people, suggesting that our fears have been shaped by natural selection. In some regards, men and women have very different sorts of sexual preferences; in other regards, we are just the same. Moral emotions, including compassion and outrage, are universal, and make society possible.

And there's more to come.

11

Social Butterflies

Let's start with ourselves, our special, special selves.

Much earlier in the book, we talked about how our self-focus leads us to overestimate the extent to which others are aware of us. This spotlight effect drives us to think that our mishaps and successes and lies are more salient to others than they really are.[1] It's hard to discount the salience of our personal experience, to recognize that we are not as special as we think we are, to appreciate that others are at the narrative center of their own stories, not ours.

This tendency to think we are special shows up in other ways. Compared with others, are you average, below average, or better than average, in:

Intelligence?
Driving skills?
Being a good friend?
Sense of humor?

If you said that you are better than average, you're in good company. Most people say the same, pretty much regardless of what you're asking about. In one survey, over eight hundred thousand high school seniors were asked to rate themselves on their ability to "get along with others." Fewer than 1 percent said that they were below average, and over half ranked themselves in the top 10 percent.[2] Now, strictly speaking, it's not impossible that most people are above average. If one person has a dime and a hundred people have one dollar, then the average is ninety-nine cents, and 99 percent of the people have more than average. But most capacities—driving, friendliness, and so on—don't have this skewed distribution, and so this above-average effect suggests that people really are overestimating themselves.

This is sometimes known as the Lake Wobegon effect, inspired by the fictional town from the radio series *A Prairie Home Companion*, where, as the creator Garrison Keillor jokingly puts it, "all the women are strong, all the men are good-looking, and all the children are above average."

You may think that the above-average effect doesn't apply to you—but this itself might reflect the above-average effect. In one study of 661 people, most thought that they were more immune than others to being biased, and only one admitted to being more biased than the average person.[3]

One interpretation of this effect is that goodness is a fuzzy notion. There are a lot of ways to be good at something and so a lot of ways that one could excel. Am I a good driver? Sure, I'm very cautious. Sure, I can drive long distances without complaining. Sure, I take risks nobody else will.

This fuzziness account cannot be entirely right, though, because we get the same effect when performance can be precisely defined. One experiment was done at a chess tournament.[4] Players were asked both about their most recent official chess rating—computed by their win-loss record and the strength of their opponents—and about what their rating should be to reflect their real current strength. One in five thought their rating was accurate, a tiny 4 percent thought they were overrated, and a whopping three quarters thought they were underrated. On average, people said that they were 99 points underrated, which means that two players of equal rank would each tend to believe that they have an excellent chance of beating the other one.

A different sort of self-enhancement bias involves how we make sense of events that involve us. Unless you're depressed (more about this in a later chapter), you will tend to view your positive accomplishments as the result of your efforts and abilities and your failures as caused by forces outside your control. I did well in this class because I worked hard; I did poorly in that class because it was held early in the morning and the professor doesn't know how to write an exam.[5]

Another form of self-enhancement is more subtle: We wish to be consistent. When our actions clash with our beliefs and preferences, it's uncomfortable, and so we are motivated to change them for the sake of consistency. This discomfort is sometimes known as "dissonance," and the general theory is known as *cognitive dissonance*. You might have heard of this; as one scholar notes, "Cognitive dissonance is the best-known psychological theory of rationalization; indeed, it is among the best-known psychological theories of anything."[6]

Here's one classic demonstration, from 1959.[7] Subjects were assigned to a very boring task—turning pegs in a pegboard for an hour. Then they were given either a dollar or twenty dollars (a lot of money at the time) to go into another room and lie to another person, to say that the task is a lot of fun. Afterward, the subjects were asked how enjoyable the task really was.

It turns out that those who got paid a small amount of money rated the task as more enjoyable. The explanation for this comes from cognitive dissonance. It makes sense to lie for a large sum of money. But it's uncomfortable to realize that you lied for a dollar, and one way to reduce this discomfort is to convince yourself that the task wasn't so bad, and so you weren't lying after all.

Some cognitive dissonance studies involve choice. Ask people to choose between two things that are valued roughly equally, and later on they will tend to like the chosen one more than when they started and the unchosen one less.[8] This effect occurs even when the choice is blind—where you don't know what you're choosing.[9] But it doesn't happen if someone else makes the choice for you, suggesting that the shift in preferences plays a psychologically palliative role—it makes you feel better about your own decisions.

Here cognitive dissonance nicely flips the usual logic of commerce:

Everyone knows that how much you value something determines whether you will choose it. The insight of cognitive dissonance is that the opposite is true as well: Whether you choose something influences the value you give to it.

Cognitive dissonance has the promise of explaining certain puzzling aspects of our lives. Take hazing. Why would groups, such as fraternities and medical schools, make new members suffer? It seems irrational; it should make people less likely to stay in the group. This is how unpleasant experiences work more generally, after all—no hotel chooses to make the beds uncomfortable and the bathrooms smelly, because customers wouldn't return. But the logic of hazing is that if you willingly go through some sort of humiliating or unpleasant experience to enter a group, that'll make you more invested. You chose to do it, so you must have had a good reason to do so, and the best reason is that it's a valuable group to belong to.

———

We judge other people. Some of this happens very quickly.[10] In one classic study, people were asked to rate teachers based on short video clips. It turns out that after watching a teacher in a six-second silent clip, subjects gave much the same teacher evaluations as those who were with that teacher for a full semester. Based on quick exposures, people are also good at detecting other seemingly invisible properties of a person, such as how extroverted they are, their sexual preference, even their political orientation.[11]

It's strange that this is possible. One recent study found that a facial-recognition algorithm, tested on images of over eight hundred thousand people on dating sites, could predict political orientation at the ridiculously high rate of 72 percent accuracy, better than the human rate of 55 percent.[12] Some find this work offensive and worry about privacy concerns. I get this, but I'm also amazed that it works at all. Doesn't this harken back to phrenology, the disproven theory we discussed earlier where the secrets of people's nature are revealed by the shape of their skulls?

It turns out that some of the accuracy has to do with demographic factors—in the United States, for instance, if you see an older white male face, it's more likely than not that you're looking at a conservative. Accuracy is still high, though, when you give the algorithm a sample of faces that are of the same age, gender, and ethnicity. Some of the accuracy has to do with how people pose—for some reason, liberals are more likely to face the camera directly and more likely to look surprised—and also with choices they make about how to present themselves, such as their facial hair and type of eyeglasses. These are all useful cues, but again, the algorithm does fine without them. It's quite the mystery what's going on here.

———

We also judge people based on what they do, and here a specific bias emerges. We've just seen that we tend to be soft on ourselves, blaming our failures on factors beyond our control. We are not so gentle with others. For them, we blame their character, not the situation. If someone is unkind to you, you might assume that they are a rude person, not that they are just having a bad day. This is sometimes known as the *fundamental attribution error.*[13]

As a classic example of this, consider the "Quiz Show" study done in 1977.[14] Three people walk into the lab and—in front of them, so it's totally clear what's going on—one of them is chosen to be a Questioner and another is chosen to be a Contestant. Questioners are asked to write hard but not impossible general knowledge questions and ask them to Contestants. Not surprisingly, Contestants do poorly; it's tough to answer random trivia. Then the third person is asked who knows more overall. They tended to say that it's the Questioner—after all, this person knew all the answers. Somehow these observers failed to consider that this difference in success is because of the roles that the other two subjects were arbitrarily assigned to.

Perhaps the weirdest reflection of the fundamental attribution error is a tendency to see actors as the characters that they play. We often see actors who play action heroes as tough and brave, and actors who play soap-opera villains as evil. Leonard Nimoy, who played the iconic role of the half human, half Vulcan in *Star Trek*, struggled with this throughout his career, writing a book called *I Am Not Spock*—and then, years later, writing another book called *I Am Spock.*

This confusion between actor and character was exploited (or joked about, I'm not sure) in a famous television commercial, where a soap-opera actor, Peter Bergman, who played Dr. Cliff Warner on a show called *All My Children*, sells cough medicine with the wonderful line: "I am not a doctor, but I play one on TV."

Why do we possess such social biases?

Let's return to Lake Wobegon. Perhaps part of the above-average effect is showing off; even in an anonymous survey we want to present ourselves at our best and impress the experimenter. Or perhaps these beliefs are genuine and arise because of the asymmetric feedback we get in everyday life. After I give a talk, people sometimes come up to me and say how much they enjoyed it. So I might think that I'm a better-than-average speaker, not realizing that most of the people who find me boring aren't going to tell me. (Perhaps I'm like a dog who is continually being told that he is a good boy, not realizing that this is just what people often say to dogs.)

At its core, though, the Lake Wobegon effect may be just one case of a more general positivity bias we have toward ourselves. We've already seen how we tend to attribute our victories to ourselves and our failures to circumstance. More than this, we are natural optimists. When people embark on diets and exercise programs, they likely already have a history of failure. And yet they plug on, believing that this time is the exception.

One example of our optimism is what behavioral economists call the planning fallacy—when people plan a future task, they are too optimistic about how long it will take. The psychologist Paul Rozin (whom we met earlier when talking about disgust) tells this story.

> Every night, I bring home a pile of work to do in the evening and early morning. I have been doing this for over 50 years. I always think I will actually get through all or most of it, and I almost

never get even half done. But I keep expecting to accomplish it all. What a fool I am.[15]

There is a more sympathetic take on what's going on. There is no way to be certain about how well our diets will work out, how long this book will take to write, and how much work Rozin will get done when he goes home at night. No matter how wise we are, much of life is simply out of our hands. And so we play the odds. For a rational actor, this involves an assessment of the chances, but it also involves a judgment of the costs of getting it wrong and the benefits of getting it right. Rozin is no fool; I bet it was no big deal for him to schlep the work back and forth, and I bet that sometimes he ended up needing the material.

Sometimes it's smart to be an optimist. Suppose a shy young college student enjoys the company of another student and is considering asking her out. There are two types of mistakes he can make. He can ask her and she says no and his feelings will be hurt. Or he can choose not to ask her but she would have said yes if he did, and he missed the opportunity of a date, and then, possibly, friendship, perhaps a passionate love affair, and then, just maybe, marriage, children, retirement, grandchildren, and an overall wonderful life together. It seems like the second mistake is worse.

One proposal is that our assessment of costs and benefits influences our judgments of the likelihood of success. This has been proposed as a theory of why heterosexual men tend to overestimate the sexual interest of women.[16] The mistake that a man makes by assuming that a woman isn't interested if she really is, is worse than the mistake he makes by assuming that she is interested if she really isn't. (It should go without saying that these are the costs and benefits to *men*; the math for those women who are the targets of men's unwelcome overtures is quite different.)

And now we're able to develop a more general theory of when it's good to have a positivity bias. As psychologists Martie Haselton and Daniel Nettle put it, "If the cost of trying and failing is low relative to the potential benefit of succeeding, then an illusional positive belief is not just better than an illusional negative one, but also better than an unbiased belief."[17] If you were building a robot to survive in the world, you might well wire in positivity biases.

But things are rarely so simple. If the advantage of a positivity bias is its motivational force—you are more likely to act if you believe, perhaps unrealistically, that your odds are good—the disadvantage is that it steers you away from accuracy. My positivity bias might make me stick to an unrealistic goal—publishing in a top journal, acting in a Broadway play, world domination—while it would be better for me, in the long run, to have more realistic aspirations. The writer and rationalist Julia Galef quotes Francis Bacon on this point: "Hope is a good breakfast but it is a bad supper."[18]

Also, there are cases where the math flips, and it's better to focus on the negatives, to have a gloomy view of the world. It would be nice to know for certain whether that movement in the branches in front of you is a tiger. But sometimes you must make an educated guess, and so, again, there are two sorts of mistakes you can make:

False alarm: You think there is a tiger and there's not.
Miss: You think there is no tiger but there is one.

The proportion of these types of errors is inversely correlated. If you are highly prone to think "Tiger!" you will have more false alarms and fewer misses; if you are less prone, you will have more misses and fewer false alarms. So which mistake is worse? Well, a false alarm is not perfect; you freeze up, burn some calories, maybe wet yourself a little. But the cost of a miss is becoming the tiger's lunch. It's best, then, to err on the side of caution: to flinch and look around when you see something and to live with an abundance of false alarms.

The psychiatrist Randolph Nesse has an engaging discussion of anxiety that builds on this point.[19] Many of us worry too much; going into a dangerous neighborhood, we are too concerned about being mugged; we overprepare for social encounters; we obsess needlessly about the safety of our children. And most of the time nothing bad happens and all this anxiety seems like a waste. But Nesse argues that it is a part of a successful life. Our anxiety motivates us to plan, to obsess, to ready ourselves for low-frequency worst-case scenarios, and the benefits outweigh the cost.

Yes, in some cases, we go too far; people do suffer from anxiety disorders. But Nesse notes that hardly anyone worries about the flip side of anxiety disorders—not enough anxiety, or *hypophobia*.

> Hypophobia is serious and potentially fatal but underrecognized and rarely treated. Hypophobics don't come to anxiety clinics. Instead, they are found in experimental aircraft, on creative frontiers, and on the front lines of battlegrounds and political movements. They are also found in prisons, hospitals, unemployment lines, bankruptcy courts, and morgues.[20]

In some circumstances, then, a positivity bias would be catastrophic, and pessimism and overcaution are wiser attitudes. In the television show *The Wire*, Omar Little evades an assassination attempt and later returns to shoot one of his assailants, taunting him as he does so: "You come at the king, you best not miss."

———

Culture might influence other biases as well. Consider again the fundamental attribution error, our tendency to explain what others do in terms of their character, not their circumstances. This is often thought of as part of our natures—it's fundamental!—but not everyone agrees. In 1943, Gustav Ichheiser, an early leader of social psychology, was clear in his assumption that this error is the product of a certain time and place: "A consistent and inevitable consequence of the social system and of the ideology of the nineteenth century, [has] led us to believe that our fate in social space depended exclusively . . . on our individual qualities—that we, as individuals, and not the prevailing social conditions, shape our lives."[21]

This focus on culture brings us back to an issue we discussed earlier, which is that a certain individualist stance on the world, something that many social psychologists take as inevitable, may instead be a product of certain WEIRD cultures—Western Educated Industrialized Rich Democracies.

When I was a graduate student, I read the anthropologist Clifford Geertz, and came across his view that the Western sense of a person as a "bounded, unique, more or less integrated motivational and cognitive universe; a dynamic center of awareness, emotion, judgment, and action" isn't universal. Instead, it is "a rather peculiar idea within the context of the world's cultures."[22]

This blew my mind. The idea that we are autonomous beings—bounded, unique, and so on—seemed so obvious, so *true*, that it was hard for me to imagine that there was any other way to think. But maybe I was wrong. Here is an expanded version of the quote I gave earlier from Joseph Henrich, from his book *The WEIRDest People in the World*, on how the WEIRD-os are different.[23]

> The world today has billions of inhabitants who have minds strikingly different from ours. Roughly, we weirdos are individualistic, think analytically, believe in free will, take personal responsibility, feel guilt when we misbehave and think nepotism is to be vigorously discouraged, if not outlawed. Right? They (the non-WEIRD majority) identify more strongly with family, tribe, clan and ethnic group, think more "holistically," take responsibility for what their group does (and publicly punish those who besmirch the group's honor), feel shame—not guilt—when they misbehave and think nepotism is a natural duty. . . .
>
> We see ourselves as unique beings, not as nodes in a social network that stretches out through space and back in time. When acting, we prefer a sense of control and the feeling of making our own choices. When reasoning, WEIRD people tend to look for universal categories and rules with which to organize the world, and mentally project straight lines to understand patterns and anticipate trends. We simplify complex phenomena by breaking them down into discrete constituents and assigning properties or abstract categories to these components—whether by imagining types of particles, pathogens, or personalities. . . .

Paradoxically, and despite our strong individualism and self-obsession, WEIRD people tend to stick to impartial rules or principles and can be quite trusting, honest, fair, and cooperative toward strangers or anonymous others. In fact, relative to most populations, we WEIRD people show relatively less favoritism toward our friends, families, co-ethnics, and local communities than other populations do. We think nepotism is wrong, and fetishize abstract principles over context, practicality, relationships, and expediency.

Henrich gives an example that illustrates the difference.[24] Please complete this sentence in ten different ways. Do it now, in your head. I'll wait.

I am ___.

If you are from a WEIRD culture, you likely answered with a word or phrase that conveys some property of yourself, such as "intellectually curious" or "hot-tempered." Or perhaps you expressed your occupation or status, like "student" or "retiree" or "billionaire tech mogul." This is the case for U.S. undergraduates, who usually talk about abilities, attributes, and aspirations.

But this is not universal. Cross-cultural studies suggest that people in rural Kenya, for instance, are more likely to talk about relationships and less likely to mention these sorts of personal traits. They are more likely to have spontaneously said something like "Mona's mother" or "the brother of Aidan."

Henrich argues that our culture's focus on people's "true nature" leads to phenomena such as the fundamental attribution error. He cites research that, among non-WEIRD cultures, this tendency is less pronounced, and concludes: "It's not that fundamental; it's WEIRD."[25]

———

These insights about cross-cultural differences are important. One of the sins of psychology—and social psychology in particular—has been to assume that our subjects are typical of the rest of the world. They are not. We must be receptive to the idea that some of the most intuitive and seemingly most natural social biases and preferences may emerge from the specific culture of the social psychologists and the people they study.

But we shouldn't take this too far. I'm often struck by the extent to which putatively non-WEIRD modes of thought live within very WEIRD people. (And to be fair, scholars like Henrich are clear that there is not an all-or-nothing contrast between WEIRD and not-WEIRD; it's a matter of degree.)

Take the "I am ___" demonstration. I was so impressed by the finding that I immediately tried it out on a friend. I asked him to complete the sentence "I am ___," and he said promptly, "I am American," naming his nationality, and ruining the point I wanted to make. Later that same day, someone on Twitter made a remark I found interesting, and so I clicked on his bio and there—on social media, in the WEIRD-est of forums, from someone who describes himself as a fourth-generation New Yorker—it said:

Sam's father. Sally's husband.

So a focus on relationships is not *that* strange to us.

Or consider the psychology of conformity and persuasion. People in WEIRD cultures are less autonomous and more swayed by convention than we are often willing to admit. An example of this comes from the old television program *Candid Camera*, which specialized in what one might call "street psychology."[26] In one episode ("Face the Rear"), they set up a scenario in which actors are standing in an elevator. Once someone walks into the elevator, all the actors turn to face the back of the car. This is a strange thing to do, but inevitably, the victim of the prank turns and faces the back as well. We don't want to be different.

It turns out that a great trick for getting people to do something, such as to vote or recycle, is to tell them that most other people are doing it. When a Holiday Inn in Arizona wanted guests to reuse their towels, they found the most effective method was to leave a card in their rooms, informing them that "seventy five percent of our guests use their towels more than once."[27] This is worth knowing if you're in the persuasion business. Sometimes people get it wrong, though. I once ate at a dining hall at the University of Chicago, and they had signs complaining about how many students were stealing cutlery and asking them to stop. This was exactly the wrong tactic: I had never thought of slipping the fork and knife into my coat pocket, but after reading the sign, the thought did come to mind. After all, this seems to be what people do at the University of Chicago. Much better for them would be a sign saying something like, "95 percent of students don't steal; be like them and respect your university."

The issue of conformity was explored in a classic study by Solomon Asch. Subjects would enter a room to participate in a psychology experiment and would sit around a table alongside people they believed were fellow participants—but who were actually confederates of the experimenter. They were given a series of tasks in which they were shown a target line and had to say which line it matched up to. The answer was obvious, and if subjects were simply asked, they would almost always get it right. But what if the subjects had to say their answers aloud, and before it was their turn, they heard every other participant give the wrong answer? Sometimes the subjects resisted the pull of the crowd, and you can see in this the force of autonomy and independence. But, still, over three quarters gave the wrong answer in at least one trial. Again, people are motivated to conform.[28]

———

Asch's experiment interested a young untenured professor at Yale, Stanley Milgram, who wondered about taking it further. In an interview, Milgram said,

> I was dissatisfied that the test of conformity was judgments about lines. I wondered whether groups would pressure a person into performing an act whose human import was more readily apparent, perhaps behaving aggressively toward another person, say, by administering increasingly severe shocks to him. But to study the group effect you would also need an experiment control; you'd have

to know how the subject performed without any group pressure. At that instant, my thought shifted, zeroing in on this experimental control: Just how far would a person go under the experimenter's orders?[29]

And from this was born one of the most famous—or infamous—experiments in all of psychology, done at Yale in 1961, in Linsly-Chittenden Hall, a neo-Gothic building a few blocks from the office I spent twenty years in.[30]

Milgram put an ad in the *New Haven Register* advertising for men between the ages of twenty and fifty to participate in a scientific study of memory and learning. When someone who answered the ad would arrive at Yale, they met an experimenter, a somber young man in a laboratory coat. Also in the room was a large friendly man of Irish descent described as a fellow participant. The two "participants" would randomly choose who would be the "teacher" and who would be the "learner." Actually, the man waiting in the room, James McDonough, was an actor, and the drawing was rigged: McDonough would always be the learner, and the real subject of the experiment was always the teacher. Then McDonough would go to an adjacent room, out of sight but close enough so that the two men could easily hear one another, and the experiment would begin.

The teacher would be told that this is a memory experiment. He was to read a string of four word pairs to the learner ("strong-arm, black-curtain . . .") and then, as a test of memory, give the first word and then four possible completions (as in "strong . . . back, arm, branch, push"). If the learner answered correctly, the teacher would say "Correct." In front of the teacher, there was a large machine with a horizontal row of thirty switches, with printed text on top, running from SLIGHT SHOCK to DANGER—SEVERE SHOCK to XXX. If the learner got an answer wrong, such as saying "back" to the prompt "strong," the teacher was supposed to say "Wrong," read out the voltage, and then throw the switch, zapping the learner, and give out the right answer ("strong-arm").

At first the learner would do fine, but then, according to a planned script, he would make mistakes. The teacher would then shock him and read out the correct answer. The teacher was instructed to increase the shocks as the mistakes continued. Soon the learner would yelp. Then he would cry in pain, and soon he would demand to be let out, saying that he has heart trouble. Then he would scream. Then he would entirely fall silent. The teachers often asked the experimenter, throughout the process, if they could stop, but the experimenter would insist, "The experiment must continue" and "Please continue."

The question is how far the teachers would go. Milgram found that about two thirds went to the highest setting, at a level that, if this wasn't rigged, would likely kill the learner. This was the main finding of one of the most surprising experiments in the history of the field—normal people are so obedient that they will kill other people when asked.

When the experiment was over, the "learner" emerged from the room, smiling and unharmed, and he and the experimenter would explain to the teacher that this was just an experiment, the machine was fake, and nobody was harmed. Nobody except for, perhaps, the subject. We appreciate now that this is a deeply unethical experiment, as it runs the risk of traumatizing the volunteers, who had been fooled into believing that they had done a terrible act. (Discussing the Milgram experiment and similar ones, a satirical *Onion* article has the title: "Report: Majority of Psychological Experiments Conducted in 1970s Just Crimes.")

The Milgram study is often cited as showing that we are prone to obey, to be conformist. But it's unlikely to be as simple as this. We disobey all the time. People don't obey speed limit signs, they cheat on their taxes; my students often don't read the syllabus. Rather, there are certain other ingredients in the Milgram study that motivated compliance, including:

1. The power of authority—these men were being told what to do on Yale campus, by a serious fellow in a white lab jacket.
2. A confusing and novel situation.
3. A technique for inducing compliance known as "foot in the door," which exploits the tendency to be more likely to agree to a request if you just agreed to a less extreme one. If subjects were asked to deliver a seemingly fatal shock at the very start, it's likely that few would, but the Milgram study was designed

so that they started small and worked their way up, and this further increased compliance.

Still, the Milgram study shows that under the right circumstances, we are subject to authority, to a far greater degree than many would have thought.

———

One theme of this book has been the impressive powers of the mind. We've seen that our capacities for language and perception and reasoning far surpass anything that a machine is capable of. Even when we get it wrong—as with visual illusions, base-rate neglect, and the planning fallacy—such errors might be the unavoidable by-products of a system evolved to do its best in an uncertain world. We are impressively rational beings.

But this is not the story that some social psychologists would tell. They will say that my review of their field missed an important discovery that shows we are nowhere near as smart as people like me think we are.

The discovery is social priming. Consider some truly striking findings. (Some of these might seem a bit crazy, but I'll explain the theory behind them in a bit.) Sitting on a wobbly workstation or standing on one foot makes people think their romantic relationships are less likely to last.[31] College students who fill out a questionnaire about their political opinions when next to a dispenser of hand sanitizer become, at least for a moment, more politically conservative.[32] Exposure to noxious smells makes people feel less warmly toward gay men.[33] If you are holding a résumé on a heavy clipboard, you will think better of the applicant.[34] If you are sitting on a soft-cushioned chair, you will be more flexible when negotiating.[35] Standing in an assertive and expansive way increases your testosterone and decreases your cortisol, making you more confident and more assertive.[36] Thinking of money makes you less caring about other people.[37] Voting in a school makes you more approving of educational policies.[38] Holding a cold object makes you feel more lonely.[39] Seeing words related to the elderly makes you walk slower.[40] You are more likely to want to wash your hands if you are feeling guilty.[41] Being surrounded by trash makes you more racist.[42] And so on and so on.

These findings are said to support a radical claim about human psychology—our conscious reasoning and motivations and goals matter a lot less than we think they do. You might think that there are good reasons for your political opinions, your feelings of loneliness, your judgment of a job applicant, and so on. But that's an illusion. You aren't judging or thinking at all; you are being *primed*.

This rejection of the centrality of consciousness might look at first like a reboot of Freud, but the proponents of social priming tend to see themselves more as bringing back behaviorism. You aren't influenced by unconscious desires, after all; rather it's the environment—what you are looking at or how you are standing or what sort of smell is in the air—that's changing how you think or act. As the authors of a classic defense of social priming put it: "Most of a person's everyday life is determined not by their conscious intentions and deliberate choices but by mental processes that are put into motion by features of the environment and that operate outside of conscious awareness and guidance."[43] Another researcher makes the connection to behaviorism more explicit: "As Skinner argued so pointedly, the more we know about the situational causes of psychological phenomena, the less need we have for postulating internal conscious mediating processes to explain these phenomena."[44]

This is the story that many social psychologists would tell, and it's sufficiently influential that this book would be incomplete without mentioning it. I'll end the chapter with why I think it's mistaken.

———

But first, as promised, I want to say a bit about why these priming effects might exist in the first place.

The British Empiricists thought the brain was an association machine, and while it's almost certainly a lot more than that, we do form associations, and these can trigger our thoughts and actions. I smell pizza and my stomach rumbles; I pick up a boomerang and have the urge to throw it; Pavlov's dog drools at the sound of the bell. From this perspective, it makes sense that words associated with the elderly might cause you to subtly act like an older person, or that standing in a way associated with dominance (a "power pose") might make you feel more powerful.

Other effects might reflect more subtle aspects of how the mind works. In language and thought, there is a universal connection between physical warmth and social contact. Just think of the meanings of "cold person" and "warm person." Perhaps this emerges in the course of development, where close body contact becomes associated with intimate connections.[45] This might make it so that physical coldness causes people to feel lonely. Similarly, moral purity is reliably connected to physical cleanliness, which is why we have phrases like "a dirty mind," and so it's not unreasonable to propose that feeling guilty might make one more likely to want to clean their body.[46] (This phenomenon has been dubbed "the Macbeth effect," based on the scene where Lady Macbeth frantically scrubs her hands after the murder of Duncan, saying, "Out, damned spot.")

Such effects, then, have some theoretical plausibility. But even if they exist, they might not make any real difference outside the laboratory. They certainly *don't* show that "most of a person's everyday life" is determined "by features of the environment and that operate outside of conscious awareness and guidance." Such a claim makes the huge jump from "X has some effect on thought or behavior" to "X is mostly responsible for thought or behavior." Suppose I discovered that people find food slightly tastier when it's served on a simple white plate (I just made this up, but who knows?). Cool finding, but I shouldn't pretend that I've shown that most of our taste experience is determined by plate color.

Or, to take a real finding from my list above, suppose your impression of the sturdiness of your romantic relationship is subtly influenced by the sturdiness of the chair you're sitting on. This is interesting from a psychological point of view. But the effect of the chair is minuscule compared to hearing a rumor that your partner is planning to run away with their yoga instructor. Maybe drinking sauerkraut juice makes an experimental subject more likely to endorse extreme right-wing policies[47] (a real finding— check the notes), but you don't want to conclude that the major factor in the rise of fascist movements is too much sauerkraut juice consumption.

We are fascinated by weird effects. These are what get into the journals, go viral, and excite all of us. It's interesting to find that how you position your body influences your confidence, that what you're smelling influences your attitudes, and so on. Such findings are particularly interesting when it comes to biases, a topic we'll turn to in the next chapter. I think it's worth knowing that bidders on eBay offer less money for a baseball card held by a Black hand than by a white one,[48] or that mock juries are influenced by the physical attractiveness of the defendant.[49]

The problem arises when people forget about the more mundane and rational considerations that are far more important, but that are too obvious to get into journals. Nobody would publish an article reporting that the bidding for baseball cards on eBay is influenced by how valuable people think the cards are, or that juries will give a harsher punishment to a murderer than to a shoplifter. But it's these more commonsensical (and more boring) influences that really make a difference in our lives.

Now, things would be different if these unconscious effects were powerful and robust. One can imagine a world where the very best way to predict a person's loneliness was to ask about how warm they are; where the best determinant of how one judges a job candidate is the physical weight of their résumé. Such discoveries really would challenge the notion that we are rational and deliberative beings.

We don't live in this bizarro world. For the sake of argument, I've been assuming that these effects are real. But social priming effects are the most controversial in psychology. We talked about the replication crisis just before this chapter, describing how certain research practices have led to false positives—findings in the literature that are not true. This crisis has hit hardest in the domain of social priming, devastating what one critic called "primeworld."[50] Indeed, one of the items on my list is clearly fraudulent—the "finding" that being surrounded by trash makes you more racist was published by Diederik Stapel, who later admitted to making it up. As far as I know, none of the other findings are fraudulent, but they are all uncertain (including one that I was coauthor of), and several of them have failed to replicate.

Again, some instances of social priming might be robust, just subtle and hard to find. Now, there's nothing wrong with subtle. I've always been fascinated with the potential connection between physical purity and moral purity,[51] and some of the demonstrations of a link might hold up. Standing like Wonder Woman might not influence hormones, but it does seem to boost your confidence.[52] Even if the effects only emerge in controlled laboratory conditions, the findings, if replicable, can tell us some interesting things about the mind.

But the subtlety of the effects suggests that they really don't play much of a role in everyday life. They certainly don't show us that consciousness is irrelevant. From a practical point of view, they might not matter at all.

There's a more general reason to be skeptical about this resurgence of a Skinnerian worldview. Fans of social priming see human behavior as if we are leaves blowing in the wind, swayed by various random forces. This just isn't so. We are certainly social beings, influenced by both our immediate environments and the practices of our communities, but there is an intelligence to our social behavior. We don't just ape those around us; we are smart about who we choose to learn from, and who we choose to model ourselves after.[53] Furthermore, even the most basic human activities require complicated planning. Consider something as simple as inviting some friends to get together at a restaurant. You need to check your calendar, send out emails, consider options of where to meet, check with people about allergies, and so on. A five-year-old can't do it. The most sophisticated AI in the world can't do it. And a mind that works through social priming can't do it. It requires the exercise of powerful and uniquely human capacities for reason.

12

Is Everyone a Little Bit Racist?

We've been talking so far about how we think about individuals—ourselves and others. But much of social psychology explores how we think about groups. And a central focus of this research is about how we think about (and feel about, and act toward) members of marginalized groups, such as ethnic and sexual minorities.

This research gets a lot of interest. I often experience the rush of seeing my colleagues' research on this topic being taken seriously by people in power. When I worked on a first draft of this chapter, Merrick Garland was being questioned as part of his confirmation hearing for becoming the attorney general of the United States, and one of the senators questioned him about the implicit bias test (which we'll talk about later). The senator asked: "Does it mean that I'm a racist . . . but I don't know I'm a racist?" This is a good question, and Garland gave a common enough response—no, everyone has biases, and this doesn't make you a racist.

But others would give a different answer. The title of this chapter is from one of the songs of the Broadway show *Avenue Q:* "Everyone's a Little Bit Racist." The provocative lyrics point out that our racism

> *Doesn't mean we go around committing*
> *Hate crimes*

But it does mean that nobody is color-blind. We all make generalizations, sometime cruel ones, based on race, and these influence our behavior in all sorts of ways.

Which answer is right? Well, it's complicated.

———

We think a lot about certain social categories, and not just for people. The journalist Masha Gessen begins an article about trans children with this story:

> Every night, when I walk my dog, several strangers, similarly tethered, will ask me the same two questions: "Boy or girl?" and "How old?" The pragmatic meaning of these questions escapes me. The answers do not inform the interactions between our dogs, nor do they tell a story. Wouldn't it be more interesting to learn whether the dog was a longtime family member or a pandemic puppy, whether it lived with other pets, how much exercise it got or desired, how it tolerated last summer's orgy of fireworks, or to learn at least the dog's name? These are the questions I usually ask other dog owners as our pets sniff each other, but in response I am still asked—hundreds of times a year—about my dog's age and gender. These categories, it seems, are so central to the way we organize the world around us that we apply them to everything, including random dogs in the night.[1]

Numerous laboratory studies find that we automatically encode three pieces of information when we meet a new person, and Gessen is right about two of them—age and gender. The third is race or ethnicity.[2]

Indeed, gender is so important that many languages, including English, force speakers to think about it when we talk, as there is often no convenient way to refer to a stranger without choosing a pronoun like "he" or "she." Many are frustrated by this, and there are movements to normalize using plural pronouns like "they" for singular reference—I've done this several times in this book—or to add a new gender-neutral singular pronoun like "zee" to English. (This might work for "they," but the prospects for "zee" are not good—new nouns and verbs pop up all the time, but new pronouns are very slow to enter a language.)

Age, gender, race. These three considerations trump the rest; they are what interests us. They are also what lingers in memory. In "memory confusion" studies, experimenters give subjects a series of pictures of people who are depicted as uttering different sentences.[3] If you have enough of these people-sentence pairs, subjects make mistakes—they misattribute who said what. And these mistakes tend to hew to how we categorize people. If there is a sentence said by a white female eight-year-old and you get the speaker wrong, you're more likely to misattribute it to another white girl than to someone else; presumably because, beyond the specifics of the person, you will have encoded the speaker as "white girl." There was an interview one can find online where someone who is interviewing Samuel L. Jackson confuses him with Laurence Fishburne, both older Black men. (Jackson is not amused.) This is the sort of mistake that's often made; few would mix up Samuel L. Jackson and Lucy Liu.

Age, gender, race—two thirds of this makes sense. Age and gender (for the purposes here, we can blur gender together with sex) matter for all sorts of things, such as who you can have babies with and who is likely to be a threat to you. It makes sense that we would naturally notice these categories, associate them with the people we meet, and take them seriously.

Race is the odd man out. The categories we call "races" correspond, very roughly, to people who descended from different parts of the world, and since our ancestors traveled by foot, their societies wouldn't include people of different races. So why would our brains be so quick to zoom in on a category that didn't used to exist?

We might solve this puzzle by appealing to the interplay of evolution and culture. As we'll discuss, we are naturally biased to carve the world into social groups (evolution), but it is our particular societies that teach us which social groups matter (culture). The idea, then, is that there might be nothing psychologically natural about the notion of race, but it just so happens to be an important way people currently divide up the social landscape—it is a basis for coalitions, it is part of social hierarchies[4]—and so we glom onto it.

This analysis makes an interesting prediction, which is that we should be less sensitive to race in cases where it is *not* a cue to coalition or hierarchy, as when diverse groups of people work together in a common cause, against a common enemy. The best example is sports. If you're watching the Celtics versus the Spurs, it gets things seriously wrong to see it as an interracial conflict. The conflict is between the teams, and perhaps in such a situation we are less sensitive to race.

This idea was tested in a clever study called "Can Race Be Erased? Coalitional Computation and Social Categorization."[5] Following standard practice in memory confusion studies, these investigators showed pictures of people associated with different sentences. The twist here was that the men were members of competing basketball teams, wearing different jerseys, with each team half white and half Black. People still made mistakes based on race, but mistakes based on *team* were made more often, because now team membership is a better cue to coalition.[6]

———

Why attend to age, gender, and race (or coalition more broadly) at all? I said above that such categories "matter," but wouldn't it be more feasible—and more moral as well—to just treat people as individuals? One way we think about groups is in terms of stereotypes, and so one way to frame this question is to ask: What are stereotypes good for?

To answer this question, we need to distinguish different senses of stereotype. Some people use the term only to refer to beliefs about groups that are false and cruel, and that are seen as applying to every member of the group. The ideas that gay people molest children or that Muslims are terrorists are stereotypes in this sense, and there is nothing positive to be said about them.

But when psychologists say that we have stereotypes of certain groups, they usually mean something different. Stereotypes here refer to characteristics that are likely to be true of members of a group. If you believe that Gen Zers tend to spend a lot of time online or that psychology professors in the United States tend to vote Democratic or that the Dutch tend to be taller than the Japanese, these are stereotypes in a psychologist's sense, even if they are not false, cruel, or indiscriminate.

In this sense of stereotype, social psychology meets with cognitive psychology—the study of mental processes such as perception and language and reasoning, the field we dealt with in the middle third of this book. It turns out that the stereotypes we form of human groups reflect a more general process of understanding and learning about the world. Bearing this in mind, I am going to do something perhaps a bit surprising. I am going to defend stereotypes. I will argue that we can't live without them.

———

Humans, and many other critters besides, sort things in the world into categories and possess mental representations of these categories—*concepts*—that help us make sense of novel experiences and act in rational ways. We are lumpers.[7]

To start with mundane examples, your brain encodes the concepts of chairs, tomatoes, and dogs. These are created through your experience with chairs, tomatoes, and dogs, as well as with information acquired through other sources, such as shopping at Ikea, gardening with your grandfather, or reading a book about the evolution of quadrupeds. And these concepts do a lot of work for you. They are connected to words, so you can understand what someone means when they say they are going to buy "a new chair" or when they offer you "a tomato" or ask whether you are afraid of "dogs." They help you recognize instances of these categories in the world—if you go into the street late at night searching for a friend's lost dog, you know what you're looking for; you won't bring back a rock or a shoe.

The psychologist Gregory Murphy begins *The Big Book of Concepts* with this:

We seldom eat the same tomato twice, and we often encounter novel objects, people, and situations. Fortunately, even novel things are usually similar to things we already know, often exemplifying a category that we are familiar with. Although I've never seen this particular tomato before, it is probably like other tomatoes I have eaten and so is edible. . . . Concepts are a kind of mental glue, then, in that they tie our past experiences to our present interactions with the world.[8]

Someone without the right concepts will starve to death surrounded by tomatoes, "because he or she has never seen *those* particular tomatoes before and so doesn't know what to do with them." Without concepts, we are helpless.

We not only group things in the world into categories such as tomatoes, chairs, and dogs; we also group people into categories based on sex, gender, age, ethnicity, occupation, religion, sexual orientation, body type, and much else—including idiosyncratic categories like assistant professor or *Star Wars* fan. Have you ever written an ad for a dating service such as OkCupid or Tinder? Here's one from the classified ads in the *New York Review of Books*:

> PAIR OF UNREPENTANT QUEERS (one pansexual Asian punk femme & one curly-haired nonbinary flâneur) found love in these pages. Seeking COVID-negative company to complete the hat trick; be enlightening, generous, flexible, spirited.

So many categories here, some I had to look up. (A "flâneur" is an idler or lounger.) We define ourselves in terms of these categories, and they shape our social preferences, our likes and dislikes. Some ads seek out political radicals, gay men, and free spirits; some say that squares and smokers need not apply.

It's not just the dating circuit. As we saw earlier with age, sex, and race, social categories really do tell you a lot about a person. If you lost the ability to make social generalizations, you would be lost indeed.

Now, for any of this to work, stereotypes should be accurate.[9] And they usually are, not just for chairs and the like, but also for kinds of people. As one example, men really are more prone to commit physical violence than women are. If you need to quickly judge the threat posed by a stranger standing at the corner of the street you're about to walk down at night, you'll probably fall back on this stereotype, and your heart might pound a little bit less if you see that it's a woman. Perhaps this is a moral mistake, an odious form of profiling—but it's not irrational.

Children rely on a similar stereotype. In one study, a stranger would approach three-to-six-year-old children and try to lure them from school, saying, "You are so adorable. I really like you. I have a gift that I want to give to you!" and "Let's go together and get it, and I will bring you back here after a while." The depressing finding here was that, despite how often children have been told not to fall for this, about half of them left with the stranger. But another finding is that they were much more likely to leave with a woman than with a man.[10]

We use categories all the time. I recently got an email from an address ending in aol.com, and drawing on stereotypes, I thought to myself, This person is going to be older than me (and he was). A friend described someone we both knew, who died too young, as "thoroughly midwestern," and I knew exactly what she was saying. But these are generalizations, not absolutes, and we know this. I can grasp the idea of some young hipster using this old-timey email address; I get that a midwesterner can be cold and unfriendly. Nobody drops dead in surprise when they see a gay conservative, and maybe that elderly woman on the dark street at night (*safe*, your stereotypes tell you) will conk you on the head and take your money. Our stereotypes guide us, but in an imperfect way.

With all of this in mind, our perspective flips: Who could object to stereotypes? We look at the world, we encounter individuals and make useful generalizations about the categories they belong to, and then use these generalizations when reasoning about new individuals. Nobody is disturbed by the idea that we carry an understanding in our minds that dogs usually bark or that one can usually sit on chairs, so what's the problem with stereotypes in the social realm?

There are many. One worry is that a statistical learning machine is only as good as its input, and a lot of the information we get about people is from a biased sample. If all you know about Italian Americans is from the TV show *The Sopranos*, that's going to have distorting effects.

Also, social stereotypes are influenced by factors other than statistics. We are bad at assessing the traits of groups we are opposed to. To take one recent example from the United States, on average, Republicans estimate that 32 percent of Democrats are gay, lesbian, or bisexual, and Democrats estimate that 38 percent of Republicans earn over a quarter of a million dollars a year. Both judgments are comically wrong—at the time of the poll, the right numbers are about 6 percent and 2 percent, respectively.[11]

Then there are the moral issues that arise. Racial profiling by police is an obvious example. Defenders of the practice claim that this is more efficient, but even if it were, it makes life worse for the people who are profiled, the vast majority of whom are innocent of any crime. I'm framing the issue here in a utilitarian way, but one might also see a practice like racial profiling as violating some fundamental human rights—people deserve to be treated as individuals, not as members of categories.

Our intuitions are complicated. We are not morally bothered by some choices based on group membership. Many people are comfortable with certain personal decisions along category lines—few think it's wrong for people to choose their partners on the basis of sex and gender. Some people (though far from all) are comfortable using social categories to further certain goals, as when companies and universities take racial and ethnic categories into account when hiring workers and accepting students, with the goal of increasing diversity. And just about everyone is fine with laws and policies that discriminate based on the category of age. This is in part because the stereotypes are so clearly rooted in facts (four-year-olds really are too young to drive); and because such policies apply to a slice of everyone's life span, they seem fairer. Sooner or later, everyone will get their chance.

———

There is a further issue that arises with social categorization, and this harkens back to the psychological essentialism we discussed in the developmental chapter. We learned then that certain categories are seen as possessing deeper properties, essences, that make them what they are.[12] A tiger isn't just an animal of a certain size and appearance; it is a creature with a certain internal structure.

The problem is that we often think the same way about human categories, including ethnic and racial ones. Now this isn't entirely wrong. The surface appearance of people, such as height and skin color, is in part the result of genetic factors. Furthermore, gene frequencies do differ slightly across groups that have different ancestries. This is why members of different groups have vulnerabilities to certain illnesses. Ashkenazi Jews, the Amish, and other groups are at risk for Tay-Sachs disease, for instance, while most other people are not. Such genetic differences are one reason why it's reasonable for members of minority groups to lobby for inclusion in medical research.

But our essentialism can make us go overboard. We tend to be too quick to assume that the racial categories of our societies capture deep discontinuities, that they have a transcendent reality. We tend to mistakenly think that the categories that are called "races" correspond to distinct genetic groups, failing to appreciate the role of social forces in the creation of these categories.

For instance, people often think there's an objective answer as to whether someone with a Jewish father and a non-Jewish mother is Jewish, or someone with a Black mother and a white father is Black or white. And this is false—these are social decisions. Some Jews believe that having a Jewish mother is essential for being Jewish; others are more mellow. In America, the practice of hypodescent—also known as the "one-drop rule"—entails that having *any* African genetic ancestry makes a person Black.[13] Over 90 percent of Black Americans have some European ancestry, which means that if the rule was flipped—if it were "one drop" of European blood that mattered—they would all become white. The current facts about who is Jewish and who is Black and who is white, then, are in part social facts, not biological reality.

Finally, our essentialism makes us prone to see distinctive properties of these human groups as reflecting deep facts about their natures.[14] This makes us too nativist; such differences are usually better understood through history and sociology than through neuroscience or evolution-ary biology. It would be absurd to explain the gross economic disparities between white people and Black people in America, for instance, without reference to the legacy of slavery and subsequent racial segregation and discrimination. To make things worse, in the world, just as in the lab (see below for examples), distinctions that start off as arbitrary can become real if enough people believe they are. This is why social differences are so slow to eradicate: they are self-perpetuating. If the world starts discrimi-nating against Capricorns, Capricorns will soon become different from everyone else.

Stereotypes and essentialism are just part of the story of how we think about members of social categories. There is also what we can call our Us/Them psychology, our *groupiness*.

Some of our negative feelings toward members of other groups make sense. If my first thought about the group down the river is that they killed my brothers, it's not surprising that I don't like them. If I believe that some group of people in my country wants to change the laws to take away my rights, and if they believe that my group wants to corrupt their children and destroy their traditions, you don't need a psychologist to tell you that there will be bad blood here.

But there is something that psychologists have discovered that might be less obvious. It turns out that we don't need ugly histories of confronta-tion to get people to break the world into Us versus Them. Some of this comes naturally, in the most minimal of circumstances.[15]

One classic study that purported to demonstrate this was led by the psychologist Muzafer Sherif. In the 1950s, Sherif got twenty-two ten-and eleven-year-old boys, all with a Protestant, two-parent background, all white and middle class, to enlist in a fake summer camp at Robbers Cave State Park in Oklahoma.[16]

Yes, fake. Social psychology experiments are often deceptive, but this one is a doozy. It's a form of paranoid delusion to imagine that your life isn't real; it's all a sham—you are actually a subject in an experiment, studied like a rat in a cage. Well, welcome to Robbers Cave, where the boys were the subjects, the counselors were trained researchers, and the janitor was none other than the leader of the project, the brilliant social psychologist Muzafer Sherif!

At the start of the experiment, the boys were split up into two groups so that neither of them knew of the other's existence. Each was housed in a different cabin, and each group gave itself a name—"the Eagles" and "the Rattlers."

Then the "counselors" arranged for the groups to meet, but under unpleasant circumstances. They set up a situation where conflicts would arise, such as having a picnic but delaying one of the groups so that when one group arrived the others had eaten all the food.

The groups started to develop distinct identities. They made their own flags. The Rattlers would swear while the Eagles took pride in their clean language. They would use racist epithets to describe the other group, though they were all white. When tested, boys from each group said that their own compatriots were stronger and faster. Conflict escalated: after the Rattlers won a competition, the Eagles stole and burned the Rattlers' flag; while the Eagles were at dinner, the Rattlers trashed the Eagles' cabin.

Later in the study, the experimenters tried to bring the groups together. Many attempts failed, such as shared meals and group movies. Success was finally attained by creating a problem that the boys had to solve together—a cut water pipe, which they were told was caused by a vandal. The factions were brought together by a common cause, and perhaps a common enemy, a discovery that has been said to have considerable relevance for how to bring conflicting groups together in the real world.

This is the textbook version of the study, and I've discussed it at length elsewhere. I've come to worry, however, that it doesn't show what it's said to show. What's less known—and wasn't known to me when I first wrote about it—was that this was Sherif's second experiment of this sort.

The first experiment was in Middle Grove, New York, and here the two groups were the Pythons and the Panthers. But the outcome was different. No matter how much Sherif and his team tried, they just couldn't get the boys to hate each other. Here's how one summary of the study describes it:

> [The attempts] included his assistants stealing items of clothing from the boys' tents and cutting the rope that held up the Panthers' homemade flag, in the hope they would blame the Pythons. One of the researchers crushed the Panthers' tent, flung their suitcases into the bushes and broke a boy's beloved ukulele. To Sherif's dismay, however, the children just couldn't be persuaded to hate each other.
>
> After losing a tug-of-war, the Pythons declared that the Panthers were in fact the better team and deserved to win. The boys concluded that the missing clothes were the result of a mix-up at the laundry. And, after each of the Pythons swore on a Bible that they didn't cut down the Panthers' flag, any conflict "fizzled." By the time of the incident with the suitcases and the ukulele, the boys had worked out that they were being manipulated. Instead of turning on each other, they helped put the tent back up and eyed their "camp counsellors" with suspicion. "Maybe you just wanted to see what our reactions would be," one of them said.[17]

Why didn't this initial version work? The problem, Sherif concluded, was that the boys started off together as a single group. So in the second, more "successful" Robbers Cave study, he put them in different cabins from the start. He may have been right, but the existence of the first study weakens any overall conclusions from this line of work. It's more theater than science.[18]

But there's better research that supports the same conclusion. Consider the "minimal group" studies done in the early 1970s by the psychologists Henri Tajfel and his colleagues.[19] In these original experiments, people were asked to rate a series of paintings and afterward were told that they were either fans of the artist Paul Klee or the artist Wassily Kandinsky (typical psychologists' trickery—what they were told was randomly determined). Then they were asked to distribute money to other people, described as either Klee lovers or Kandinsky lovers.

Tajfel and colleagues found that these seemingly meaningless categorizations really mattered. People would give more to the group that they themselves were associated with. Klee lovers favored other Klee lovers; Kandinsky lovers favored Kandinsky lovers. Other studies find that you can get the same effect by handing out red T-shirts and blue T-shirts, or even by categorizing people on the epitome of a random procedure—the flip of a coin.[20]

This effect isn't limited to adults. Young children placed into arbitrary groups will give more money to their own group, they will predict that their own group will be better people, and they will be more likely to remember good acts by their own group members and bad acts done by members of the other group.[21] The power of what the developmental psychologist Yarrow Dunham calls "mere membership" is disturbingly powerful.

———

Let's stick with children for a moment and focus on real-world differences. The psychologist Katherine Kinzler and her colleagues explored a question that takes us back to the issues raised at the start of the chapter: When dealing with individuals in the world, what social categories do children care about?[22]

Two considerations are not going to be surprising, as they are the same ones that Masha Gessen pointed out—age and gender. When three-year-olds get to choose who to accept an object from or engage in an activity with, children tend to choose a child over an adult, and boys tend to favor boys while girls tend to favor girls.

A third might be more surprising. Children also tend to take language seriously. In one experiment, run in Boston and in Paris, ten-month-olds got to listen to both an English speaker and a French speaker. Then each speaker held out a toy and the babies had to choose who to interact with. Those babies raised in an English community tended to go for the English speaker; those raised in a French community tended to go for the French speaker.

Does this language bias reduce to a reasonable preference to associate with those whom one can easily communicate with? Not entirely—other studies find that children are sensitive to accent. English-speaking American children prefer to interact with individuals who speak American English versus French-accented English, even if both are perfectly understandable. Children tend to view unaccented individuals as more trustworthy and prefer to have them as friends. Early on, it seems, we use language as a cue to who is Us and who is Them. If you don't speak as I speak, you probably aren't a member of my group.

And what about race? I've often heard it argued that babies are little racists (and sometimes seen my own research cited, mistakenly, as evidence for this). Early on, they do distinguish faces of different human groups and prefer to look at faces that look like those who are around them most of the time—a familiarity effect.[23] But race doesn't seem to influence their preferences of children under the age of about five.[24] When children start to take race into account, it's less important to them than language. Kinzler and her colleagues find when white five-year-olds are asked to choose between a white child and a Black child as a friend, they tend to prefer the white child—but when asked to choose between a white child with an accent and a Black child without one, they tend to choose the Black child.[25]

———

We've been talking so far about attitudes toward groups that are conscious and explicit. Subjects from Sherif's "camps" could tell you what they think about the Rattlers and the Eagles, and subjects from Tajfel's "minimal groups" could tell you what they think about the Klee fans and the Kandinsky fans. Once they're old enough to talk, children can tell you their views about Black people and white people, English speakers and French speakers, boys and girls. Sometimes, as adults, we keep quiet about our opinions—certain beliefs are taboo to express. But we have them nonetheless.

Social psychologists have long been interested in other sorts of mental representations of social groups, those that are unconscious or implicit and are said to influence us in important ways.

As an example—and sorry if you've heard this one before—check out this riddle:

A father and son are in a car accident. The father dies. The son is rushed to the hospital, he's about to get an operation, and the surgeon says, "I can't operate—that boy is my son!" How is this possible?

Answer: The surgeon is the boy's mother. Well, duh. When I first heard the riddle, it was a real head-scratcher, but that was a while ago, and I would have thought it would no longer work now. Everyone knows that women can be surgeons. But a study done in 2021 found that only about a third of college students who had never heard this riddle before gave that answer.[26] The students were either stumped or gave other answers. The surgeon could be the boy's stepfather, adoptive father, or second father in a same-sex marriage. Other answers included "It could be a dream and not reality," "While the son was in the operating room he died, and when he saw the surgeon, it was his dad's ghost," and "The 'father' killed could be a priest, because priests are referred to as fathers and members of the church are 'sons.'"

Presumably, it's not that university students in 2021 didn't know that surgeons could be women. And yet, when they thought of a surgeon, they thought of a man, so much so that they entertained outlandish alternatives. This hints at a certain sort of duality between our considered views ("Of course women can be surgeons") and what naturally comes to mind when given a riddle like this one ("Surgeons are men").

Such implicit notions sway us in ways that matter. As one example, consider the study, mentioned earlier, in which psychologists put baseball cards up for sale on eBay.[27] Sometimes the cards were held by a light-skinned hand and sometimes they were held by a dark-skinned hand. This has an effect: People's maximum bids dropped about 20 percent for the darker hands. Other studies have been done in which résumés that are identical are sent out to different individuals, except that one is John Smith and the other is Jane Smith, or one is a member of the Young Republicans and the other a member of the Young Democrats, or one is white and the other Black, and so on. If people were just responding to the merits of the candidates, and ignoring factors such as sex, politics, and race, they should, on average, respond identically to these résumés. But they don't; they are influenced by biases, in all sorts of ways, even when they believe themselves to be fair and egalitarian.[28]

———

I said earlier in this book that many of the great discoveries in psychology involve novel methods. This area of social psychology is no exception, and there are clever techniques used to tap implicit attitudes or associations.

The most-used method brings together social psychology and cognitive psychology. It was developed by the psychologists Anthony Greenwald and Mahzarin Banaji, with their first article published in 1988. This is the Implicit Associations Test—the IAT.[29]

To get a sense of it, I encourage you to go online and try it out yourself (see the notes for the link).[30] If you aren't able to do this right now, here's how it goes, taking as an example a test that's developed to explore implicit attitudes toward the young and the elderly. The subject watches a screen as either words or pictures flash by. The pictures are of either old faces or young faces, and the words are either positive (like "pleasant") or negative (like "poison"). Then, for one set of trials, subjects are asked to press one key for either a young face or a positive word and another key for either an old face or a negative word. For another set of trials, it's reversed: one key for a young face or a negative word and another key for an old face or a positive word.

The logic here is that if you have a positive association with youth and a negative one with the elderly, then performance on young-positive/old-negative trials will be quicker and have fewer errors than young-negative/old-positive. And, in fact, people do find it more natural to associate young with positive and old with negative than the other way around.

Such studies have been done with millions of people and have found the same pattern of negative associations when tested on attitudes toward Black people, gay people, the overweight, the disabled,[31] and others. These effects are present even when questions about explicit attitudes find no bias and are often present even in subjects who belong to the group that is less favored. The elderly, for instance, also have negative associations with the elderly and positive associations with the young.

The IAT can tell us some very interesting things. In one study, Tessa Charlesworth and Mahzarin Banaji used online data from 4.4 million IATs from 2007 to 2016, focusing on attitudes about sexual orientation, race, skin tone, age, disability, and body weight.[32] They also looked at explicit judgments along these dimensions, asking people to rate on a scale their agreement with statements such as "I strongly prefer young people to old people."

Looking at explicit attitudes, they found that people have become less biased over time. This makes sense. We know that there have been radical changes. When I was a boy, for instance, many people I knew were openly hateful of gay people; now both anonymous polls and everyday experience show a real transformation in attitudes. Similarly, studies find that between 1990 and 2017, there has been a steady decline in the proportion of white Americans who say that Black people work less and are less intelligent than white people, with the numbers today dwindling to virtually zero.[33]

The results with implicit attitudes were more subtle. There was a decline in negative implicit attitudes in the domains of race, skin tone, and sexual orientation. But there is less of a decline for age and disability and a *worsening* of attitudes in the domain of body weight, suggesting that, on an implicit level, we've become harsher to those who deviate from "ideal" bodies.

The IAT can be a useful tool, then, when it comes to exploring population changes in bias over time, assessing changes that would be hard to see if you asked people directly.

The IAT can also be used to explore the relative strength of different biases. In a study done in 2012, the researchers explored racial bias, using the standard racial IAT, and political bias, looking at the association with the symbols of Democrats and Republicans and positive words (Wonderful, Best, Superb, Excellent) and negative words (Terrible, Awful, Worst, Horrible).[34] In the racial IAT, white people favored white people, and Black people favored Black people; in the political IAT, Democrats favored Democrats and Republicans favored Republicans. But the extent of the bias was much stronger for politics than for race.

———————

This line of research has gotten a lot of press and has shaped how many people think about sexism and racism and other forms of bias. A lot of this is to the good. The IAT and related measures do capture much of value. They show many people have, at some implicit level, certain biases, and we've seen that they provide a tool to study how the biases change over time and to compare their relative force.

More generally, it's worth knowing that someone might not want to be biased, might wish to treat people equally, but nonetheless be influenced by psychological forces that are beyond their control. I would certainly like to think that when I'm bidding for baseball cards (or whatever), I don't take the race of the seller into account—but based on the psychological research, I know that I can't be certain.

But there have been some criticisms about these measures over the last few decades, and it's clear that certain claims about them have been overblown. Here are some qualifications to keep in mind:

1. Some believe that you can get someone's score on the IAT and then know how biased they personally are. This isn't true. The test is too noisy. If you do the IAT online and don't like how it turned out, take it again—your score will change, sometimes by a lot. When we discuss individual differences later on, we'll see that some psychological tests, such as IQ tests or the "Big Five" personality tests, have pretty strong test-retest correlations (reliability); your score won't change that much over time. But the IAT has lousy reliability.[35] Your score on the same test taken at different times can vary wildly, and so it's not precise enough to use as a measure of how individuals differ from one another.

2. Related to this, there is reason to doubt that your score on the IAT can predict much about your actual behavior. One meta-analysis finds that one's score on the IAT explains only a tiny pro-

portion of the variance between people on a behavioral measure.[36] This is no surprise given the reliability problem; if your IAT score bounces around depending on when you take the test, how can it do a good job at capturing your behavior in the real world?

3. Our biases might be unconscious in the sense that we don't know how or when we are influenced by them,[37] but it's not like people don't know that they are biased. When I list groups— Black people, gay people, the overweight, the disabled, and the elderly—you don't have any problem figuring out which way your positive or negative evaluation is likely to go.

4. The biases may not have anything to do with animus or dislike. When you associate the elderly with negativity, for instance, it shouldn't be taken as showing any personal bad feelings toward the elderly. As an illustration of this point, consider an experiment in which people were taught about novel groups, which were called Noffians and Fasties.[38] Some people were told that the Noffians were oppressed and the Fasties were privileged, and some groups were told the opposite. Then they were given an IAT. As predicted, subjects had negative associations with Noffians in the condition where they were told that they were oppressed. But plainly they themselves didn't have bad feelings about this imaginary group; they were just responding to what they were told.

So how should we think about implicit biases? One promising approach is defended by the psychologists Keith Payne and Jason Hannay, and it brings us back to the issue of stereotypes and the acquisition of information in the world.[39] They argue that measures such as the IAT tap our appreciation of regularities in the environment, including regularities in how people think. In other words, tests like the IAT don't measure attitudes, let alone bad attitudes—they pick up associations.

Such associations are everywhere: Given the environment I was raised in, I associate peanut butter with jelly, Ringo with George, the song "O Canada" with hockey games. I also associate doctors with men and nurses with women. And I associate some groups, like the young, with good things and other groups, like the elderly, with bad things. If my world were different, I would have different associations.

Payne and Hannay argue that these are the sorts of associations captured by an IAT. Countries with citizens that associate men and science on the IAT tend to have greater achievement gaps between men and women in science and math.[40] Differences in racial IAT scores correspond, across communities, to the extent of real disparities between Black people and white people.[41] Based on those studies and others, and focusing on the race IAT, Payne and Hannay conclude that we should think of implicit biases as "the natural outcome of a mind that generates associations based on statistical regularities, whenever that mind is immersed in an environment of systemic racism."

———

This book has been filled with dichotomies. Dualism versus materialism. Soups versus Sparks (or more technically, the neurotransmitters theory versus the idea that neurons communicate through electricity). Nature and nurture. Top-down and bottom-up processes. System 1 reasoning versus System 2 reasoning. Sometimes the dichotomies are sharp, as with Soups/Sparks; sometimes, as with nature versus nurture, they exist, but have fuzzy boundaries.

When we think about our groupish natures, we end up with another dichotomy. This is between how we believe we should act and how we actually act. The first arises through reflection and is our considered view as to how we should treat people. The second is influenced by all sorts of forces, including a strong bias to favor our own group and all the associations, explicit and implicit, we carry about in our heads.

For some people, there is no clash at all.[42] If informed that they are bidding more for a baseball card held by a white hand than for one held by a Black hand, or that they are more likely to give a scholarship to someone who is a member of their political party, they will shrug and say that this is fine. But some of us are at war with ourselves. For at least some cases, we don't want to be swayed by stereotypes and we don't want to be influenced by ingroup-outgroup bias (even if we do have bad feelings about members of the opposing group). We want to be fair, and we see this as requiring that we treat people as individuals and ignore the categories they fall into.

You might think that the solution here is to try hard to be unbiased. Perhaps learning about and thinking about our biases can help us override them, just through force of will. Unfortunately, the evidence suggests otherwise. We are good at self-justification. We make choices that are shaped by prejudice and bias and convince ourselves that we are fair and impartial.

My own view is that we do better when we construct procedures that override the biases we don't want to have. If you're choosing who to hire and don't think that race or gender should matter, set up the situation in such a way that you don't have this information about the people you are judging. This is the logic of procedures such as blind auditions. Or, from a different moral viewpoint, set up diversity requirements that explicitly take into account factors such as race and gender so as to override the prejudices you're trying to overcome. These are different solutions—and people have strong views about which is preferable—but the impetus is the same, to engineer processes to eradicate bias where we think that bias is wrong.

This is how moral progress happens more generally. We don't typically become better merely through good intentions and force of will, just as we don't usually lose weight or give up smoking just by wanting to and trying hard. But we are smart critters, and we can use our intelligence to manage our information and constrain our options, allowing our better selves to overcome those gut feelings and appetites that we understand we would be better off without.

DIFFERENCES

13

Uniquely You

In our everyday lives, we care a lot about differences. If I'm telling you about my sister, I wouldn't say, "She has short-term memory and long-term memory, and can understand sentences she's never heard before." If I wanted to introduce you to a potential date I wouldn't say, "He enjoys eating when hungry and gets angry when treated unfairly." Tell me the interesting stuff, you'd say. Tell me what makes these people special.

The linguist Noam Chomsky once playfully made the same point about frogs: "No doubt, they would not be interested in what makes them frogs, but in what makes them different from one another: whether one jumps further, etc.; anything that makes a frog remarkable for other frogs. The frogs assume that it is perfectly natural to be a frog. They are not preoccupied by 'frogness.'"[1] Only scientists, thinking as scientists, care about universals. Most of the time we are interested in how individuals are different.

And there are so many ways in which people—and in particular, people's minds—differ. Just to take a domain that most everyone cares about, some of us identify as male, others as female, and others in ways that don't fall into this binary classification. Typically, this self-categorization—often called gender identity—matches those traits (chromosomes and genitals) that constitute natal sex, the sort of traits that people are talking about when they look at a baby and say "It's a boy" or "It's a girl." But not always—gender and sex can go in different directions.

Then there is sexual desire: Most people have a primary sexual attraction to those of the opposite sex—the biggest psychological sex difference is that men tend to be primarily sexually attracted to women while women tend to be primarily sexually attracted to men. But plainly, some people are gay or bisexual; and others have desires that, again, don't fall easily into any of the traditional categories.

Also, not all our preferences hew to the categories of sex and gender. You might be drawn to dominant older people, or to furries (anthropomorphic animals). Perhaps you have the hots for smart people, and then you would be *sapiosexual*—not really a scientific term, but a fun one. Sometimes our focus is narrow; a good friend only dates female mental health professionals (except when he was in high school, then he only dated the daughters of mental health professionals).

Other differences have to do with styles of decision making. Take the contrast between Maximizers and Satisficers, between those who seek out the best and those who go for what's good enough. Here are some descriptions that fit Maximizers:[2]

> Whenever I'm faced with a choice, I try to imagine what all the other possibilities are, even ones that aren't present at the moment.
> I often find it difficult to shop for a gift for a friend.
> Selecting movies to watch is really difficult. I'm always struggling to pick the best one.

I'm in the Satisficer range myself. My superpower is that I can glance at a menu and pick a main course in under sixty seconds. The perfect is the enemy of the good. But I am married to a Maximizer, someone who once needed a chair for her study and spent six months looking for the one that perfectly met her many desiderata. Maximizers tend to do better at life in objective ways, such as getting better jobs, though there is some evidence that they are overall less happy, perhaps because they end up in situations where they spend six months without a chair.

These are all differences that fall in a normal range, no call for a therapist or priest or constable, just the everyday variation that makes humanity so very interesting. But some differences bleed into pathology, criminality, and immorality. There are a variety of possible sexual desires that are compatible with a flourishing life, but if you are turned on by pre-pubescent children, that's a problem. There are a range of personal styles when dealing with other people, each with strengths and weaknesses, but if you are choked with rage if someone mildly disagrees with you, or so worried about embarrassing yourself that you can't leave the house, this may fall into the category of mental illness, and we will get to some of these cases in the next chapter.

———

Consider now one of the major ways in which humans differ—*personality*, which refers to people's style of dealing with the world, most notably with other people.

Intuitively, we have a sense of what it is to have certain personality traits. You likely see yourself as shy or moody or short-tempered or goofy or introspective, or whatever. You think of your friends in certain ways as well, and you can easily describe the personalities of famous people you don't know personally, such as Donald Trump or Beyoncé, or even fictional people like Homer Simpson or Anna Karenina or Wonder Woman.

This might all seem obvious—of course people have different personalities! But there's long been a debate in psychology over how much personality matters when it comes to explaining what people do.[3] To take an extreme example, if you wake up surrounded by smoke and fire, you're likely to be anxious, but it would be a mistake to say that this is due to something about you in particular—it's better explained by the environment. Just about anyone would have been anxious in that situation. Maybe the same point holds for less extreme cases. As we saw earlier, in our discussion of the fundamental attribution error, when explaining someone's action we are often too quick to credit the person's character instead of the situation that the person finds themself in. Perhaps we take personality too seriously.

This tendency is particularly strong when we talk about those who act cruelly to others. We often ask, *What's wrong with these people?* and there's some comfort in the answer that they are sadists and psychopaths, warped creatures devoid of human feeling. Now, I do think that there are some such monsters that walk among us, but they are rare, far more common in literature, movies, and television than in the real world. Much evil is caused by people just like us.[4] Remember the Milgram study on obedience? This is one of its lessons.

The writer Ian Parker, in his discussion of Milgram, summarizes the view of prominent social psychologists:

> The Obedience Experiments point us towards a great social psychological truth, perhaps the great truth, which is this: people tend to do things because of where they are, not who they are."[5]

Even if there are enduring traits that distinguish us from one another, the skeptic might wonder how general they are. Is there really such a thing as an aggressive or timid or adventurous person in general? Surely, someone might be a fearless driver, but nervous when making financial choices; reserved in their everyday life, but bold and creative when it comes to sex (I believe the phrase is: Geek in the streets, freak in the sheets). We all know people who embody such contradictions, and we are sometimes urged to embrace them ourselves: Gustave Flaubert advised artists, "Be regular and orderly in your life, so that you may be violent and original in your work."

These are valid points about an overemphasis on personality, and we should be careful to consider the power of the situation and to accept that some traits are narrow. But it's clear as well that personality is real and helps explain human behavior. My aunt is very extroverted, and I can predict how she'll deal with people in all sorts of situations. A friend is anxious, and in just about any social situation you can count on him to be, well, anxious. We'll see in a bit that these impressions are backed up by research; how people's scores on tests of traits like extraversion and anxiety really do predict how they behave.

As for the sorts of studies done by Milgram and Asch, that argument cuts both ways. Yes, these unusual situations do make people, in the aggregate, act in certain ways. But these studies also put a spotlight on human difference, as some people resist the demands to kill (Milgram), and some resist the pull of conformity (Asch).

———

Before we talk about what we know about personality differences, we need to say a few things about what makes for a good psychological measure.

When psychologists talk about the merits of psychological tests, they focus on two criteria, reliability and validity. *Reliability* involves lack of measurement error. If somebody steps onto a bathroom scale and it indicates that they are 160 pounds and they step off and they step back on and it's 160 pounds again, it suggests that it is a reliable scale. If they step off and then step on and it reads that they are 20 pounds lighter, the scale isn't reliable. To return to an issue raised in the previous chapter, if your test of my implicit bias has me as "highly biased" on Tuesday and "the impartiality of a saint" on Wednesday, it too has problems with reliability.

Validity is about whether a test measures what it's supposed to measure. If I'm a world champion sumo wrestler and the scale has me at 130 pounds, then I don't care how reliable it is, it's a bad scale. There are astrologers who claim to gain insights by your astrological sign, but such tests are not valid because they don't measure your personality or anything else. (Yes, I know, this is exactly what a closed-minded traditional Capricorn like me *would* say.)

There are a lot of bogus personality tests online. Recently, I took the test "Which superhero are you," and it said that I am Batman. It told me: "You are dark, love gadgets, and have vowed to help the innocent not suffer the pain you have endured." Sadly, this test is neither reliable nor valid. It's not reliable, because when I first took it, I came out as the Hulk, then when I took it again, I was Wonder Woman. I was unhappy with these answers, and so, like a bad social psychologist in the 1990s, I kept redoing the study until I got the answer I wanted—the Dark Knight. And it's not valid because I'm not actually much like Batman.

Some other tests don't fare any better. One well-known personality test is the Rorschach Inkblot, where a person describes what they see in an ambiguous figure like the one below. (Fun fact: the inventor of this test, Hermann Rorschach,[6] was obsessed with inkblots; his nickname in school, in Switzerland, was . . . *Klex*, or "inkblot"!)

This is a test that's been used for psychiatric diagnosis, criminal cases, custody hearings, and other high-stakes decisions. But many psychologists and psychiatrists believe that it has little or no validity.[7] It does remarkably poorly at predicting any outcomes that matter.

You might have heard of the Myers-Briggs Type Indicator, or just the Myers-Briggs. Developed from the work of Carl Jung, it is based on the idea that people experience the world in four main ways—sensation, intuition, feeling, and thinking—and people vary in the extent that they fall into these categories. (I've received emails from students who want to work with me that begin by telling me their Myers-Briggs type.) But this popular test also lacks validity.[8] Among other flaws, it puts people into distinct bins, seriously distorting the real variability of human behavior. Someone (ahem, me) might fall into the INTP category—introverted, intuitive, thinking, and perceiving—but all these categories fall on a continuum. It's not that some people are introverts and others are not; rather, some people are more introverted than others. It would be like a basketball scout observing players and categorizing them into Tall and Not-Tall, Fast and Not-Fast.

One of the pioneers of personality science was Gordon Allport. In the 1920s, as a young man, he visited Freud. (Every famous psychologist of a certain generation seems to have a good story about when they met Freud.) Allport was nervous, and Freud was totally silent. So Allport started to talk, a bit desperately, the way one does in these situations, and he told Freud a story about how on the train to meet him he saw a small boy who was very worried about dirt. Allport noticed that the boy's mother was dominant, and suggested that this supported Freud's theory.

And Freud sat silently through this then looked at Allport and said, "And that little boy was you?"[9]

End of Freud Story. Allport came to endorse a highly analytic (one might say anal) approach to the study of personality, one based on the words that we use to talk about people. In one study, he and a colleague went through a dictionary and came up with 4,500 terms that refer to personality traits. In subsequent work, Allport put these traits into different categories, distinguishing them on the basis of their importance and their centrality.[10]

Later theorists worked to shrink Allport's list. After all, many words are just different ways of saying the same thing. Saying someone is friendly, sociable, and welcoming, for instance, is just using three words to capture a single characteristic. The project that occupied the personality psychologists who followed concerned just how much one can narrow down the number of traits. To put it differently, if you wanted to sum up personality in a series of numbers, each capturing a point on a continuum for a trait, how many numbers would you need?

Hans Eysenck proposed a minimal conception—two dimensions in which people vary:[11]

Introverted—extroverted
Neurotic—stable

He later added a third:

Introverted—extroverted
Neurotic—stable
Psychotic—non-psychotic

Maybe there are more. Raymond Cattell broke it down to sixteen different dimensions, having to do with warmth, reasoning, emotional stability, dominance, liveliness, rule-consciousness, social boldness, sensitivity, vigilance, abstractedness, privateness, apprehension, openness to change, self-reliance, perfectionism, and tension.[12]

Many contemporary psychologists have come to the consensus that sixteen is too many, and two to three are too few. The right number is . . . five, and these are called the Big Five.[13] They are:

Openness:	Open to new experiences—Not open to new experiences
Conscientiousness:	Conscientious—Undirected
Extraversion:	Extraverted—Introverted
Agreeability:	Agreeable—Antagonistic
Neuroticism:	Neurotic—Stable

The idea here is if you want to know everything about a person's personality, these five parameters will do the trick. Lose one, you miss out on something important. Add one and you're just wasting everyone's time. Fortunately for Introduction to Psychology students who are studying for my tests, when you put these in the right order you get the acronym OCEAN. (Though I've met renegade psychologists who insist on shuffling the items so that they form the word CANOE.)

I don't want to overstate the dominance of the Big Five; there are other scales that capture things in a different way. The HEXACO model, for instance, adds a parameter for honesty-humility.[14] But the Big Five is currently the major framework in personality theory, and so this is what I'll stick to.

What makes this better than what-superhero-are-you and inkblots and Myers-Briggs? For one thing, it is reliable. There is a stability over the course of a lifetime, although there are also some general shifts in personality with age (more about this later).

It is also valid—it turns out that the Big Five captures the sorts of traits that connect to the right sorts of real-world behaviors. People who are high in openness to new experiences change their jobs more,[15] are more creative and liberal,[16] and have more body piercings.[17] Those who are high in conscientiousness are more faithful in romantic relationships[18] and do well in jobs and at school;[19] those low in conscientiousness are less successful and more likely to smoke and eat unhealthy foods.[20] Extroverts have more friends.[21] Agreeableness predicts success at relationships, while neuroticism shows the opposite pattern.[22]

———

Armed with the Big Five, we can explore differences among groups of people.

Talk of group differences is fraught, often for good reasons. Many claims about differences are ugly and unfounded. It makes sense to distrust those who make sweeping claims about how some other group differs from their own, particularly when the comparison makes them look good. It's not just that they get things wrong; they are wrong in ways that motivate and excuse brutal mistreatment, unfairness, and cruelty. Examples are easy to find; here's Gustave Le Bon, a pioneering scholar of group behavior, writing in 1879:

> In the most intelligent races, as among the Parisians, there are a
> large number of women whose brains are closer in size to those of
> gorillas than to the most developed male brain. . . . Without doubt

there exist some distinguished women, very superior to the average man, but they are as exceptional as the birth of any monstrosity, as, for example, of a gorilla with two heads.[23]

Not an example one wants to emulate! So why wade into this issue of group differences at all?

One answer is that it's of scientific interest. Like many of my colleagues, I am primarily interested in universals, in our evolved nature, in how the human mind works—and so looking at differences is less of a priority. But a true science of the mind can't duck the issue of variance, including group variance. A complete theory of psychology should be able to explain why the elderly are more vulnerable to Parkinson's disease than the young, why Norwegians tend to be happier than Americans, or why men commit so many more rapes than women.

Further, the study of how we differ often has the potential of informing us about what we all share. It's only by studying different languages, for instance, that linguists and psycholinguists can make strong claims about what aspects of language are universal. The same holds for memory, perception, emotions, and just about everything else that psychologists study.

Finally, some groups do worse than others. The poor, for instance, are particularly vulnerable to all sorts of mental illness; people living in some countries are less happy than others; suicide rates differ radically based on age and gender, and so on. Putting aside the theoretical issues here, studying what explains these differences is of vital human importance, as it can lead to interventions that improve people's lives.

———

One problem in discussing these issues is that people are often confused about what claims about group differences mean. To see this, take an anodyne example. Suppose it turns out that Canadians are, on average, more polite than Americans. That's the stereotype; but just suppose it's true. What *doesn't* follow from this?

First, it does not mean that all Canadians are politer than all Americans. You can't show me a rude Canadian or a polite American and tell me that you've refuted the claim. This is easy enough to appreciate with physical differences between groups. Men are taller than women, on average, and everyone appreciates that this can be true even if there are men who are shorter than some women. We appreciate that older people are more likely to get cancer, but that some young people get cancer and many older people don't.

Second, to say that there is a difference says nothing about the cause of the difference. If Canadians are, on average, more polite than Americans, this might be due to cultural factors, or genetic factors, or the politeness-inducing effects of high maple syrup consumption, or whatever. It's often devilishly hard to figure out the cause of real-world differences. As an example, taller men have higher salaries.[24] Is this because there is a societal prejudice in favor of the tall? Maybe, but perhaps the genes for tallness also code for other attributes that correlate with higher pay, like intelligence and aggression. Or maybe tall men dominate in school, which expands their confidence and self-esteem and improves their performance on the job market, even if the situations in which they dominate are no longer present. Or maybe it's not height at all, but racial bias—maybe the ethnicities that tend to be tall also tend to be favored. Or perhaps it's class; maybe rich kids tend to be taller and rich kids tend to end up with better jobs. I'm just getting started. My point is that it's a lot easier to find a difference than to explain it.

Third, to say that there are differences is not to say that they are immutable. Women in the United States used to be less educated than men—in 1970, men earned almost 60 percent of college and university degrees. Now the number has flipped, and men are less educated than women. One day, perhaps, Canadians will be less polite than Americans.

Fourth, differences are facts about the world; to assume that they are inherently good is to commit a blunder we discussed earlier in the book when we talked about the evolution of the emotions—the naturalistic fallacy. To say (on average, always on average) that Canadians are politer than Americans, or tall men have higher salaries, is not to say that these are good things. Now, some differences might be good. If it turns out that kind people do well in life, this would be wonderful. I'm comfortable with the meek inheriting the Earth. (I guess I am really a Canadian.) On the other hand, the fact that people with darker skin tend to get more severe prison sentences than those with lighter skin who are charged with identical crimes is a deeply immoral difference that we should work to get rid of. You can't make a blanket statement about the goodness or badness of differences in general.

Having gotten this out of the way, we are prepared to talk about some group differences in personality.

―――――

We'll start with age. Our personality shifts somewhat during our lifetime. Studies of over a million people looking at dozens of different countries find that, from early to middle adulthood, we get more agreeable, more conscientious, and less neurotic.[25] This jibes with another finding, which we'll talk about at the very end of the book, which is that the elderly tend to be pretty happy.[26]

Let's turn now to cultural differences. One study looked at fifty-one cultures across six continents.[27] They collected data in different ways, including asking people to think of a typical person in their society and then rate them on a large personality test that matches the Big Five.

One finding is that the five-factor model works—these factors nicely fit people's intuitions of how individuals differ across the many cultures of the world. (It's more controversial, though, whether they hold for small-scale societies, of the sort that aren't usually studied in cross-cultural research.[28]) A second finding is that the above generalization about personality changes in age match people's intuitions across cultures—older people were thought of as less neurotic and more agreeable and conscientious.

The third finding is that there are real cultural differences in personality and these correspond to broader facts about the lives of people in these societies. For instance, cultures whose members score low in openness, extraversion, and agreeableness are more likely to place greater emphasis on agriculture instead of industry, and cultures high in openness are less likely to be religious. (Of course, cause and correlation are, as usual, hard to disentangle. Is it that some communities, for whatever reason, start with people who are high in openness and then become more secular, or does living in a secular society influence the personalities of the citizens? Or is there some third factor that influences both personality and the broader culture?)

What about political differences? Here the findings can be summarized in one sentence: Liberals score higher on openness and conservatives score higher on conscientiousness.[29] (These are both seen as good traits, and so this is perhaps an outcome that both liberals and conservatives will be pleased with.)

Let's end with sex differences. Even if you never read a psychology study and never heard of evolutionary theory and knew nothing about the neuroscience data (based on brain scans, one can tell whether a brain is male or female with over 90 percent accuracy, even once you factor out head size, though there is no single brain area where there is a clear difference[30]), you won't be surprised to hear that there exist psychological sex differences.

The obvious one was mentioned at the start of the chapter, having to do with sexual partner preference. Another is that, by a large margin, men are more likely to commit violent crimes, like murder and assault and rape, than women. This is true everywhere in the world. If I tell you that someone is running through the streets shooting people, and you say, "I hope they catch him," it would be persnickety for me to respond, "Why are you so quick to assume that it's a *him*?"

What about the Big Five? Across most nations, women have higher levels of neuroticism, extraversion, agreeableness, and conscientiousness than men.[31] (Openness to experience is more of a toss-up.) These differences aren't enormous, but they are large enough that if you took a random person around the world and gave him or her a series of personality tests, you would have about an 85 percent success rate at telling whether the person you are testing is a man or a woman.[32]

These differences get larger when you look at specific subscales. For instance, men are, on average, more assertive than women (one form of extraversion), while women are, on average, more sociable and friendly (another form of extraversion). There are universal differences between men and women in traits such as aggression, risk-taking, dominance, co-operating, and nurturance.[33] We see here that the largest and most reliable differences between men and women are those that follow from the evolutionary logic outlined in a previous chapter, where we talked about differential parental investment (in a species like ours, females have relatively more of an investment in offspring than males, and males compete with one another for access to females).

One thing to keep in mind when thinking about human variability more generally is that even small differences in group averages can have large real-world effects, because they lead to bigger differences at the extremes. Think about sex differences in height. The average American man is about five feet nine inches and the average American woman is about five feet four inches, but it's not hard to find men who are shorter than most women and women who are taller than most men. But now consider the extremes. When you get to very tall people, say, there are many more men than women. In the United States, the top 1 percent for men is six feet four inches and over; only about one in a hundred thousand women is that tall.

We see the same difference at the extremes for psychological traits as well. For instance, the average sex difference for the personality trait of Agreeableness isn't much, but there are far more women than men who are extremely agreeable and far more men than women who are super disagreeable. This is valuable to keep in mind when thinking about the real world. When you notice a large group difference at the tail ends—as with the generalization that violent criminals are much more likely to be male—we should be careful to realize that is compatible with average group differences that are not so large after all.

———

That's personality. Another way in which we differ is intelligence, and this brings us to IQ tests.

This is a hard topic to think straight about. Some associate the study of IQ—intelligent quotient—with racist and far-right movements. Others just see it as silly and might agree with Christopher Hitchens' remark that there is "an unusually high and consistent correlation between the stupidity of a given person and [their] propensity to be impressed by the measurement of IQ."[34] Some don't believe the tests measure intelligence; others see intelligence as overrated—what really matters, they'll tell you, is grit, or creativity, or emotional intelligence, or street smarts, or common sense.

At the same time, many of the same people who make these arguments are obsessed with intelligence.[35] Take my own community of university professors. When talking about our colleagues, potential hires, graduate students, and undergraduates, there is continuous reference to smarts or lack of them. For better or worse, the magical word "brilliant" is what many look for when searching for new colleagues or deciding whom to grant tenure.

Academics are fascinated by geniuses for whom everything comes easy. In the field of psychology, someone who is often nominated for genius status is the late Amos Tversky, whom we met earlier when we talked about his pioneering work on rationality with Daniel Kahneman. In Michael Lewis's book *The Undoing Project*, he tells the story about this collaboration and recounts a famous psychologist's one-line intelligence test: "The sooner you figure out Amos is smarter than you, the smarter you are."[36]

Indeed, academics are devotees of what the writer and education scholar Fredrik DeBoer calls "the cult of smart," whose members have the notion that intelligence is what matters the most.[37] DeBoer tells this story of a party he went to as a graduate student.

> I was chatting with a PhD student from another department, the spouse of one of the many grad students from China who attended Purdue. She was talking about her older son with obvious pride, describing his achievements in his robotics club, how well he always did in math. And then her younger son ran by, and she said, offhand, "This one, he is maybe not smart."

DeBoer was shocked. Parents just don't say this about their children. Clumsy, bad at music, awful at drawing, socially awkward—sure. But what sort of parent would casually tell someone that their kid is not so smart? Intelligence is too important to say something like that.

———

There were many motivations for the creation and development of intelligence tests, some rotten, some enlightened.[38] A considerable role in the development of the test was played by Francis Galton, a half cousin of Charles Darwin. Galton was by any account a brilliant scholar—he invented central statistical concepts, pioneered the science of fingerprints, drew the first weather map, and much else. He also came up with the term "eugenics" and was himself a proud eugenicist. This is the sort of history that critics of the test think about.

Intelligence tests have another origin story too. At the start of 1890, a test was developed by Alfred Binet as a tool to find children's academic difficulties—so that the children could be given special assistance. Even at the onset of testing, then, at least some people were wise enough to appreciate that one's intelligence is not necessarily immutable, and that measurement can be consistent with egalitarian goals. This applies right now: anyone who cares about helping children with cognitive difficulties should support testing to identify the difficulties in the first place.

While we're still on history, I'll add that the Nazis hated intelligence testing—in large part because they worried that it would favor the Jews. They preferred the study and measurement of traits they believed Germans would fare better in, such as what they described as "practical intelligence."[39]

———

As intelligence tests were developed, a discovery was made.

If you have ever taken an intelligence test, you'll appreciate that it is really a series of different tests. For instance, an IQ test might include a measure of vocabulary, a test of spatial skills, and an assessment of one's ability to appreciate analogies. Because these are different tasks, it's perfectly conceivable that the scores on these measures might have nothing to do with one another. But this is not the case.

The discovery was that there is a common factor across all those different tests. To get a sense of what this common factor means, forget for the moment about cognitive differences, and imagine testing individuals on a series of athletic tests, such as running, gymnastics, and weight lifting. People would score differently on these tests. Someone with a lot of upper-body strength might be great at weight lifting but not very flexible. A long-distance runner might not have much upper-body strength. But there would also be some correlation across these tests. Somebody who is a trained athlete will likely do well in all of them; somebody who never exercises will do poorly. So, you could measure being good at weight lifting and being good at running, but you could also have a general factor derived from these subtests that corresponds to a person's overall fitness.

It turns out that this is true for psychological measures as well. We can talk about performance in math and performance in vocabulary—and these are separate; one can do well at one and poorly at another—but there's also such a thing as a general factor here, what one can call general intelligence, or g.

In a thoughtful discussion, the psychologist Stuart Ritchie pointed out that the existence of this common factor is a surprise.[40] It didn't have to be that way. It could have been that these different abilities have nothing to do with one another, or the abilities might have turned out to be negatively correlated—it could have been that someone who is good at math will tend to be poor at verbal skills, and vice versa. This intuition would be consistent with some sort of limited-resources theory, where we each have a fixed amount, and more ability in one domain comes at the cost of less ability in another.

It doesn't work that way, though, at least not for the different manifestations of intelligence. Weirdly, your skill at tasks involving figuring out which three-dimensional objects match up when they are shown in different orientations is related to how many words you know, which is related to other seemingly different manifestations of cognitive skills. Scores at these tests, then, can allow you to extract a single number—

your IQ score. A single number! (At least with personality tests, there were five of them.)

Before getting to what that number really means, keep in mind that the average score on an IQ test is 100. This isn't due to some magical coincidence; it's because the test is calibrated that way. The scores are further tweaked to correspond to what statisticians call a normal distribution, where about two thirds of the people have an intelligence between 85 and 115.

———

We said that the two criteria for assessing psychological tests are reliability and validity. IQ tests satisfy the first hurdle; the numbers don't move around that much over time. As one extreme example of this, a psychologist gave a group of Scottish ninety-year-olds the very same intelligence test that they took as eleven-year-olds, in 1921.[41] The results were highly correlated; preteens who scored high turned into old people who scored high.

Validity is harder to assess. Do these tests really measure what they are supposed to measure?

To answer that, we need some idea of what intelligence is. Discussions of the topic usually start by citing the following definition, written many years ago by top researchers in the field, including much of the editorial board of the journal appropriately enough named *Intelligence*:

> Intelligence is a very general mental capability that, among other things, involves the ability to reason, plan, solve problems, think abstractly, comprehend complex ideas, learn quickly, and learn from experience. It is not merely book-learning, a narrow academic skill, or test-taking smarts. Rather, it reflects a broader and deeper capability for comprehending our surroundings, "catching on," "making sense" of things, or "figuring out" what to do.[42]

If IQ tests measure intelligence in this broad sense—and not merely "book-learning, a narrow academic skill, or test-taking smarts"—then one's score on these tests should have real-world implications. Does it? I'm fond of this answer, from an excellent textbook:

> You've probably heard people say things like, "IQ tests don't predict anything important." That's absolutely true—so long as you don't think grades, jobs, money, health, or longevity are important. The fact is that intelligence test scores are highly correlated with just about every outcome that human beings care about.[43]

It's hard to find anything valuable that doesn't correlate with IQ. Higher-IQ people tend to have more satisfying relationships and better mental health. They tend to be less racist, less sexist, less likely to commit crimes, more attentive to the benefits of long-term cooperation, and less tempted to exploit others. They tend to be healthier and live longer, which may be in part because they tend to make wiser lifestyle choices; they eat better, exercise more, and are less likely to smoke.

As best anyone knows, there is only one bad trait associated with high IQ—bad eyesight.[44]

Now, there are presumably diminishing returns here; increases in intelligence matter less as intelligence goes up. People with very low IQs have difficulty making their way through the world. There is a big difference between 80 and 100. But what about going from 100 to 120? Or from 120 to 140? Perhaps after a certain point, more isn't better.

We can test this claim. In one classic decade-long study, psychologists took 320 children who were tested before the age of thirteen as having high mathematical or verbal skills—at the top 1 in 10,000.[45] Then they looked at the same individuals twenty years later to see how well they did in life, focusing on their occupations. And it turned out that their lives were awful because they were total nerds.

No, I'm just kidding. Actually, they tended to do very well at life, becoming scientists, journalists, politicians, CEOs, leaders in society. A similar study, this time with a larger sample, looked at thirteen-year-olds who were in the top 1 percent in mathematical reasoning, and then studied them four decades later.[46] Again, members of this group were overrepresented as attorneys, top executives, and tenured professors; they wrote more books, secured more patents, and so on.

In our society right now (and we'll see this has an important qualification), if you had to give a child one test to predict his or her fate in life, you couldn't go wrong with an IQ test.

———

There's much more to a successful life than intelligence, so let's step back and return to personality and consider a trait known as *grit*. This is a concept made popular through the work of the psychologist Angela Duckworth and her colleagues, and it refers to "a combination of passion and perseverance for a singularly important goal."[47] Grit is not as important as IQ when it comes to life success, but once you are dealing with communities where most everyone has a high IQ, it can be an excellent predictor of performance.

Qualities like grit and the related capacities of conscientiousness and self-control are different from intelligence, but they naturally go together. They work together in good decision making—intelligence because it leads to rational choices, and grit, conscientiousness, and self-control because they let you translate good decisions into good actions, working toward long-term consequences and abstaining from tempting but less optimal choices.

We've seen that there is a relationship between IQ and various forms of kindness; this is true as well for these other traits.[48] Lack of self-control is associated with a tendency to engage in all sorts of cruel acts, which is why drugs like alcohol, which loosen our inhibitions, are so often involved in the commission of violent crime.[49] Psychopathy is associated with high impulsivity, while studies of extraordinary altruists, such as those who donate their kidneys to strangers, find that they are less impulsive than the rest of us.[50]

These points were anticipated by the philosopher Adam Smith long ago. In his book *The Theory of Moral Sentiments*, published in 1759, Smith discusses the qualities that are most useful to a person and comes up with the following (note that "superior reasoning and understanding" is what we are calling intelligence, and "self-command" is what we now call self-control). Smith writes:

> The qualities most useful to ourselves are, first of all, superior reason and understanding, by which we are capable of discerning the remote consequences of all our actions, and of foreseeing the advantage or detriment which is likely to result from them: and secondly, self-command, by which we are enabled to abstain from present pleasure or to endure present pain, in order to obtain a greater pleasure or to avoid a greater pain in some future time. In the union of those two qualities consists the virtue of prudence, of all the virtues that which is most useful to the individual.[51]

We've talked about how we are different, focusing on personality and intelligence. Now let's look at why we are different.

One consideration has to do with our genes, and when we talk about the genetic contribution to human differences, we are talking about *heritability*. This can be a confusing concept. It's important to stress that heritability is not about how much of a trait is due to your genes. Rather, it refers to the proportion of the difference within a population that is due to genetic variation. To see the difference, consider the trait of having a brain. This is surely due to genes, if anything is. But the notion of heritability doesn't apply here, because there are no differences—everyone has a brain.

In contrast, consider the trait of being good at basketball. This isn't encoded in the genes in any direct sense; basketball is a modern invention. But some of the traits that make one good at basketball, like height, are partially genetic, and in our society, these translate into being good at basketball. And so basketball skill is heritable.

There are genes, and then there's the environment, which can be divided into two parts.

One is *shared environment*; this is the environment that family members have in common. If a father is violent and hits his children, this is part of the shared environment of his children. If the family lives above a toxic waste dump, or is rich, or is in a good school district, this is also shared environment.

The other is *nonshared environment*, which is everything else. Meeting a special someone at a party, getting the flu on the day of an important exam, winning the lottery—that's nonshared environment.

As an illustration of the difference, imagine identical twins raised in the same family. They have virtually the same genes and have the same shared environment. But they do not turn out the same later in life, and this is because of nonshared environment. Maybe they were in different positions in their mother's womb, or one got food poisoning on a school trip, or they had different friends. We know these events must matter: If nonshared environment was irrelevant, then identical twins would turn out psychologically identical, and they plainly don't.

Together, this is it; everything is either heredity, shared environment, or nonshared environment.

Psychologists—and everyone else—are fascinated by the extent to which different forces make us what we are. Some people are more anxious than average, say. To what extent is this because of their genes, to what extent is it due to shared environment (being raised in poverty, having authoritarian parents, not being hugged enough as a baby), and to what extent is it due to none of the above but to nonshared environment (being the victim of a violent mugging, say)?

In the real world, it's hard to pull these apart. Genes and shared environment are particularly intertwined. Everyone knows how similar family members are. They share physical traits, like being tall, and they share psychological traits, like enjoying music, or hating to travel, or extraversion. But it's not obvious how much of these commonalities is due to heredity or to shared environment. To find out, you need to apply the methods of behavior genetics.

One such method is to look at adopted children. If a couple adopts a newborn baby into their family, it will have, by definition, the same shared environment as the couple's biological children, but it won't share the same genes. Suppose the biological children and their adopted siblings tend to have the same IQ. That would suggest that the family environment is shaping their IQs. If adopted children end up with wildly different IQs than their siblings, this suggests more of a role of the genes.

Or consider twins. Monozygotic twins are essentially clones; they share (almost) all their genes, as opposed to fraternal twins (also known as dizygotic twins, since they come from two zygotes), who are just regular old siblings and have an approximately 50 percent overlap in genes. One might ask, then, for any given trait, are monozygotic twins more similar to one another than are dizygotic twins? If so, this might suggest there's a role of genes—heritability. If not, it would suggest that the genes don't matter that much.

The most exotic case is twins raised apart. If environment is all that matters, they should be as different as any two strangers. If they are more alike than two people plucked from the population at random, this suggests a role of genes.

These methods are still in use when exploring the role of genes and environment. There are questions they address that can't be answered otherwise.[52] But we now have more direct ways of exploring the relative contribution of genes and environment, including genome-wide complex trait analysis (GCTA), where you can take large samples of people (sometimes hundreds of thousands of people; sometimes more than a million) and analyze their DNA and find parts of the genome that vary across individuals. Then you can test for the correlation between genetic differences and something of interest, such as diseases like diabetes or schizophrenia, or traits such as height or educational attainment.[53]

GCTA has its limitations—for instance, it only looks at some sources of genetic variation—but it also has certain advantages over previous methods. Just as one example of a problem with the methods above, if you find that monozygotic twins are a lot more similar than dizygotic twins, you can conclude that this is due to the greater genetic similarity of monozygotic twins—but you might also wonder if they are treated more similarly than dizygotic twins, both within the family and outside the family, because they look more alike. GCTA gets around many of these problems.

The findings from twin studies and adoption studies and GCTA converge, and they support one of the great discoveries of our field, what the psychologist Eric Turkheimer, in 2000, dubbed the First Law of Behavior Genetics:

First Law: All human behavioral traits are heritable.[54]

And not just heritable to a trivial degree. It turns out that for just about every trait, a large proportion of the variation across individuals—often in the neighborhood of 50 percent—is due to genes. This is true for intelligence, for the Big Five of personality, for religiosity, for many types of mental illness, for happiness, and much else.

This finding came as a big surprise to psychologists, who tend to be more sympathetic to environmental explanations for human differences. But perhaps it's not surprising for everyone. In one study, people were asked to estimate the contribution of genes to human differences in a variety of traits and their estimates matched well with published estimates of heritability.[55] They appreciated, for instance, that schizophrenia, bipolar disorder, and intelligence have a relatively high level of heritability, more so than alcoholism, political beliefs, and musical talent.

———

What do findings about heredity tell us about group differences? Many people are quick to make a connection here. Those who are eager to believe in the genetic basis of ethnic and racial differences will take Turkheimer's First Law as support for this view; those who reject the notion of innate group differences will often dismiss discussion of heredity altogether because they see it as opening the door to scientific racism.

This sort of inference is mistaken. Heredity within a group—the sort of thing we've been talking about so far, the focus of the First Law—does not entail that differences between groups are genetic.

The geneticist Richard C. Lewontin had a well-known example that makes this point.[56] Someone picks up a bag of corn seeds, randomly takes one handful of seeds, and puts them in a garden plot with great soil and bright sunshine. Then they take another handful and put them in a plot with lousy soil and little light. Within each plot, the plants will vary in size, and this variance is entirely due to genetic differences—heredity. But the plants in the good garden plot will grow bigger, on average, than the plants in the bad plot. This difference—which might be much larger than any within-plot difference—is entirely due to the environments in which the seeds are growing.

It's not just seeds; it's also people. South Koreans are taller than North Koreans. Why? Well, for any Korean, like any other person on Earth, height is partially predicted by the height of one's biological parents—it's heritable. But the North-South difference in height is environmental (presumably having to do with diet); the country split in the 1940s, nowhere near long enough for any genetic difference to exist between citizens of the two countries. This is another illustration of how differences within groups can be influenced by genes, while a large effect between groups may have nothing to do with genes.

And now let's turn to a question that is often asked. Can differences between human groups in IQ be due to environment, similar to what we see in Lewontin's seed example?

We know they can, and I can illustrate this with a true story: There exist two groups, each composed of many millions of people. They live in the same country but they have a staggering difference in IQ; one group has an average IQ of 100; the other group has an average IQ of about 82. These groups have no major genetic difference from one another; in fact, they are drawn from the very same families and communities.

So what happened to cause this difference? Time. These groups are people who took IQ tests in 2002 and those who took IQ tests in 1948. IQ scores have been gradually going up since they were invented early in the last century. The gradual increase in IQ is called the Flynn effect, named after its discoverer, the political scientist James Flynn.[57]

Flynn discovered it by asking a clever question. The average IQ for any year is 100, because that's how the test makers construct the test. But Flynn wondered: What would the numbers be if the tests weren't tweaked? And he found that we've been getting better and better at the tests, and so the same performance that would get a genius score many years ago would be merely average today.

The Flynn effect has been found in dozens of countries. It's not a growth in every aspect of intelligence. There hasn't been much improvement on the parts of the test that measure vocabulary or mathematics. Rather, we've been getting better at tests that tap the capacity for abstraction, including, to use some easy examples, similarities ("How are morning and afternoon similar?") and analogies ("Book is to reading as fork is to . . .").[58]

Nobody knows for sure why the Flynn effect exists—hypotheses include improvements in schooling, an increase in how stimulating the environment is, and better nutrition. Nobody knows why the effect has flattened and possibly even reversed in some countries.[59] We do know, however, that a few generations isn't enough time for radical genetic change. As such, the Flynn effect provides an existence proof of large group-level intelligence differences that are entirely environmental. This supports the claim that the differences we find in existing ethnic groups, much smaller than those that hold across time, could be environmental as well.

———

One qualification about heredity is that it says nothing about the role of genes in between-group differences. Here's another: Heredity is not constant. There is no absolute answer to the question of how heritable a trait is, only an answer that is relative to a specific environment.

For one thing, traits tend to be more heritable in environments where people are allowed to flourish and less heritable in more impoverished environments. To see how this could work, consider a country where everyone has access to an excellent mathematical education, and there is no discrimination based on race, gender, income, or anything else. Compare this to another country where women and minorities are discouraged from learning math, and where entrance to the best graduate institutions is determined by wealth and family connections. In the first country, everyone has a chance to flourish, and so presumably the differences you find among individuals will reflect, to the greatest extent possible, their genetic potential. Heredity will be on the high end. In the second country, your mother and father can be math geniuses, and you could have a genetic predisposition toward giftedness at math, but if you're the wrong race or sex, or come from the wrong side of town, you'll never see the inside of a math class. Your genetic potential will go to waste, and more generally, heredity will be on the low end.

One way to think about this is that in good environments, where our gifts are given free rein to flourish, our genetic capacities can "max out." In bad environments, such gifts are stanched—people are poisoned by lead, deprived of the ability to learn and practice, discriminated against in all sorts of ways. I talked about seeds before and described how, when in the same environment, their flourishing is determined by their genes. But if their garden plot is always in darkness and you forget to water it—now all the seeds have a dismal fate, and genes hardly matter.

It's not surprising, then, that in the United States, the heritability of intelligence for rich children is higher than the heritability of intelligence for poor children.[60] While moving children from one reasonably adequate environment to another doesn't seem to make a difference in their intelligence, moving them from an impoverished environment to a better one leads to an IQ bump of over ten points, a powerful effect.[61]

———

One final, but critical, qualification about heredity is that knowing how heritable a trait is tells you nothing about the mechanism through which genes influence that trait.[62]

To take an important example, educational attainment is partially heritable. If you have data about someone's genetic makeup, you can predict, to a reasonable extent, whether they will complete college or university. There is a temptation, upon hearing this, to conclude that certain genetic variants influence people's brains; they make some people smart—or perhaps give them more grit or ambition or some such personality trait.

This might be so, but it's not the only possibility. To modify an example from the sociologist Christopher Jencks, consider a country where there's discrimination against red-haired children; they have a more difficult time getting accepted into good schools.[63] Since red hair is genetic, this would mean that certain genes would be reliably correlated with a lack of educational attainment, even if these genes had nothing to do with brains. If hair color sounds like a fanciful example, think about physical attractiveness and skin color, both influenced by the genes, and both of which are powerfully related to how well you do in life.

When we talk about the heritability of intelligence, extraversion, alcoholism, and so on, we must appreciate that the environment itself influences the causal link between genes and traits. In a country where nobody cares about the color of your hair, genes involved in hair color no longer relate to how long you stay in school. If the best colleges decide to choose students in part on their athletic ability or musical skills or how good they are in face-to-face interviews, then the genes that influence these capacities will relate to success in life; when they stop doing so, they will matter less.

We've talked about Turkheimer's First Law—"All human behavioral traits are heritable"—and then added various qualifications. Now here are the Second and Third Laws of Behavior Genetics:[64]

Second Law: The effect of being raised in the same family is smaller than the effect of the genes.

Third Law: A substantial portion of the variation in complex human behavioral traits is not accounted for by the effects of genes or families.

What arises from all of this is that, if the environment is good enough (always an important qualification), it's genes and nonshared environment that have the most effect on how people are different. The research suggests that what doesn't seem to matter very much, at least for the traits of personality and intelligence, is family environment. To put this in a more radical way, once the moment of conception is over, and the parental genes have been fused into the zygote, then for certain important aspects of how children turn out, parents just don't matter that much.[65]

You might be thinking that this can't be right—there must be an effect of shared environment. After all, there are a million studies—as well as common sense—showing that there's a high correlation between parent and child for everything. Parents who read a lot tend to have children who read a lot. Religious parents are more likely to have religious children, extroverts are more likely to have extroverted children, and so on. These are obvious and much replicated findings.

The problem is, they are consistent with explanations other than parenting effects. As one example among many, children who are hit by their parents tend to be more violent.[66] The immediate conclusion that many draw from this is that parental hitting has a bad effect on children. But it's equally consistent here that the genes that make parents likely to hit (those related to low impulse control, say) are shared by adult and child.

(There is also a third possibility. Such correlations might be due to the child influencing the parent, not the other way around. Perhaps it's not that reading to children makes them interested in books, but rather that if parents have a child who is interested in books, they are more likely to read to him or her. Developmental psychologists call this a *child effect*.)

For the longest time, trained research psychologists, who really should know better, have thought of everything in terms of the effects of parenting and ignored the alternative explanations, particularly the genetic ones. The psychologist Kathryn Paige Harden observes that you can open a prominent journal in developmental psychology and find claims about all sorts of connections between what parents do and how children turn out—maternal depression is related to child depression, parental drinking is related to child sleep habits, maternal language is related to child self-control, and so on.[67] Then she points out that "all of these studies use data from genetic relatives with only a cursory mention, at best, of the possibility that observed associations might be due to genetic inheritance." I don't want to be too harsh here, as the data for these studies can be of interest for other reasons. But to take them as evidence for the power of the environment is a terrible mistake, and many such studies are, as Harden puts it, "an enormous waste of scientific resources."

The same problem of disentangling genes and environment arises when we tell stories about our own lives. I am neurotic because Mom never loved me—or maybe I'm neurotic because she showered me with too much love for my own good. I am a good father to my children because Dad was such a good father to me . . . or I'm a lousy father because I decided that I could never live up to his example. See how easy storytelling is? We love personal narratives, and the best stories involve other people, usually the people who raised us. The research, however, suggests that such stories are often false, or at least unsupported.

I am aware how unintuitive much of this is. If someone tells you that psychology is just common sense, show them this chapter.

Let's end this section with some qualifications. I said that parenting doesn't matter that much when it comes to explaining variation in certain traits. But of course, parenting matters in all sorts of other ways. Parents have extraordinary power over their children's lives. The right parents, biological or otherwise, will treat their children with respect and love, and do their best to give them a happy and fulfilling childhood; the wrong parents, through abuse, neglect, or pure sadism, can make their children miserable.

How mothers and fathers treat their children also influences how children think of their parents and whether they want to associate with them later in life, when they have a choice. In this regard, the parent-child relationship is like any other relationship; how you treat the other person influences what they think of you and whether they want to be with you.

Parents might matter as well in countless ways that don't show up in psychological tests or other such assessments. Parents usually spend more time with their children than anyone else, and so of course they have an influence. If your dad was obsessively into Pokémon, and now you devote hours a day to Pokémon, it makes sense to credit him for this aspect of your psychology—there may be genes for obsession, but there are no genes for Pokémon. Children often choose the same occupations as their parents. I know two families where both the father and the child are famous philosophers, and another where father and son are both butchers. While genes may have some role here, it's likely that there are environmental influences at play as well.

Finally, there is room for skepticism. One possibility is that parenting has effects in the domains we've been talking about, like intelligence and personality—but it's hidden. For half of children, say, being raised by an overprotective parent makes them more introverted as adults; for the other half of children—who are different in some subtle way in their genes and their past experiences—this makes them less introverted as adults. Each person would be entirely justified in saying that the parent had an influence on their personality (if they had different parents, they would have turned out differently), but a psychologist looking at a million people to see if there's any relationship between having an overprotective parent and being introverted later in life would find nothing, because the influences perfectly cancel one another out. While this sort of example is unlikely (it would be a strange coincidence if things worked out in a way that so mischievously hid the effect), the possibility remains open that some of the environmental effects on our psychologies involve a complex interaction of parents and other factors that we just don't fully understand yet.

———

We've been talking so far about the way things are; let's turn now to how we can change them. Suppose that we want to make people better—smarter, kinder, more resilient, whatever. What do these findings about individual differences tell us about the prospects for doing so?

We know that heredity is important, so why not fiddle with the genes? If parents don't influence their children through how they raise them, can they make up for this through a touch of genetic engineering?

A full treatment of the ethical issues here would take us beyond the scope of this book, but I do think people are too quick to get upset at this prospect. Suppose there was an environmental intervention that would improve children's intelligence—for instance, what if we discovered that smaller class sizes in preschool and elementary school did the trick? Is it morally wrong if the government worked to have lower class sizes? Suppose pregnant mothers could change their diet to help raise their children's IQ. Should they be discouraged from doing so? Most people don't find such interventions upsetting. Is genetic manipulation really so different?

There are arguments on both sides here, but maybe it doesn't matter. Before we get to the moral and political problems with fiddling with genes, there's a high technological hurdle. Traits like intelligence turn out to be highly polygenic—they arise through a combination of thousands of genetic variations. To put it simply, there is no "gene for intelligence." The same can be said about grit, kindness, extraversion, and most everything else interesting about people—including physical traits like height. Accordingly, psychologists have coined a Fourth Law of Behavior Genetics:

> **Fourth Law:** A typical human behavioral trait is associated with very many genetic variants, each of which accounts for a very small percentage of the behavioral variability.[68]

It might be possible, one day, to do some sort of widespread genetic improvement, but for now it remains science fiction.

The good news is that we don't need to fiddle with the genes. The existence of genetic causes for normal variation is fully compatible with successful environmental interventions. The obvious example here is eyesight. Myopia is partially heritable, but fortunately, it can usually be handled with technological fixes such as glasses and contact lenses and surgery.

Intelligence and personality don't have the same obvious fixes, but a similar logic applies. Recall Lewontin's example about the seeds. Among other things, this illustrates how the most heritable of traits can be radically influenced by changes in the environment. For seeds, you can move them to a better garden plot. For people, you can get rid of lead in the environment and increase iodine supplementation, two interventions that likely have profound effects.[69] The existence of the Flynn effect suggests that changes in society, perhaps in schooling, can raise intelligence to a striking degree, and I don't doubt that similar changes in the environment can modify personality. Trust in others, for instance, is a trait that varies a lot across societies, allowing for the possibility that low-trust societies can transform into high ones.

The point here is that these two claims . . .

1. Much of the variation that we see in a trait is due to genetic factors.
2. We can successfully modify this trait in some nongenetic way.

. . . are perfectly compatible.

———

This is not a book on moral philosophy or public policy, but I will end with one point that ties the psychological research to some broader issues. Here I'm drawing upon two books that came out as I was working on this chapter, Fredrik DeBoer's *The Cult of Smart*[70] and Kathryn Paige Harden's *The Genetic Lottery*.[71] DeBoer and Harden accept the fact of human differences that are to some extent genetic, but they both argue, from somewhat different directions, that it is up to us as a society to determine how much this variation matters.

Take IQ. We've seen that IQ correlates with good things—in fact, it correlates with just about all good things. But this is in part because many societies, including the United States, are constituted so that it works out that way. If the best jobs go to graduates of top universities and the system works so that high IQ (as measured by tests that strongly correlate with IQ, such as the SAT and the LSAT) is needed to get into top universities, we shouldn't be surprised to find that IQ correlates with getting the best jobs. That's just how we set things up. To go back to Christopher Jencks' analogy and modify it a bit more, if we gave slots at top universities to candidates with red hair, we would quickly live in a world in which being a redhead correlated with high income, better health, and more political power—and then psychologists like me would write books that talk about the importance of red hair.

It's not a perfect analogy, of course. Hair color, in itself, has no intrinsic value, while intelligence does. It's no accident that institutions as different as the Yale Philosophy Department and the Coast Guard both take intelligence seriously when considering candidates.

But—and here I'm expressing my own moral and political intuitions; your mileage may vary—it doesn't have to be *so* important.[72] In the United

States, for instance, there is an enormous social and financial advantage to those highly gifted students who graduate from a small cohort of Ivy League schools. A society that puts so much value on a certain specific set of cognitive capacities is exaggerating them beyond all rational proportion. A good society should respect other traits such as kindness and tenacity and courage. Indeed, a good society should make space for individuals who, simply through bad luck, aren't particularly gifted at all. Many of us would agree that it would be unfair to so sharply limit the possibilities of someone based on the color of their skin or where they were born or their sexual orientation or gender identity. Why, then, would we be comfortable with societies that offer so little to those who are below average in intelligence?

14

Suffering Minds

When asked what I do for a living, I used to answer, "I'm a psychologist." But this gave the wrong impression; people thought that I was a therapist. So now I add: ". . . the kind that does research." One friend has a better answer: "I'm a psychologist but not the sort that helps people."

These qualifications are helpful, because the treatment of mental illness is so often seen as what psychology is about. If you've made it this far through the book, you know that there's so much more. But still, clinical psychology is central to our field, and if you wanted to argue that it's the most important part, I wouldn't push back. If psychologists could cure depression or schizophrenia, we would have earned our keep.

The usual approach to introducing the domain of clinical psychology is to list the major mental illnesses, talk about their symptoms, describe theories of their causes, and then go on to discuss treatment—and I'll get to all this soon enough. But there are more general questions that we should grapple with first, starting with: What do we mean by mental illness?

———

To get a list of mental illnesses, you might go to the *Diagnostic and Statistical Manual of Mental Disorders*. Mostly used within the United States, this is the bible of psychiatry and clinical psychology, used for diagnosis, research, and treatment. It is revised every several years; as I write this we are on the fifth edition—the *DSM-5*. Alternatively, you might check out the *International Classification of Diseases*, which is now on the eleventh edition—the *ICD-11*. This is maintained by the World Health Organization, and is more commonly used in other countries.

What is it about certain patterns of thought and behavior that made psychologists and psychiatrists include them in these big books?

Much of these books is devoted to problems that are plainly severe. Children with social phobias so bad they can't leave the house. Young people with schizophrenia who act in disordered and seemingly irrational ways. The elderly with diseases like Alzheimer's that disrupt their memory and reasoning.

Then there are the harder-to-classify cases—and, indeed, decisions about what gets into these books is often controversial. Some of you reading this are addicted to drugs or alcohol—do you have a mental illness? The *DSM* says: Yes. What about online gaming addiction? The *DSM* says: Not sure—it's categorized as a "condition for future study."[1]

One long-standing debate concerns personality disorders, including the subtypes of histrionic personality disorder (exaggerated emotional responses, inappropriate attention seeking), antisocial personality disorder (lack of concern for others, failure to conform to social norms—this is related to what laypeople call psychopathy or sociopathy), and narcissistic personality disorder (an inflated sense of self-importance, a need for attention and admiration). Are these real illnesses or just personalities that most of us don't like? There are some successful people, after all, who look like they have narcissistic personality disorder. Is it reasonable to point to someone who is thriving—maybe doing better in every tangible way than the psychologist or psychiatrist who is evaluating them—and insist that they have a problem and need treatment? Perhaps yes, but then the problem might be more of a moral one, rooted in ideas about what a good life should be and how people should treat others. Should psychologists and psychiatrists be in the business of making these sorts of judgments?

Such questions of diagnosis aren't just airy theoretical problems. Insurance companies and government health-care programs pay for certain treatments, but unless they are to pay for everything, they must choose what count as an illness. I find a glass of whiskey to be just the thing after a stressful day, but I won't have much luck getting my insurance company to reimburse me for my Jameson as the treatment for Stressful Day Syndrome.

Or suppose I'm teaching a class and a student emails me with, "Dear Prof. Bloom—Sorry about getting in my paper late. I am in treatment for bipolar disorder and I had a manic episode." I would respond with sympathy. But if the student said that they didn't work on the paper because they have a sex addiction and had no time left for study, I would probably have a different, less forgiving response. But should I? (Sex addiction is not in the *DSM*, by the way; *hypersexual disorder* was proposed for inclusion, but got voted down.[2])

———

Some argue that there is no such thing as mental illness. They hold that there are diseases of the body, but mental illness is a metaphor, and a bad one at that. This view was championed by the psychiatrist Thomas Szasz in his book *The Myth of Mental Illness*.[3] In addition to specific criticisms about the coercive nature of psychiatric practice and the vagueness of psychiatric diagnoses, Szasz argues that the label of mental illness deprives individuals of responsibility and dismisses their legitimate concerns. People are not ill; rather they are "disabled by living," struggling with the problems of everyday existence.

Nobody could doubt that some psychiatric diagnoses have been illegitimate. In what used to be the Soviet Union, people who rebelled against the repressive government were put in insane asylums. In the 1800s in the United States, enslaved people who escaped were proposed to be suffering from *drapetomania*—an insane compulsion to wander. More recently, the diagnosis of homosexuality as a mental illness was broadly accepted by many psychologists and psychiatrists. It was only by the *DSM-III*, published in 1980, that this was changed after a heated battle.

Here's a description of how different things once were.[4] In the 1972 meeting of the American Psychiatric Association, there was a panel discussion called "Psychiatry: Friend or Foe to Homosexuals?" One of the panelists announced, "I am a homosexual. I am a psychiatrist. I, like most of you in this room, am a member of the A.P.A. and am proud to be a member." He then talked about how psychiatrists should use their "skills and wisdom to help themselves . . . grow to be comfortable with that little piece of humanity called homosexuality." To hide his identity, he wore a mask and a wig, and spoke through a voice-distorting microphone.

As I am writing this, the *DSM-5* includes *gender dysphoria*, referring to the distress people feel when their gender identity doesn't match their natal sex. Some trans activists feel that this should be jettisoned just as homosexuality was; they worry about the stigma of being categorized as mentally ill.[5] There are mixed feelings, though, because being in the *DSM-5* allows insurers to pay for various treatments, such as hormone therapy.

As another critique of current psychiatric practice, proponents of "neurodiversity" argue that certain so-called mental illnesses are better seen as different styles of thinking, distinct ways of making sense of the world and dealing with it. Perhaps what we are calling mental illnesses can even lead to advantages in certain aspects of life.

A further concern is that the typical psychiatric approach misunderstands the source of the problem.[6] Imagine that you are suffering from low mood, inability to experience pleasure, poor sleep, fatigue, difficulty in concentrating, and loss of interest in eating and sex. You have been in this state for a few weeks. You fit the criteria for major depressive disorder. But suppose that three weeks ago your young child got hit by a car and died. You might still benefit from treatment, even medication, but your depression shouldn't properly be seen as an illness. It is a normal response to a terrible event.

Now, psychologists and psychiatrists are no dummies. In an earlier version of the *Diagnostic and Statistical Manual*, there was an explicit "bereavement exclusion" to the diagnosis of mental illness, and the newest version of the *DSM*, while it no longer has this, is careful to note that we shouldn't confuse ordinary grief with a major depressive episode.[7]

But Johann Hari, in his book *Lost Connections*, objects that this doesn't go far enough. Once you admit that depression is rational when someone you love dies, why stop there?

> Why is death the only event that can happen in life where depression is a reasonable response? Why not if your husband has left you after thirty years of marriage? Why not if you are trapped for the next thirty years in a meaningless job you hate? Why not if you have ended up homeless and you are living under a bridge? If it's reasonable to use one set of circumstances, could there be other circumstances where it is also reasonable?[8]

Hari goes on: "What if depression is, in fact, a form of grief—for our own lives not being as they should? What if it is a form of grief for the connections we have lost, yet still need?" And he has a corresponding skepticism about biological treatments (although he is careful not to entirely reject them). For him, they focus on the wrong problem. Don't blame the mind, blame the world.

———

These points are worth taking seriously. But nothing we just talked about should lead us to abandon the view that mental illnesses really exist.

Let's return to neurodiversity, and take autism as an example, since this is often where the case is made. Some argue that we should see autism as reflecting a certain style of processing information, one that is different from that of neurotypical individuals, but not inferior.[9] The difficulties that many individuals suffer should be understood as reflecting the intolerance that a society of mostly neurotypical individuals has toward those who think in different ways.

In certain cases, this analysis seems exactly right. In their discussion of how to find talented people, Daniel Gross and Tyler Cowen write, "By now it is well known that many autistic individuals are highly skilled at programming, mathematics, and other technical subjects; this has virtually become a cliché. There are whole companies, typically in the tech area, that specialize in hiring autistic people for jobs of this kind."[10] They go on to note that other capacities associated with autism, such as decreased susceptibility to reasoning biases, might lead to advantages in other domains as well. And there are many individuals with autistic traits, such as the climate activist Greta Thunberg and the Nobel Prize–winning economist Vernon Smith, who do exceptionally well in life.

But autism—or more precisely, autism spectrum disorder—is a broad category. In an early discussion of the topic, Oliver Sacks notes that when most people and, indeed, most physicians think of someone with autism, they imagine "a profoundly disabled child, with stereotyped movements, perhaps head-banging, rudimentary language, almost inaccessible: a creature for whom very little future lies in store."[11] This is no longer a popular stereotype, but such children do exist, and they need constant care. It is not stigmatizing or social control to treat such a disorder as something to be treated. It is compassion.

The same holds for those who suffer from schizophrenia, or who are severely depressed, or who have serious anxiety disorders such as panic attacks. Such people suffer, and a cure for their suffering would be a godsend. The notion of neurodiversity, and the concerns of critics like Szasz, are relevant for many cases—but not for these.

What about Hari's concerns? He's right to point out the importance of the environment, but surely the individual also matters. As we'll see, genetic predispositions and early experiences play a strong role in determining who suffers from mental illness. Some who are, say, trapped in meaningless jobs fall into deep depression, but others don't, and find meaning and joy elsewhere.

Also, even if this weren't true—even if mental illness was solely a product of the immediate environment—it wouldn't be a knockdown critique of the medical approach. Environmental causes don't mean environmental treatments. An analogy with cancer might be helpful here. There are carcinogens in the world, and this is well worth knowing and acting on. It doesn't follow that the only way to treat cancer is to change the world, and it doesn't follow that cancer isn't a disease. One can make precisely the same argument for disorders such as depression.

So, mindful of these concerns, I'm going to talk in terms of illness, symptoms, underlying causes, and treatments. But then I'll come back at the end and return to some of these skeptical worries, because I'm in some ways skeptical myself.

———

Schizophrenia is not the most common mental illness. Only about 1 percent of the world's population suffers from it. But it might well be the most terrible one. The World Health Organization classifies it as one of the ten worst diseases on Earth in terms of its economic impact.[12] The life expectancy for someone with schizophrenia is more than fifteen years less than normal, due to suicide, illnesses, and accidents,[13] and over a third of people in mental hospitals are there due to diagnoses of schizophrenia.

The onset of schizophrenia occurs somewhere between the late teens and the midthirties, a bit later in women than men. It is more common in men.[14] To be diagnosed, two or more of the following symptoms must be present for a significant amount of time:[15]

1. **Hallucinations:** You experience things that don't really happen. Typically, these are auditory—you hear voices. These might be from God, from the devil, or from sinister people berating you or demanding you do terrible things.
2. **Delusions:** A delusion is an irrational belief, one that is very hard to change. Maybe the government is tracking you, or you are Jesus Christ, or aliens are reading your mind, and so on.
3. **Disorganized speech:** This refers to odd incoherent language. In extreme cases, those who suffer from schizophrenia might produce gibberish, sometimes called "word salad."
4. **Disorganized or catatonic behavior:** This refers to both odd and inappropriate actions, such as inappropriate giggling, wearing strange clothing, and either a decrease in movement, sometimes becoming frozen in place, or sometimes the opposite—excessive and purposeless movement.

The symptoms above are *positive* symptoms, in which someone with schizophrenia will think and act in unusual ways. But there are also *negative* symptoms, where certain normal aspects of thoughts and behavior are lacking.

5. **An absence of cognition or emotion:** This can include a certain flatness of emotional expression, loss of motivation, loss of pleasure, and withdrawal from the world.

Many people get their impression of schizophrenia from the movies, and sometimes movies get it wrong. *A Beautiful Mind* is based on the life of John Nash, a brilliant mathematician known for his work on game theory. It's a fine movie but a lousy depiction of schizophrenia. Just as one complaint, for most of the movie, we see Nash with a friend who lives with him and advises him. And then there's a twist. The friend doesn't really exist; he's a hallucination. But this isn't how hallucinations usually work; they tend to be auditory, not visual, and rarely appear as persisting lifelike characters. More importantly, the movie takes seriously the notion that the more serious symptoms of schizophrenia can go away just by sheer force of will, without medication and without therapy. And this, most experts in the field would agree, is mistaken.

A better film in this regard is *The Soloist*, which was also based on a real person: Nathaniel Ayers, a musician who developed schizophrenia. Jamie Foxx, playing Ayers, does a terrific job at depicting what it's like to have this condition.

In popular culture, schizophrenia is sometimes confused with what's colloquially called "split personalities"—actually "dissociative identity disorder," which we will soon turn to. The confusion might have had its origin because "schizophrenia" comes from the Greek words meaning "split" (*schizo*) and "mind" (*phren*). But in the case of schizophrenia, the split refers to a loss of touch with reality. This is reflected in emotions, motivations, and cognition; individuals with schizophrenia have problems with learning, with memory, with processing speed, with attention, with just about every aspect of mental life.

To make things worse, the problems associated with schizophrenia make it difficult for sufferers to deal with others, and this leads to social isolation. This in turn has consequences for mental health—delusions may be seen in part as an elaboration of the private world of a person who's lost contact with other people. We have a vicious circle here: This isolation can make the disorder worse, which leads to further isolation, and so on.

What is the cause of schizophrenia? It used to be thought that the culprit was a brain chemical called dopamine—receptors to this neurotransmitter became overly sensitive. In support of this, some antipsychotic drugs work by blocking dopamine reception at the synapse, which usually helps with symptoms. Also, drugs that boost dopamine—amphetamines—can sometimes cause a form of psychosis that looks a lot like schizophrenia.

But the dopamine theory cannot be entirely right. If the benefit of antipsychotics has to do with dopamine, dopamine-blocking drugs should work right away, because they have immediate effects on neurotransmitters—but they are slow to take effect. Furthermore, there are problems with the brains of those who suffer from schizophrenia that have nothing to do with dopamine, including structural abnormalities such as reduction of gray matter (the bodies of neurons, dendrites, and the terminals of axons) and enlarged ventricles, which are fluid-filled cavities. These abnormalities are not subtle; they are large enough to be seen on an MRI scan.[16] Also, antipsychotic medications that don't involve dopamine have some success at alleviating symptoms of schizophrenia. All of this suggests a more general problem. Perhaps, as one group of scholars proposed, schizophrenia is "fundamentally a disorder of disrupted neural connectivity."[17]

Schizophrenia has a genetic basis. If a close biological relative has schizophrenia, there is a considerably higher likelihood of you having it yourself. If your identical twin has it, you're about 50 percent likely to have it, while if your fraternal twin has it, you are about an eighth as likely.[18]

However, there is no such thing as a "gene for schizophrenia." Rather, hundreds or thousands of genes work together to predispose people to this disorder.[19] Remember the Fourth Law of Behavior Genetics discussed in the last chapter:

A typical human behavioral trait is associated with very many genetic variants, each of which accounts for a very small percentage of the behavioral variability.[20]

The genetic data reveal an important role of the environment. If schizophrenia were entirely genetic, then if your twin had it, you would have it too, and this isn't the case.

What's genetic is not schizophrenia itself, then, but a vulnerability to schizophrenia. This is an example of the "diathesis-stress" model of mental illness, where getting the illness is a combination of a vulnerability (unhelpfully called "diathesis," a word derived from the Greek term for "disposition"), plus some stressful event that serves as a trigger.

But what's the trigger? One theory is that it is a difficult family environment during childhood. And indeed, when adults who suffer from schizophrenia report on their childhood, they tend to report unusually bad early life experiences.[21] But it's difficult to pull apart cause and effect here—people with mental illness might be more likely to see their childhood in negative ways. Still, in support of the early environment account, there is a strong relationship between schizophrenia and certain life stressors—children born into poor families are more likely to develop schizophrenia as adults.[22]

It's possible as well that the trigger isn't the sort of usual trauma or bad environment that we tend to think about. It might be a difficult birth, maternal malnutrition, or a viral infection.[23] In this regard, there is a similarity with autism, which is also associated with problems with birth,[24] infections,[25] and more likelihood of being conceived in the winter months.[26]

Finally, schizophrenia emerges in early adulthood, but precursors emerge in childhood. In a clever study, psychologists looked at home movies taken of adults with schizophrenia when they were five years old or younger, where they appeared with siblings who didn't develop schizophrenia later in life.[27] Observers who didn't know who was who could guess, better than chance, who in the movies will grow up to suffer from schizophrenia—they behave more oddly and have fewer positive and more negative facial expressions.

———

The hallucination and delusions of schizophrenia—the split from reality—is known as psychosis. This is not unique to schizophrenia; psychosis can also emerge in bipolar disorder and certain extreme drug reactions. It is unfamiliar to most people, though perhaps some of us experience a much milder form of it. There's a story about James Joyce talking to the psychologist Carl Jung about Joyce's daughter, Lucia, who suffered from schizophrenia. Jung was describing the unusual mental processes characteristic of Lucia's condition and Joyce observed that the same sort of strange associations occurred in his own writing. He said, "Doctor Jung, have you noticed that my daughter seems to be submerged in the same waters as me?" And Jung answered, "Yes, but where you swim, she drowns."[28]

Other disorders are more familiar. Have you ever been sad? Really sad? Didn't want to get out of bed, didn't want to eat, life had no joy for you? Multiply this many times and you may have some inkling of what major depression feels like.[29] Here is a description from the writer Andrew Solomon:

> At about this time, night terrors began. My book was coming out in the United States, and a friend threw a party on October 11. . . . I was too lackluster to invite many people, was too tired to stand up much during the party, and sweated horribly all night. The event lives in my mind in ghostly outlines and washed-out colors. When I got home, terror seized me. I lay in bed, not sleeping and hugging my pillow for comfort. Two weeks later—the day before my thirty-first birthday—I left the house once, to buy groceries; petrified for no reason, I suddenly lost bowel control and soiled myself. I ran home, shaking, and went to bed, but I did not sleep, and could not get up the following day. I wanted to call people to cancel birthday plans, but I couldn't. I lay very still and thought about speaking,

trying to figure out how I moved my tongue, but there were no sounds. I had forgotten how to talk. Then I began to cry without tears. I was on my back. I wanted to turn over, but couldn't remember how to do that, either. I guessed that perhaps I'd had a stroke. At about three that afternoon, I managed to get up and go to the bathroom. I returned to bed shivering. Fortunately, my father, who lived uptown, called about then. "Cancel tonight," I said, struggling with the strange words. "What's wrong?" he kept asking, but I didn't know.[30]

Depression is common, with a prevalence of about 15 percent in men and 26 percent in women.[31] Like schizophrenia, it is heritable. If your biological relatives have it, you are more likely to have it. But, also like schizophrenia, it's not perfectly heritable, far from it, and so environment must play a role. (In fact, I am going to stop saying this—all the mental disorders we will discuss are partially heritable.) The stressors that trigger depression include, as you might expect, difficult life events, like a bad end of a relationship or the loss of a job. While someone who is living what seems to be an objectively great life might get major depression, still, people in worse circumstances are more vulnerable—and as with schizophrenia, depression is more common in the poor.[32]

One theory of depression is that it's due to low levels of neurotransmitters such as serotonin. In support of this, when people are given medication that is believed to increase the level of serotonin in the brain, including popular ones that are sold under the names of Prozac, Zoloft, and Paxil, they often have a beneficial effect.

But, again just as with schizophrenia, it can't be as simple as this. The drugs influence neurotransmitters right away, but they take weeks to take effect. And the evidence that serotonin is implicated in both the cause of depression and its treatment is surprisingly weak, as one commentator said: "To put it bluntly, there is no decisive evidence that low mood is caused by low serotonin levels."[33] The drugs often work, but we don't know why they work. A quite different theory is that depression is related to a more general lack of plasticity in the brain, where one loses the capacity to make appropriate restructuring in response to the environment—though there is hardly a consensus here.[34]

At a more cognitive level, depression is associated with (though perhaps not caused by—the usual problems with cause and effect apply) certain patterns of thought. These include a tendency to ruminate on one's problems.[35] The Ruminative Response Scale asks people how often they do the following when they feel sad, blue, or depressed:

> I think "Why do I react this way?"
> I think about how hard it is to concentrate.
> I think "I won't be able to do my job if I don't snap out of this."

Responses are related to the likelihood of getting depressed, though not the duration of the depression. The psychologist Susan Nolen-Hoeksema proposed that this may be why depression is more common in women; women ruminate more.[36]

Depression is also associated with a certain negative style of thinking about one's life. We talked earlier about how people often see themselves and their situations as better than they really are. One can think of depression as involving the opposite bias. If nondepressed students do poorly in a course, they might think that the teacher was incompetent and they'll do better in the future; depressed students might think that they are stupid and this poor grade will haunt them for the rest of their life. As one psychiatrist notes, "Every psychotherapist is familiar with patients who have high-paying prestigious socially useful jobs, happy families, are beloved by everyone in their community—and who sit down on the couch and say, 'I guess I'm a total failure, I haven't accomplished anything with my life and nobody likes me.'"[37]

Put this way, depression doesn't make you stupid; you're as clever as you've always been, but your intelligence is turned against you. Solomon writes, "You don't think in depression that you've put on a grey veil and are seeing the world through the haze of a bad mood. You think that the veil has been taken away, the veil of happiness, and that now you're seeing truly."[38]

Of the mood disorders, major depressive disorder is the most common. I'll quickly mention a less common one, which about 4 percent of people experience in their lifetime—bipolar disorder. This can involve a swinging back and forth between mania and depression. Mania is an extremely high mood that often comes with exaggerated patterns of thought, such as inflated self-esteem and erratic behavior. Sometimes mania feels great, sometimes it involves anger, and sometimes it blends into psychosis.

Bipolar disorder is more heritable than major depression. It also seems to have a different brain basis than depression and is best treated with different drugs; if you have one of these disorders you are more likely to have family members with the same disorder, not the other. (Indeed, bipolar disorder has a stronger link with schizophrenia than with major depression.[39])

Many creative people, such as Vincent van Gogh and Virginia Woolf, suffered from bipolar disorder, and some have argued that their disorder has led to the creation of great work.[40] Even if that were true, the terrible cost of depression likely outweighs the advantages of mania; great artists and writers would probably have done better work if they were psychologically whole.

Like sadness, anxiety is a part of life. Everybody experiences it some of the time. As discussed in earlier chapters, if you had no fear and dread, no anxiety, you might not have lived long enough to be reading this book. But anxiety can become a problem when it's irrational, uncontrollable, or disruptive. Here we get to anxiety disorders.[41] These are the most common of all psychological disorders, averaging about 30 percent lifetime prevalence,[42] and like mood disorders are more common in women than men.

One specific type is *generalized anxiety disorder*—essentially the problem of worrying about things all the time. About one in every twenty people get this at some point in their lives.[43] The worry interferes with functioning and gives rise to physical symptoms like headaches, stomach-aches, and muscle tension.

A more specific disorder is that of *phobias*—intense irrational fears. Discussed in detail earlier in this book, they tend to emerge during child-hood.[44]

A related problem is *obsessive-compulsive disorder*. (This used to be categorized as an anxiety disorder in earlier versions of the *DSM* and it plainly does involve anxiety, but it has been moved to its own special cate-gory in *DSM-5*.) Obsessions are irrational and disturbing thoughts. These give rise to compulsions, which are repetitive actions or mental rituals such as counting that serve to alleviate the obsessions. The descriptions of people suffering from this disorder—about 1 percent of men and 1.5 per-cent of women[45]—are heartbreaking:

> When I was six years old I started picking up things with my el-bows because I thought it would get my hands dirty if I picked things up with my hands. By the time I was seven I was wash-ing my hands 35 times a day. . . . When I swallowed saliva I had to crouch down and touch the ground. I didn't want to lose any saliva—for a bit I had to sweep the ground with my hand—and later I had to blink my eyes if I swallowed. . . . Each time I swal-lowed I had to do something. For a while I had to touch my shoul-ders to my chin. I didn't know why. I had no reason. I was afraid. It was just so unpleasant if I didn't.[46]

I've given mood disorders and anxiety disorders separate sections, and this might give the impression that they are two different problems. This isn't so; they blur together. People with one often have the other, and some of the drugs that sufferers take work for both. Perhaps they are best seen as different manifestations of the same underlying disorder. As Hari puts it, they are "cover versions of the same song by different bands. Depression is a cover version by a downbeat emo band, and anxiety is a cover version by a screaming heavy metal group, but the underlying sheet music is the same."[47]

I'll end this menagerie of troubles and torments with perhaps the most controversial category of mental illnesses—the dissociative disorders.[48] According to the *DSM-5*, dissociation is defined as a disruption or discontinuity in the normal integration of mental life, in our consciousness, memory, identity, and the like.

Two forms of this are *depersonalization/derealization disorder*, where you feel detached from yourself or your environment, and *dissociative amnesia*, where you lose the ability to remember important information about yourself. Such experiences are familiar to some of us, particularly if we overindulge in alcohol and drugs, or have sleep problems. Under certain unusual circumstances, you might know what it's like to feel like you're outside of your body, or to feel like a robot, devoid of feeling, or to experience the temporary loss of important memories. The clinical versions of dissociative disorders are of this sort, only they are more serious and last longer.

The dissociative disorder that has spawned the most interest and the most debate is *dissociative identity disorder*, or DID, known best by its earlier name: *multiple personality disorder*. People with this disorder are said to possess distinct personalities, selves, or alters that might have different ages, genders, and styles of interacting with the world. Often these personalities are seen as taking turns in controlling the body; one personality may have no memory of the actions of another. This might be familiar from popular books and movies that are said to depict real cases, such as the movie *The Three Faces of Eve*, and the novel and movie *Sybil*, as well as from florid fictional recreations such as M. Night Shyamalan's movies *Split* and *Glass*. There is now a large online community, on TikTok and other sites, where millions of viewers follow "systems"—single bodies that are said to possess multiple selves.

The standard theory of the origin of this disorder, proposed in the late 1800s, is that it's the consequence of childhood trauma. A child, usually a girl, experiences terrible abuse, and, as a way to cope or escape, she *splits*—creating different selves. In support of this theory, there is a significant relationship between suffering from DID and reports of childhood abuse—though a substantial minority of those with DID report no trauma and neglect.

An alternative theory is that this disorder is in large part the product of social forces.[49] Under this view, the popular books and movies about DID don't merely depict it, they cause it. Therapists are seen as shouldering a lot of the blame here. Convinced of the reality of this disorder, psychologists and psychiatrists got their patients to construe their problems in terms of multiple personalities. They would encourage their patients to playact and reenact experiences in different voices and different "selves," ultimately giving rise to DID in people who didn't start off with it.

In support of this second account, diagnoses used to be rare, then the books and movies came out, and then thousands and then tens of thousands of patients, mostly women, were said to have DID. It's not just the overall number of cases that have shot up, but the number of personalities that people are said to possess have also grown—it used to be two or three; now the average is fifteen, with some reporting hundreds.[50]

There is also controversy over what DID really is. Nobody denies that people who are diagnosed with DID feel distress and act in bizarre ways. Nobody is saying that they are just faking. But are there really multiple personalities—more than one person in a single head (whatever, precisely, this means)? Those who endorse the trauma explanation often say yes, but proponents of social explanations are skeptical. One research team concludes, "People with DID are not, in reality, a conglomeration of indwelling entities, despite their subjective conviction that this is so. That is, individuals with DID hold *the mistaken belief* that they house separate selves."[51] From this perspective, there is no such thing as people with multiple personalities. Rather, there are people who think they have multiple personalities. Put this way, DID is fundamentally "a disorder of self-understanding."

Regardless of how one comes down on these debates, there remains the question of why some people and not others develop this disorder. One intriguing theory is that it is related to problems with sleep. People who have problems such as sleep paralysis or hallucinations while going to sleep are more prone to suffer from dissociative disorders.[52] Indeed, the correlation with DID and sleep disorders is higher than the correlation with DID and trauma. It might be that a poor distinction between sleep and waking life makes one vulnerable to various dissociative experiences, including, perhaps, those that predispose you to becoming (or thinking that you have become) more than one person.

––––––

Many people urgently want to know how to best treat these problems. Many of the readers of this book either suffer from them, have suffered from them, or have people in their lives who suffer from them. I often get emails from people who see my Introduction to Psychology course online, and while they sometimes ask me about theory or research, the most common question concerns their own psychological issues.

I'm not in any way qualified to give specific advice to these people who write me—I'm a research psychologist, not a psychiatrist or clinician. What I do instead is suggest that if the problem is serious enough, they should seek out therapy.

There are many forms of therapy.[53] One is psychodynamic, built on the ideas of Freud that we discussed earlier. The idea of psychodynamic therapy is that the problems that people struggle with are mere symptoms; proper treatment focuses on discovering and addressing the underlying cause. Classical Freudian *psychoanalysis* is a lengthy process, often going on for years, involving many sessions a week, and using techniques like free association and dream analysis.

There's *behavior therapy*, motivated by ideas of B. F. Skinner. Some of the treatments for phobias, for instance, use the techniques of classical conditioning.

There is *cognitive therapy*, where the goal is to identify and correct distorted ways of thinking. Someone with a depressive style of thought, for instance, might come to realize that they are exaggerating the importance of bad events.

Perhaps the most popular therapy is a blend of the last two—*cognitive-behavioral therapy*, or *CBT*. This involves working on patterns of thought, but also on skills and behaviors, which are seen as fundamentally connected to mental life.

There are medications. These include antipsychotics for schizophrenia, antianxiety drugs for people with panic disorder, and, of course, antidepressants. Most people who are treated for depression in the United States are given such medications.

There is *electroconvulsive therapy*, or *ECT*, in which electric current is run through your brain. This is used for major depression. It's nobody's first resort, but if talk therapy and medication fail on their own, electroconvulsive therapy can save people's lives. As a milder variant, there is a lot of interest in *transcranial magnetic stimulation*, a new technology involving sending highly localized pulses of electrical energy into the brain.

I'm dividing therapies into strict categories, but it doesn't really work this way. Most therapists are eclectic. You might get one who talks to you about the underlying causes of your depression *and* assigns homework exercises that improve your interactions with other people *and* addresses your maladaptive patterns of thought *and*—if you are working with a psychiatrist—prescribes some sort of medication like Wellbutrin.

———

I recommend therapy because the evidence suggests that it is, on average, better than no therapy.

It's not just that people report feeling better after therapy. This doesn't show much—maybe they would have gotten better anyway. To test the efficacy of therapy, one needs a control group, one that includes the same sort of people as the treatment group, so that it's blind luck who gets therapy and who doesn't. For instance, you might have one hundred people who are looking to get some sort of therapy—fifty of them get treatment, and fifty of them are put on a waiting list. When you do this, in study after study, those who get the therapy tend to do better later on than people who didn't.[54]

Therapists know some surprising things about mental illness. One psychiatrist, Randolph Neese, writes:

Recognizing common patterns can make even ordinary clinicians seem like mind readers. Asking a patient who reports lacking hope, energy, and interest, "Does your food taste like cardboard, and do you awaken at four a.m.?" is likely to elicit "Yes, both! How did you know?" Patients who report excessive hand washing are astounded when you guess correctly, "Do you ever drive around the block to see if you might have hit someone?" If a student has weight loss and a fear of obesity, she will likely be astonished when asked, "You get all A's, right?" Clinicians recognize these clusters of symptoms as syndromes: major depression, obsessive-compulsive disorder, and anorexia nervosa. After seeing thousands of patients, expert clinicians recognize different syndromes as readily as botanists recognize different species of plants.[55]

Therapists also often know how to make people better. Certain specific therapies do well for anxiety disorders, for instance. Nesse writes: "Patients with panic disorder get better so reliably that treating them would be boring if it were not for the satisfaction of watching them return to living full lives."[56]

In addition, there seems to be a general benefit to therapy regardless of type.[57] In 1936, this was playfully dubbed the Dodo Bird effect, from a line in *Alice's Adventures in Wonderland*, where the Dodo Bird ran a competition in which many of the characters who had gotten wet had to run around the lake until they were dry. When the Dodo Bird was asked who had won, he thought and said, "Everybody has won, and all must have prizes."[58]

This verdict implies that there is something that all (or just about all) therapies do right. The main one, perhaps, is the bond that the client has with the therapist—the "therapeutic alliance."[59] There is support; you're dealing with a person who accepts you, who feels compassion for you, who encourages you, who guides you, who is on your side. And then there's hope. When you go into therapy it is usually with some faith that this will work. You wouldn't be there if you didn't think it had some chance of success. This sort of hope could be a self-fulfilling prophecy. Believing something will work is often an excellent first step to making it work.

I cannot stress this enough—if you are in distress, seek treatment. It works. But, and I can't stress this enough either, the treatment of mental illness is at a primitive stage.

Things once looked rosy. In the late 1800s, one of the major mental illnesses, "general paralysis of the insane," was found to be caused by a syphilis infection and could therefore be treated as one treats syphilis. The hope at the time was that other mental disorders would come to be diagnosed and cured in a similar way.[60]

When I was in graduate school, it was another time for optimism. People mocked Thomas Szasz and his dismissal of psychiatry. We were confident that mental illnesses are diseases of the brain, soon to be diagnosed through blood tests and brain scans, dealt with the same way as other physical diseases.

Yet another wave of enthusiasm came many years later, when Thomas Insel, the director of the National Institute of Mental Health, announced that the institute would be moving away from the *DSM*.[61] He complained that the *DSM* criteria just weren't scientific enough: "Unlike our definitions of ischemic heart disease, lymphoma, or AIDS, the *DSM* diagnoses are based on a consensus about clusters of clinical symptoms, not any objective laboratory measure. In the rest of medicine, this would be equivalent to creating diagnostic systems based on the nature of chest pain or the quality of fever." The NIMH would instead pursue a more scientific taxonomy that is grounded in neuroscience and genetics, and this would lead to a revolution in treatment.

How did that all work out? Here is Insel's grim summary years later.

I spent 13 years at NIMH really pushing on the neuroscience and genetics of mental disorders, and when I look back on that I realize that while I think I succeeded at getting lots of really cool papers published by cool scientists at fairly large costs—I think $20 billion—I don't think we moved the needle in reducing suicide, reducing hospitalizations, improving recovery for the tens of millions of people who have mental illness.[62]

Insel's predecessor at NIMH, Steven Hyman,[63] recently chimed in with a similar sentiment: "No new drug targets or therapeutic mechanisms of real significance have been developed for more than four decades."

There is a lot we understand about the diagnosis and treatment of mental illness, but it's fair to say that progress here has been slower than hoped. The notion that we would treat depression like diabetes or schizophrenia like cancer has not panned out.

You would know this if you have ever sought treatment for depression or anxiety. Talk therapy—where you meet with a therapist and discuss your problems—often works, but nobody quite knows why, and it's frustrating that the specific person you are talking to makes more of a difference than any techniques that they are using. If there is a secret sauce, we don't know what it is.

Then there are drugs. Again, these do often work, but it's not like taking antibiotics for an infection. Psychiatrists often go through a lengthy period of trial and error to find the right drug for a given patient. Sometimes one will work for a while and then stop, and the process has to start again. Side effects occur—some of them serious, involving weight gain, insomnia, and sexual problems—and so many people who would be helped by medication give up on it. Furthermore, creating new drugs is expensive and time-consuming—they take a long time to test and there are considerable regulatory hurdles—and many drug companies have seemingly lost interest.[64]

Perhaps a revolution is in sight. Maybe it will come from neural stimulation methods, or from drugs now used for pleasure and exploration, such as ketamine, psilocybin, LSD, and ecstasy. Such approaches have enthusiastic proponents, and we see now the usual rash of seems-to-be-too-good-to-be-true findings. Maybe in twenty years, those troubled with terrible sadness or crippling panic attacks will go to their psychiatrists and have a helmet put on their heads that zaps them in just the right way in just the right place. Or maybe a sufferer will be prescribed a course of microdoses of LSD, and this will lead to a profound increase in happiness and flourishing. It would be great if it were so, but it's just too early to tell.

———

Why do problems like schizophrenia, anxiety disorders, and depression exist in the first place? We've seen that there are genes that predispose us to develop these disorders, or at least that fail to confer a resilience to them that others possess. So why hasn't natural selection weeded these genes out? This question isn't unique to psychology. We would ask it as well for diseases like cancer and diabetes.

There are several candidate theories.[65] Some disorders might not influence reproductive success. This could be because their effects don't get in the way of survival and reproduction—it's not clear that unusually high anxiety, say, would have negative consequences in the environment in which most of human evolution took place—or because their effects primarily strike the elderly, who are out of the mating game.

In other cases, the genes that would confer resistance to a disorder just don't exist. Evolution isn't magic; natural selection can only work if the raw material is present. A predator that can turn invisible or move at the speed of sound would be really something, but there's no mutation or combination of genes that will get you there. Similarly, perhaps there just isn't the right assembly of genes that could give you a liver that never gets cancer—or a brain that never succumbs to schizophrenia.

Another proposal is that genes that contribute to some disorders might confer other benefits. The idea *isn't* that schizophrenia and bipolar disorder are themselves beneficial for those who have them, but rather that some of the genes that predispose us to have these disorders might be advantageous. As an analogy to this, carriers of sickle cell disease have greater resistance to malaria and so the genes stick around, despite the serious consequences when two individuals carrying the gene have a child together. Similarly, it's been proposed that some of the genes that predispose people to schizophrenia and bipolar disorder are related to creativity, supporting Joyce's speculation about the relationship between his gifts and the suffering of his daughter.[66]

———

There's a certain tension that arises in these discussions. At times we talk about mental illnesses as akin to cancer or AIDS or the flu. We say that someone is suffering from schizophrenia, or bipolar disorder, or panic disorder, and while these disorders might fall into subtypes and admit of degree—still, you either have them or you don't.

This isn't how they are diagnosed, though. To focus on the most common problems, therapists and researchers appreciate that sadness and anxiety are normal parts of life—indeed we saw earlier that at least some degree of anxiety is essential for survival. To be diagnosed with a mood disorder or an anxiety disorder means that you have *too much* of a certain sort of sadness or anxiety, and this involves all sorts of judgment calls. To be diagnosed with major depression, for instance, involves meeting criteria such as low mood, diminished interest in pleasure, significant weight loss or gain, fatigue, and the like, but therapists might disagree on *how much* low mood, *how much* diminished interest in pleasure, and so on, counts as clinically significant. And some of the criteria are themselves partially arbitrary. You have major depression, for instance, if your symptoms persist for two weeks. But there's nothing magic about this cutoff. It could have been thirteen days, which would make the category more inclusive, or fifteen days, which would make the category smaller.

Or take psychopathy.[67]. This is not a diagnosis in the *DSM-5*, which instead includes antisocial personality disorder, something that is subtly different. But many do consider psychopathy to be a specific mental illness, involving lack of remorse or guilt, pathological lying, callousness, lack of empathy, and need for stimulation—and it is a classification in the other major coding system, the *ICD-11*. There is a commonly used test for it—the Hare Psychopathy Checklist-Revised or PCL-R, developed by Robert Hare.[68] The scale was developed for trained clinicians, but it forms the basis for a simple self-report measure that people can fill out themselves, which is often used in research.

Some of the findings from this research are interesting. You won't be surprised to hear that people in prison score higher on the psychopath test than students in universities, or that men score higher than women.[69]

But what about occupations? One study done online with 5,400 people found that in the United Kingdom the professions that had the lowest psychopathy scores were:

social-health assistants, nurses, therapists, artisans, stylists, charity workers, teachers, creative artists, physicians, and accountants.

And the ten professions with the highest scores were:

CEOs, lawyers, radio or television characters, salespersons, surgeons, journalists, priests, police officers, chefs, and civil servants.[70]

In 2021, a paper was published that reviewed the literature and asked: What is the prevalence of psychopathy in the general adult population?[71] The answer, based on data collected from over eleven thousand people, is: 4.5 percent.

As the authors of the study admit, though, this number can move up and down if you want it to. The Hare scale provides a score from 0 to 40; in the United States, the standard cutoff for psychopathy is 30 or more. With that score, you're a psychopath; less than that, you are not. If you wanted to say that there are more psychopaths in the population, you could drop the cutoff to 25, as is often done outside the United States. Want to have fewer? Kick it up to 32, as done in other studies. The question "What is the prevalence of psychopathy in the general adult population?" might not be a scientifically meaningful one; it's a lot like the question "What proportion of the population is really tall?" It all depends on what you mean by "really tall."

None of this is an argument against categorization. We put people into discrete categories for all sorts of reasons. Maturity is on a continuum, but we have strict age cutoffs for driving, enlisting in the military, and drinking in bars. For tax reasons, one is single or married, even if, in fact, there are plenty of sort-of-married couples in the world. It's hard to see how a health-care system could work if there weren't clean categories such as social anxiety disorder or schizophrenia.

Also, critically, it's possible that these clinical categories might in fact be all or none, even though our ways of diagnosing them are fuzzy and often arbitrary. COVID-19 isn't just a name we give to people who show a sufficiently high degree of fever, body aches, and breathing problems. It's a distinct category. We have tests for it. Perhaps psychopathy and major depression are like that, regardless of how we currently diagnose them— the correct test just hasn't been developed yet.

However, this is probably not the case. The data so far suggest that most disorders are continuous. Just like introversion is a matter of degree (and so we are better off, at least as scientists, talking about the extent of introversion rather than about a certain special sort of person who is "an introvert"), this is true as well for many of the disorders we've been discussing. One recent meta-analysis on the topic concludes by saying that the data from multiple studies "support the conclusion that the great majority of psychological differences between people is latently continuous, and that psychopathology is no exception."[72]

This continuous view, should it be right, has implications for some of the more foundational issues we began with, such as neurodiversity and the very definition of mental illness. When thinking about disorders like depression, anxiety, addiction, autism spectrum disorder, and schizophrenia, we can no longer simply view them as problems to be solved. Rather, they are the names we give to certain extremes of human variation. What counts as extreme enough to warrant treatment isn't just a psychological question. It is a moral and political one.

15

The Good Life

Psychologists have always been interested in when things go badly, and there's good reason for this. Mental illness is a terrible curse, and the study of schizophrenia, mood disorders, anxiety disorders, and all the rest is a pursuit of obvious value. But what about the bright side of life?

Over the last few decades, some psychologists have worried that there has been too much focus on the negative. We haven't been interested enough in when things go well. We haven't done enough research into what goes into a pleasant and meaningful and satisfying life. Now, human thriving hasn't been entirely ignored—it's long been part of the humanist movement in psychology, led by scholars such as Abraham Maslow and Carl Rogers[1]—but it's never been anywhere near as focal as human misery. A new movement, known as positive psychology, hopes to change all this.[2]

I want to end the book on this topic. But I'll warn you—positive psychology is a messy field. There's a lot of money and fame in the business of telling people how to be happy. If you want to get onstage and stay there, best to tell a story that's confident and clear. If you hem and haw and say that there is so much we don't know, nobody will invite you to give a TED talk and you won't get to give lucrative speeches at corporate retreats. While there are a lot of great scholars in this area—and actually some excellent TED talks[3]—there is no shortage of hucksters.[4] A lot of what you'll read in popular books and see in popular online videos is filled with exaggerations, mistakes, and occasional flat-out lies. If we ever get to meet and have a beer together, Dear Reader, I'll give you some examples of this.

To make things messier, the question of the good life relates to morality and politics in a way you don't see in other fields of psychology. It's not essential that your preferred view of how things should be—your views on issues such as religion, marriage, children, and income inequality—is supported by research about what makes people most happy, but it sure is nice when it happens. This does not encourage good science.

Finally, positive psychology is blessed with an abundance of data. Some of the findings I will discuss are based on results from studies of millions of people. But for reasons we'll see, these data are often hard to interpret, and so well-intentioned people can look at the same studies and draw opposing conclusions, even about the most important things, such as whether getting married makes you happy.

For all these reasons, you should be cautious about what psychologists have to say about human flourishing. Still, the field has real and robust findings, some of which will surprise you. A little while ago, I spoke to a group of retirees and asked: Do you think rich people are happier than poor people? In a room of about forty people, only two hands went up. But this is mistaken. We'll see that the relationship between money and happiness is one of the more robust findings in the field of positive psychology. And we're just getting started.

———

What is the good life? Some, following the philosopher Epicurus, say that the good life is nothing more than a life filled with pleasure. Others are concerned with the avoidance of suffering, which isn't quite the same thing. Still others propose that a good life is one in which a person is happy or satisfied, feeling that their life in general is good, separate from how much pleasure and pain they experience from moment to moment.

Then there are loftier concerns; maybe a good life is one that is filled with meaning (whatever that is), or one where you make a positive difference in the world, or one that has some sort of transcendent or spiritual quality.

Certain aspects of a good life don't have anything particular to do with psychological states.[5] Consider longevity. If life is not a slog and burden, then it's better to have more of it than less. A longer life means more happiness, more opportunity for pleasure, more helping others, more meaning, more of whatever is good. Perhaps there is a limit here; some worry that immortality would be bad for us. But still, holding psychological happiness constant, a life of eighty good years is better than one of fifty good years, let alone five good years. When we compare countries on how well they provide conditions for flourishing, then, we miss something when we ignore life span. If there are two countries, and the citizens of these countries claim to be equally happy and satisfied on all measures, and one has a life expectancy of fifty and the other has a life expectancy of eighty, then, holding everything else constant, the second is doing better.

My own view is that the question "What is a good life?" won't have a one-word answer. All of us want pleasure, to be sure, but we also want to engage in interesting activities, to have satisfying relationships, to find meaning, and much else. My view jibes with that of the economist Tyler Cowen, who wrote:

> What's good about an individual human life can't be boiled down to any single value. It's not all about beauty or all about justice or all about happiness. Pluralist theories are more plausible, postulating a variety of relevant values, including human well-being, justice, fairness, beauty, the artistic peaks of human achievement, the quality of mercy, and the many different and, indeed, sometimes contrasting kinds of happiness. Life is complicated![6]

I've defended this sort of pluralism earlier in this book, when talking about motivation more generally. I pointed out that there are many psychologists who think that people just want one thing (though they disagree what this one thing is), and then argued that it's more plausible to talk about many distinct drives—sex, food, social interaction, morality, and so on. I think "motivational pluralism" is good philosophy and good psychology, and I wrote a book defending it.[7]

But maybe you are unconvinced. I can hear you saying, *Forget about meaning, purpose, morality, and all that. I just want to be happy!* I feel you. Even pluralists shouldn't doubt the importance of happiness. Polls find that in just about every nation, if you ask people what they want most in life, this is where they land. They'll tell you that they want to be happy.[8]

But what do we mean by "happy"? There are subtly different meanings of the term and, accordingly, different ways to measure it. One interpretation has to do with day-to-day experience. In some studies, psychologists get people to carry around a smartphone with an app that randomly goes off and prompts them to answer a string of questions, including about how happy they are. Their overall happiness is then calculated as the average of these assessments. Call this *experienced happiness*. Another interpretation involves one's thoughtful assessment about how well one's life has been going, extending over a long period of time, separate from any specific experiences. Here, a happy life is one where someone honestly says, "My life is going great!" Call this *remembered happiness*.[9]

The happiness levels you get from these kinds of measures are correlated—but they are not identical. Someone can judge their life as happy even if their day-to-day experiences aren't so enjoyable, or, conversely, have overall positive experiences but not be so pleased by their life as a whole.[10] There is quite a bit of debate as to which of these two measures is the one to shoot for, with pluralists like me saying that they are both worthy goals.[11]

Enough futzing around. Let's talk about a study that speaks to a question that I bet you would like to be answered. What do we know about the lives of the happiest people on Earth?

To explore this, psychologists took a survey of 1.5 million people, covering 166 nations over a span of 10 years, and looked at the people who reported themselves as happy in every way that their scales measured—on a scale from 0 to 10, they all evaluated their lives as a 9 or 10; they all reported experiencing enjoyment, and said that they had no worries, sadness, or anger in the previous day.[12]

The big finding is that the very happy have rich social relationships. They spend a lot of time with family and friends. Almost all of them—94 percent—said that they have someone that they can count on to help when they are in trouble. Almost all of them—98 percent—said that they were treated with respect on the previous day.

Relative to other individuals, the happy elite make more money, are in better physical shape, have fewer health problems, exercise more, smoke less, feel more well rested, and have less stress. They have more hours of free time, more freedom to choose their activities, and are more likely to say that they learned something new on the previous day.

Now, importantly, some people who are moderately happy and even some who are unhappy are also treated with respect, spend time with family and friends, make a good living, and so on. Such considerations are not sufficient for being very happy, then. But they do appear to be close to necessary; without them, you are very unlikely to be one of the happiest people in the world.

I love this study, but it's important not to get carried away here. If someone were to look at these findings and tell you that the trick to being happy is: exercise, smoke less, get good sleep, learn a lot, make money, and, most of all, have deep and satisfying social relationships, you should respond with: Not so fast.

The problem—and I know that I must have said this a hundred times so far in the book—is that it's difficult to pull apart cause and effect.

Take having friends. Very happy people have friends. So perhaps friends make us happy, a reasonable claim. But perhaps being happy leads to having friends, which also makes sense. There's something appealing about good cheer, at least in proper doses. It's harder for someone who is sullen and withdrawn to be popular. Or perhaps happiness and friendship have no direct relationship, but some third factor independently influences them both. Perhaps some people are gregarious and agreeable and extroverted. Such traits might make them more likely to be happy, and might also help them have more friends, but their happiness and their popularity might themselves have no direct connection.

We don't have to throw up our hands here; careful studies can pull apart cause and effect. To see this, let's take the question of the relationship between happiness and helping others. One of the first thing people ask about happiness is "Does being kind make you happier?" and there is good data about causality here.

Consider first one form of kindness—volunteering. In study after study, research finds that volunteering is associated with greater life satisfaction and lower rates of depression. This relationship holds for both poor countries and wealthy countries.[13]

So good, if you want to be happy, volunteer, right? It's unclear. There are several experiments that have randomly assigned people to volunteer or not to volunteer and they find no difference in mood, self-esteem, and so on.[14] But a recent longitudinal study, looking at changes in happiness before and after volunteering, does find a substantial positive effect.[15] It's not yet known, then, whether happy people volunteer more (for whatever reason), whether volunteering makes you happy, or both.

What about giving to charity? Again, we get a correlation. Studies of over a million people find that charitable donation is one of the predictors of life satisfaction. And this time the evidence for a causal connection is clearer.[16] There are several experiments that involve handing people money, like a twenty-dollar bill, and telling them that they either must spend it on themselves or spend it on another person. People are happier if they are told to be altruistic. These studies typically find an immediate boost in mood for the people who give the money away. It is less clear whether giving increases lifelong happiness, but still, these initial results are suggestive.

———

One of the main findings in happiness research is stability. There are a couple of considerations that work to keep one's happiness pretty much the same over time.

Just as with physical characteristics like height and weight and psychological traits such as intelligence and openness to experience, happiness is heritable.[17] To put it more casually, some people are naturally cheerful, while others are by nature glum.

One bestselling book claims that 50 percent of the variation in happiness is due to genes.[18] For all the reasons discussed in the last chapter, I think we should be wary of precise numbers here. Heritability varies in different places and different times. We should keep in mind as well that moving to a new environment might change how the same genes would be expressed, and this leaves open the possibility, for almost all of us, of lives of great happiness or of terrible misery. Still, the findings from behavior genetics do suggest that there are powerful influences on us even before we're born.

Also, there are several studies that illustrate what the psychologist Daniel Gilbert has described as failures of "affective forecasting."[19] We are often poor at predicting our happiness. Most relevant to the issue of stability, we tend to think life events matter more than they will. We fail to appreciate that we are fairly resilient to bad experiences and, unfortunately, fairly unaffected by good ones as well.

In one study, college students got to provide their university with a ranked list of which dormitories they wanted to live in and were asked to predict how happy they would be if they were assigned their top choice or their bottom choice. They predicted a substantial effect, but once they were in their dorms, there was no difference at all.[20] In another study, sports fans were asked how they would feel after a pivotal soccer championship game. It was made clear to them that they were being asked about their overall feelings, not their feelings when they were thinking about the game, but still, fans of the winning team overestimated how happy they would be, and fans of the losing team overestimated how sad they would be.[21] After the 2000 election, when George Bush was elected president, Bush supporters were not as happy as they thought they would be, and Gore supporters were not as unhappy as they thought they would be.[22] Indeed, even for more severe events, such as the dissolution of a romantic relationship, people tend to overestimate how much their happiness will be affected.[23]

Why are we so poor at predicting our future happiness? One consideration is that when you make the prediction, you think about the event, and this act of thinking leads one to overestimate how much you'll think about it in the future. When I'm asked to imagine my favorite candidate losing the election, I feel crushed, and when making predictions about how I'll feel after the election, I might say that I'll be crushed, failing to appreciate that most of the time I won't be thinking about politics at all. Thinking about winning a large prize makes me flush with excitement, and so I predict this will make me happy in the future, not realizing that for the vast proportion of my future life my mind will be elsewhere. As Daniel Kahneman and Richard Thaler put it, nothing in life matters quite as much as you think it does when you are thinking of it.[24]

Finally, we often don't appreciate how good we are at telling stories and reinterpreting information in ways that preserve our good feelings about ourselves, our group, and the views we most cherish. Gilbert and his colleagues nicely put it as follows (you know who Freud is; Leon Festinger developed the cognitive dissonance work described in the earlier chapter on social psychology):

> Psychologists from Freud to Festinger have described the artful methods by which the human mind ignores, augments, transforms, and rearranges information in its unending battle against the affective consequences of negative events. Some of these methods are quite simple (e.g., dismissing as a rule all remarks that begin with "You drooling imbecile"), and some are more complicated (e.g., finding four good reasons why one didn't really want to win the lottery in the first place); taken in sum, however, they seem to constitute a psychological immune system that serves to protect the individual from an overdose of gloom. As Vaillant noted: "Defense mechanisms are for the mind what the immune system is for the body."[25]

We've seen that roughly half of the difference between individuals' happiness is due to heredity. This could be direct—one's genes could directly influence one's happiness—but more likely the causal chain is complex. The genes could influence your personality, for instance, or your capacity to form relationships or to succeed in various parts at life, and this in turn could influence your happiness.

Still, if 50 percent of the variation is genetic (and keep in mind that this is a crude estimate), this means that 50 percent is not. A lot of your happiness is determined by your environment. So what sort of environment matters the most?

One consistent finding is that people in some countries are, on average, happier than people in others.[26] These differences are not subtle. In the World Happiness Report[27] of 2020, for instance, the top country (Finland) had an average of 7.9, on a scale going from 0 to 10, and the bottom country (Moldova) had an average of 5.8, a staggering difference. Indeed, besides genes, where you live might be the largest determinant of your happiness.

Now, you might be skeptical about how much these scores really tell us, and you might worry in particular about cultural differences in how people think about happiness and use words like "happy."[28] These are real concerns, but the precise way of computing the numbers and the exact wording of the question don't seem to make a difference for how the results turn out. While some studies, like the World Values Survey, ask about "happiness," others use different methods, such as asking people to rank their lives on a score from 0 (the worst possible life) to 10 (the best possible life). And the results tend to cohere, regardless of the questions. The same countries—which include the Nordic nations of Norway, Finland, Denmark, and Sweden, as well as Switzerland, Germany, Iceland, the Netherlands, Canada, New Zealand, and Australia[29]—tend to come out on top.

The factors associated with happy countries are what we would expect. Happy countries are high in average income—there is a robust connection between the GDP of a country and how happy it is. They have high life expectancy and strong social support. The citizens of these societies perceive high levels of freedom, trust, and generosity. Progressive taxation and a strong welfare state predict happiness. So does some degree of economic competition (communist countries are unhappy countries).[30]

Is it fair to call this an effect of the environment? Perhaps it's not actually living in the country that influences one's happiness; Finns, say, might be happy because of Finnish genes, and they would be just as happy if they moved to Moldova. But this isn't so. Several studies show that, although there is some influence of country of origin, immigrants tend to be roughly as happy as native-born citizens of the country they live in.[31] It's the country, not the genes.

———

What else influences happiness?

Money.[32] It's not just that people in rich countries are happier than people in poor countries; rich people are happier than poor people in the same country. Of course, there is truth to the cliché that money can't buy happiness; there are plenty of sad rich people. But the idea that money is unrelated to happiness is just wrong; it was always more of a moral claim than an empirical one, more about how many people think the world should work than how it actually does.

And wouldn't it be strange indeed if improvements in our lives— the sorts of things that money can buy—had no effect on our happiness? Money buys better health care, better education for oneself and one's children, a better living space, better and healthier food, more leisure time, more travel, and many other opportunities. And if you don't like where you live or where you work, money can buy freedom. The advantages of money vary from society to society—in some places, even the very poor get good health care; in others, they are left to suffer—but such advantages always exist, and so it's hardly surprising that no matter how you slice it, the poor are less happy than the rich.

There are diminishing returns here. If you are making thirty thousand dollars a year, another thousand dollars might be a big deal, but if you are making a hundred thousand dollars, you'll hardly notice the extra 1 percent. Some have proposed that the returns might dwindle to nothing; once you reach a certain level of safety and security, more money might not help. Fitting with this, some studies do find a satiation point— roughly at the income level of one hundred thousand dollars, though it varies from country to country—and some studies even find evidence for a *decline* at higher incomes.[33] Perhaps extreme wealth estranges one from social relationships and somehow has a corrosive effect?

Still, other studies find that money never stops buying happiness. One 2019 poll found strong differences in life satisfaction at higher levels of income—of those who make over $500,000 per year (the top 1 percent in the United States), 90 percent were "very" or "completely" satisfied with their lives, which was substantially more than those who made over $250,000 but under $500,000.[34] Another study looked at the superrich and found that people with over $10 million in wealth were more satisfied with their lives (though just by a bit) than those with a mere $1 million to $2 million.[35]

Now, the problems that we discussed earlier about correlation and causation weigh heavily here. Maybe the sort of person who is rich is different in some ways from the rest of us and would have been happy even if all their money was taken away.

This can potentially be studied in a simple experiment: Give people a lot of money and see what happens—does this make them happier in the long run? There is just one small practical problem, though. Not many psychologists can afford to give a million dollars each to a large number of people to see what happens.

Still, there are so-called natural experiments, where luck gives rise to the random assignment associated with a proper experimental design. Such natural experiments take advantage of a certain type of real-world situation in which some random people get a lot of money and others don't, entirely by chance. These situations are called lotteries. (Though, as an important caveat, lotteries only give money to people who buy lottery tickets, and you might worry that this is an unusual group.)

A well-known study done in 1978 compared twenty-two lottery winners from the Illinois State Lottery with twenty-two randomly chosen non–lottery winners and found no significant bump in happiness.[36] But this was a small sample and other studies with larger samples do find an overall improvement in the happiness of the winners,[37] and evidence as well for an improvement due to another sort of windfall—those who inherit a substantial amount of money.[38]

Health is also associated with happiness. Not surprisingly, illness, disease, and injury, particularly when there is an effect on mobility as with strokes and spinal cord injuries, lead to long-lasting drops in happiness.[39]

Age is associated with happiness—in a strange way. As we get older, we are, on average, less strong, less mobile, more vulnerable to injury, and more likely to get certain serious diseases (most people with cancer are over sixty-five). And so, given what I just said about health, you might expect a decline in happiness as one ages. There is also a loss of status—a common complaint by the elderly is that they feel irrelevant and are treated with less respect—and this might also lead to a happiness dip.

But there is no decline in happiness with age. Rather, happiness maps onto age in what appears to be a U-shaped curve.[40] Eighteen-year-olds are relatively happy, then average happiness gradually drops until the early fifties, and then it rises again, so that the eighties are, on average, the happiest time of people's lives.

Religion relates to happiness too, but in a complicated way. Within a country, religious people tend to be happy, particularly in countries that are poor. But people living in religious countries are, on average, less happy than those living in less religious countries.[41]

What about gender? Even in the most egalitarian societies, the lives of men and women differ in all sorts of ways—income, familial roles, jobs, and so on. Women are more prone to depression, which is obviously related to unhappiness. Then again, men are more vulnerable to alcoholism and other addictions and are more likely to be both the perpetrators and victims of violent crime. Men also die much younger than women. Perhaps it all washes out, because, surprisingly, there is no interesting effect of gender—men and women experience about the same amount of happiness.[42]

Let's end our list of factors that influence happiness by returning to relationships.

We saw that the happiest people in the world tended to have good close relationships. Take marriage. My nana would have told you that to have a fulfilled life, you need a partner. Every pot needs a cover, she would loudly remind my (then) unmarried sister at family gatherings, which I found very funny, though my sister did not. Now Nana was phrasing things a bit too strongly—there are plenty of happy people who have no partners—but she was right on average; married people do tend to be happier.[43]

But why? Here we are presented with our usual trio of options when thinking about the relationship between two factors. A can cause B; B can cause A; or a third factor, C, can cause both A and B, but A and B might not be otherwise related. It might be that marriage makes you happy (plausible enough), or that being happy makes you more likely to attract people who want to marry you (also plausible), or that there is some third factor—wealth, money, a certain personality type—that leads to both marriage and happiness (also plausible).

One way to pull apart these possibilities is to look at timing. If your happiness goes up once you get married, it suggests that marriage does play a causal role. And this does seem to happen, for both heterosexual and homosexual couples, at least for the first few years of marriage.[44] But then your happiness drops back to where you started.[45] Careful studies looking at long-term consequences of marriage find subtle effects that vary due to all sorts of factors, such as whether you are a man or a woman, how old you are, and what country you're from.[46] The question of whether marriage makes you happy does not have an easy answer.

What, finally, about the effects of having children? There is enormous debate about this, and no wonder—it's a decision that many people struggle with.[47]

For a while, there was a clear answer. In a study that has been very often cited (sometimes with glee by those without children), Daniel Kahneman and his colleagues got about nine hundred employed women to report, at the end of each day, each of their activities and how happy they were when they did them.[48] It turns out that childcare is no fun—the women recalled being with their children as less enjoyable than many other activities, such as watching TV, shopping, or preparing food. Other studies find that when a child is born, there is a decrease in happiness for parents that doesn't go away for a long time,[49] along with a drop in marital satisfaction that only recovers once the children leave the house.[50] As the psychologist Dan Gilbert puts it, "The only symptom of empty nest syndrome is increased smiling."[51]

This rings true: raising children is *hard*, particularly when the children are young. It's stressful, expensive, worrying, often dull, and often associated with sleep deprivation, which is itself associated with depression. For a single parent, it can be overwhelming; for couples, it can put a strain on relationships. As the writer Jennifer Senior notes, children provoke a couple's most frequent arguments: "More than money, more than work, more than in-laws, more than annoying personal habits, communication styles, leisure activities, commitment issues, bothersome friends, sex."[52]

But as often happens, the initial studies provide a clear picture and then other research makes things more complicated. The negative effects of children hit mothers more than fathers, young parents more than older ones, and single parents hardest of all.[53] The effects are influenced by the country and whether there are policies such as paid parental leave and accessible and cheap childcare.[54] Parents from Norway, for instance, are happier than childless couples, while the greatest hit from having children is borne by Americans.

Still, children really are often associated with a decline in overall happiness, daily pleasures, and marital satisfaction. This raises a puzzle. I don't know of any scientific research on this, but from my own experience and my own interactions with parents, there are few people who claim to regret having children. Actually, many parents (including me) describe parenthood as central to their lives, a huge positive. Are we just deluded?

Some happiness researchers would say yes. Maybe it's cognitive dissonance; the psychological immune system at work—once we had children and devoted our lives to caring for them, we can't bear to admit that we messed up our lives, so we fool ourselves into thinking that we are the wise ones.

Or maybe it's the product of memory distortion. As we'll discuss in a bit, we tend to remember the peaks and endings of past experiences and forget the mundane awfulness in between. Jennifer Senior again: "Our experiencing selves tell researchers that we prefer doing the dishes—or napping, or shopping, or answering emails—to spending time with our kids. . . . But our remembering selves tell researchers that no one—and nothing—provides us with so much joy as our children. It may not be the happiness we live day-to-day, but it's the happiness we think about, the happiness we summon and remember, the stuff that makes up our life-tales."[55]

Consistent with this, one study looked at millions of people from 160 countries and found that, overall, parents are less happy than nonparents—lower life satisfaction and more daily stress.[56] But they also experience more daily joy. So perhaps we remember the highs? I had a Zoom call with a journalist during the pandemic and he mentioned that he had a four-year-old and two-year-old twins. He said that parenthood was like heroin—the lows are horrific, but the highs are wondrous. The writer Zadie Smith described it in similar terms: having a child as "a strange admixture of terror, pain, and delight."[57]

I think this remember-the-peak explanation is plausible enough, but there is another consideration to throw in the mix. Let's oversimplify here and assume that there really is a drop in pleasure and happiness when you have kids. But now I want to return to the pluralism introduced at the beginning of this chapter and suggest that it might still be worthwhile. When I say that raising my sons is a source of immense satisfaction for me, I don't mean that being a father upped my overall pleasure. I'm talking about something deeper, having to do with satisfaction, purpose, and meaning.

———

Why can't we just become happier by wanting to, in the same way that we can conjure up an image of a duck or think of the number seven? Plainly, our minds don't work that way—but why not?

This question forces us to confront what happiness is for. A plausible enough evolutionary account is given by Steven Pinker:[58]

We can see happiness as the output of an ancient biological feedback system that tracks our progress in pursuing auspicious signs of fitness in a natural environment. We are happier, in general, when we are healthy, comfortable, safe, provisioned, socially connected, sexual, and loved. The function of happiness is to goad us into seeking the keys to fitness: when we are unhappy, we scramble for things that would improve our lot; when we are happy, we cherish the status quo.

From this perspective, happiness evolved as an indicator that our lives are going well. This explains why it's impossible to simply choose to be happy. Imagine buying a car where the gas gauge was rigged so that it would always read FULL. This would be very pleasing in the short term—you would never worry about running out of fuel. But it's a bad idea for the long term because the tank would soon empty and the car would stop running. Similarly, if I'm starving and could just choose to ignore my hunger, or even to feel as if I just had a satisfying meal, it would be nice in the short term, but I'd lose the motivation to get food and then I wouldn't be around for much longer.

If I choose to always be happy, I would miss out on the information provided by being unhappy, the information that life is going poorly and that something needs to be done. Unhappiness—like a gas gauge telling you that you are approaching empty, like an aching empty stomach—is a *gift*.

———

One conclusion here is that you should just try to live the best life. If you succeed, you'll be happy; if not, your unhappiness will motivate you to keep on trying. But suppose you wanted to ignore this annoying piece of advice or are unable or unwilling to improve your life. Is there any way to hack the system?

There are pills. Some of the medication currently used for depression has been argued to not only make the depressed be less depressed, but also help the normally unhappy to be happy, and, perhaps, the happy to be very happy.[59] There are nonmedicinal quick fixes that have been proposed to bump up one's happiness level, such as certain dietary supplements, gratitude exercises, mindfulness meditation, exposure to sunshine (especially during the winter), and various forms of physical exertion. Some of these might work in the short run, though I think we should be skeptical about long-term effects.

One somewhat paradoxical piece of advice holds for people living in more individualistic societies: If you want to be happy, you probably shouldn't try too hard to be happy. Those who are most likely to say that happiness is a priority for them are more likely to be depressed and lonely, and their lives go less well in general.[60]

Now, concerns about cause and effect should come naturally by now. Perhaps having happiness as a priority is not what's bad for you; it's that when your life is going badly happiness becomes more of a priority.

Still, there are a few reasons why trying to pursue happiness is the wrong way to get happy.[61] It might lead to unrealistically high expectations. It might make you mull over how unhappy you are, which isn't itself a very happy pursuit. And you might go about being happy in the wrong way. It turns out that pursuing goals such as making money and building up status is accounted with unhappiness, and is linked as well with more depression, anxiety, and mental illness.[62] Trying to make money is a particular problem. One meta-analysis concluded: "Respondents report less happiness and life satisfaction, lower levels of vitality and self-actualization, and more depression, anxiety, and general psychopathology to the extent that they believe that the acquisition of money and possessions is important and key to happiness and success in life."[63]

Wait, wait—didn't we just establish that money is associated with happiness? It is. Money does help make you happy. It's just that, apparently, *trying to make money* doesn't make you happy. (The wise strategy here is to be born rich.)

What if you try to be happy in a different way? There is evidence that in collectivist societies, such as parts of East Asia, there actually is a positive relationship between trying to be happy and being happy.[64] This might be because, in these societies, people strive to be happy by being more connected with friends and family, which seems to be a better way to go about it.

———

We've been talking so far about happiness—or subjective well-being— usually assessed by asking people how happy they are with their lives. We also have seen that there are other facets of life that we wish to maximize. Some are deeper, like meaning and purpose; this might be what motivates us to run marathons or have children. One might want to live a moral life or a rich and interesting one.[65]

Other goals are more immediate. We often seek pleasure and avoid pain. While happiness and meaning are long term, pleasure and pain are qualities of specific experiences. I want to end this chapter—and this book—with a set of discoveries that I find particularly gripping about how these qualities are thought of and remembered.

Consider the relationship between the hedonic nature of an experience and the memory of the experience later on. It would make sense for them to match up, that we would remember an experience with two hours of pain as worse than an experience with one hour of pain. It would make sense that an hour of pleasure followed by an hour of pain should be thought of as just the same as an hour of pain followed by an hour of pleasure.

But Daniel Kahneman and his colleagues find neither of these claims are true.[66] The length of an experience has little effect on how we re- member it.[67] In reality, two enjoyable weeks of vacation has more pleasure than one enjoyable week of vacation. Indeed, assuming that the pleasure stays constant and you don't get bored, it is twice as much pleasure. But we remember them as identical. *That was a fun vacation.* Or consider our memories of negative experiences. Imagine being on a cramped middle seat of the plane, with no in-flight entertainment, nothing to read, and your neighbor at the window constantly getting up to pee. Which is worse, doing this for four hours or eight hours? This is not a hard ques- tion, I hope, but when you think back on the experience, you will pretty much remember the four-hour experience and the eight-hour experience as equally negative. *What an awful flight.*

It turns out that when we think about past events, we tend to focus on two things—the peak (the most intense moment) and the ending. One study provides a dramatic illustration of this. Experimenters exposed subjects to differing levels of pain—by having them immerse their hands in freezing water—for different periods of time.[68] Here were the trials:

A. 60 seconds of moderate pain.
B. 60 seconds of exactly the same moderate pain. Then, for 30 more seconds, the temperature is raised a bit, still painful, but less so.

Which did subjects prefer? You might think A because, duh, less pain. But, no, they preferred B because of the better ending. In another study,[69] Kahneman and his colleagues went outside of the lab. They tested volunteers who were undergoing a colonoscopy procedure (this was done at a time when these procedures were substantially more unpleasant), and, for half of the people, artificially prolonged the procedure, by leaving the scope inside them for an extra three minutes, not moving, which was uncomfortable but not painful. These people with the additional discomfort rated their experience as overall less unpleasant, just because it ended on a less painful note.

This might be familiar: Imagine going to a party that is mostly enjoyable, but suppose there is an unpleasant minute or so, suppose you have an embarrassing interaction, saying something awkward. When would you want to make that remark—at the start of the party, the middle, or the very ending? Which event is most likely to ruin the party for you? Here, common sense gives the same answer as in the Kahneman studies: Endings matter.

————

And here we are, at the end.

If you've read this whole book, you know a lot about psychology. You appreciate why scientists believe that the brain is the source of mental life. You understand the theories of Freud and Skinner. You have a grasp of what we know—and don't know—about consciousness, child development, language, memory, perception, rationality, emotions, motivations, social behavior, prejudice, individual differences, mental illness, and much else. And you are now familiar with the current discoveries about happiness and pleasure. Some of this might be useful to your everyday life. The rest, I hope, you will find interesting and important, regardless of any practical relevance.

I began *Psych* with the story of how I was entranced by a book on the origins of the universe. But once I put the book down, I never read anything on cosmology again. Perhaps this is your last book on psychology. And so I'll end by recommending two attitudes toward the field I have devoted my life to.

The first attitude is humility. There are some basic questions that nobody knows the answer to yet. We know that the brain is the source of mental life, but we don't know how a physical object, a lump of meat, can give rise to conscious experience. We know that much of the variation in personality, intelligence, happiness, and other psychological traits is heritable, but we don't know how genes have their effects and have little understanding of the role of life experiences in shaping how we are. We know about the power of social influence, but we are far from being able to predict and control what people do. We have some understanding of mental illness, but our diagnoses and treatments are, to put it bluntly, primitive. For every topic of psychology, genuine puzzles remain.

The second attitude is optimism. I believe that the methods of scientific psychology will eventually triumph. In the end, the most important and intimate aspects of ourselves, our beliefs and emotions, the capacity to make decisions, our sense of right and wrong, even our conscious experience, will be explained through constructing and testing scientific hypotheses. We've made progress; we will make much more.

Some people find this a scary prospect. They worry that a scientific approach to the mind takes a specialness away from people, it diminishes us somehow. But I have the opposite reaction—one I hope you share now that you've read this book: The more you look at the mind and how it works from a serious scientific point of view, the more you appreciate its complexity, its uniqueness, and its beauty.

Acknowledgments

P *sych* is based in part on my Introduction to Psychology course, but it also brings together many of the ideas that I've developed over my career as a research psychologist and writer of six books. So, in a sense, I've been working on this project for my whole professional life. It would be impossible to acknowledge all the friends, colleagues, and students who have influenced me on this journey, and I'm not going to try. But I do owe some special thanks to those who helped me turn all of this into a book.

Before I started, I reached out to some top Introduction to Psychology professors for inspiration and guidance, and I'm grateful to Marvin Chun, Chaz Firestone, David Pizarro, and Nicholas Turk-Browne for providing me with materials from their own courses. And I benefited from two superb introductory textbooks—*Psychology* by Daniel Wegner, Daniel Gilbert, Daniel Schacter, and Matthew Nock, and the less Daniel-heavy *Interactive Psychology: People in Perspective* by James J. Gross, Toni Schmader, Bridgette Martin Hard, Adam K. Anderson, and Beth Morling.

During the writing of *Psych*, I frequently went on social media to ask for assistance (Who originally thought of this theory? What's the experimental evidence for this effect?) and was moved by the generosity and wisdom of both perfect strangers and good friends—especially Fiery Cushman, who came through multiple times. David Pizarro and Tamler Sommers had me on their podcast, *Very Bad Wizards*, to talk about

William James, and this led me to include more of James in this book than I originally planned.

I am grateful to the authors of three excellent books that had a big influence on me as I worked on the manuscript: *The Idea of the Brain: The Past and Future of Neuroscience* by Matthew Cobb; *The Genetic Lottery: Why DNA Matters for Social Equality* by Kathryn Paige Harden; and *Freud: Inventor of the Modern Mind* by Peter D. Kramer. I am grateful as well to other scholars whose work I went back to over and over again, including Susan Carey, Frank Keil, and Elizabeth Spelke (on the topic of child development), Edward Diener (happiness), Julia Galef (rationality), Daniel Gilbert (social psychology), Joseph Henrich (culture), Steven Pinker (language, rationality), Stuart Ritchie (intelligence), and Scott Alexander (perception, clinical psychology). I know that some readers will want to know where to go to learn more about psychology—this paragraph is my answer.

Numerous experts were kind enough to provide comments on specific chapters, and I'm grateful to Arielle Baskin-Sommers, Amy Finn, Chaz Firestone, Rebecca Fortgang, Yoel Inbar, Michael Inzlicht, Peter Kramer, Stuart Ritchie, and Katherine Vasquez. No doubt there are still mistakes in this book; these likely correspond to exactly those points where I decided to ignore these experts' advice. And my biggest debt is to the four psychologists who read the whole manuscript and provided me with extensive and extremely helpful comments: Frank Keil, Gregory Murphy, Christina Starmans, and Matti Wilks.

As always, my agent and friend Katinka Matson provided wise counsel every step of the way. I started on *Psych* with Denise Oswald, a wonderful editor who worked with me on my last two books, and was sad when she left Ecco before I sent in my first draft. (Editors move around *a lot*.) But I was lucky to have Sarah Murphy take over. Sarah has been a delight to work with and has a strong background in psychological science, making her the perfect editor for this project. Also at Ecco, Norma Barksdale provided superb comments on the near-final version of this manuscript, and support in other ways as well.

I also thank Ashima Kaura for her careful work on the references,

Hendri Maulana for the wonderful drawings, and Jane Cavolina for her sharp eye as copyeditor.

Most of all, I thank Christina Starmans, who has given me more help on this book than everyone else combined. Christina is, among many other things, the best person in the world to talk ideas with. I look forward to many future discussions—I sent in the final version of this manuscript the day before we got married.

Notes

Prologue

1. John D. Barrow, *The Origin of the Universe: Science Masters Series* (New York: Basic Books, 1997).
2. To access the free course, go to https://www.coursera.org/learn/introduction-psychology.
3. John Updike, *Rabbit at Rest* (New York: Alfred A. Knopf, 1990), 206.

1. "Brain Makes Thought"

1. Malcolm Macmillan, "Phineas Gage—Unravelling the Myth," *Psychologist* 21 (2008): 828–31.
2. Oliver Sacks, "The Last Hippie," *New York Review of Books*, March 26, 1992, 53–62.
3. Francis Crick, *The Astonishing Hypothesis: The Scientific Search for the Soul* (New York: Scribner, 1994), 3.
4. Matthew Cobb, *The Idea of the Brain: The Past and Future of Neuroscience* (London: Profile Books, 2020), 104.
5. Paul Bloom, *Descartes' Baby: How the Science of Child Development Explains What Makes Us Human* (New York: Random House, 2005).
6. Owen J. Flanagan, *The Science of the Mind* (Cambridge, MA: MIT Press, 1991), 1.
7. Cited by Flanagan, *The Science of the Mind*, 2.
8. René Descartes, *Descartes: Philosophical Essays and Correspondence*, ed. Roger Ariew (Indianapolis, IN: Hackett Publishing Company, 2000), 61.
9. Bloom, *Descartes' Baby*. For a critique of this "intuitive dualism" theory, see Michael Barlev and Andrew Shtulman, "Minds, Bodies, Spirits, and Gods: Does Widespread Belief in Disembodied Beings Imply That We Are Inherent Dualists?," *Psychological Review* 128, no. 6 (2021): 1007–21.
10. Steven Pinker, *How the Mind Works* (New York: W. W. Norton, 1997), 64.

11. Cobb, *The Idea of the Brain*, 42.

12. Cobb, *The Idea of the Brain*, 43.

13. Pinker, *How the Mind Works*, 64.

14. Tomoyasu Horikawa et al., "Neural Decoding of Visual Imagery During Sleep," *Science* 340, no. 6132 (2013): 639–42.

15. For review, see Howard Robinson, "Dualism," Stanford Encyclopedia of Philosophy Archive (Fall 2020 Edition), ed. Edward N. Zalta, https://plato.stanford.edu /archives/fall2020/entries/dualism/.

16. Cobb, *The Idea of the Brain*, 20.

17. Bertossa Franco et al., "Point Zero: A Phenomenological Inquiry into the Subjective Physical Location of Consciousness," *Perceptual and Motor Skills* 107 (2008): 323–35.

18. Christina Starmans and Paul Bloom, "Windows to the Soul: Children and Adults See the Eyes as the Location of the Self," *Cognition* 123, no. 2 (2012): 313–18.

19. Cobb, *The Idea of the Brain*, 32.

20. Terry Bisson, "They're Made Out of Meat," *VOICES-NEW YORK* 39, no. 1 (2003): 66–68.

21. For accessible reviews of basic brain anatomy and physiology, see Diane Beck and Evelina Tapia, "The Brain," in *Noba Textbook Series: Psychology*, eds. Robert Biswas-Diener and Ed Diener (Champaign, IL: DEF Publishers, 2022), http:// noba.to/jx7268sd; Sharon Furtak, "Neurons," in *Noba Textbook Series: Psychology*, http://noba.to/s678why4; and Uta Frith, Chris Frith, and Alex Frith, *Two Heads: A Graphic Exploration of How Our Brains Work with Other Brains* (New York: Scribner, 2022).

22. Cobb, *The Idea of the Brain*.

23. Frederico A. C. Azevedo et al., "Equal Numbers of Neuronal and Nonneuronal Cells Make the Human Brain an Isometrically Scaled-Up Primate Brain," *Journal of Comparative Neurology* 513, no. 5 (2009): 532–41.

24. William James, *The Principles of Psychology* (New York: Henry Holt, 1908), 6–7.

25. Cited by Cobb, *The Idea of the Brain*, 110–11.

26. Andrew Hodges, "Alan Turing," Stanford Encyclopedia of Philosophy (Winter 2019 Edition), ed. Edward N. Zalta, https://plato.stanford.edu/archives/win2019 /entries/turing/.

27. Elizabeth F. Loftus, "Leading Questions and the Eyewitness Report," *Cognitive Psychology* 7, no. 4 (1975): 560–72.

28. Stephen R. Anderson and David W. Lightfoot, "The Human Language Faculty as an Organ," *Annual Review of Physiology* 62, no. 1 (2000): 697–722.

29. Cobb, *The Idea of the Brain*, 80.

30. Cobb, *The Idea of the Brain*, 40.

31. Oliver Sacks, *The Man Who Mistook His Wife for a Hat and Other Clinical Tales* (Toronto: Vintage Canada, 2021).

32. Stanislas Dehaene et al., "How Learning to Read Changes the Cortical Networks for Vision and Language," *Science* 330, no. 6009 (2010): 1359–64.

33. Iftah Biran and Anjan Chatterjee, "Alien Hand Syndrome," *Archives of Neurology* 61, no. 2 (2004): 292–94.
34. Edward H. F. de Haan et al., "Split-Brain: What We Know Now and Why This Is Important for Understanding Consciousness," *Neuropsychology Review* 30, no. 2 (2020): 224–33.
35. Jerry Fodor, "Diary: Why the Brain?," *London Review of Books* 21, no. 19 (1999): 68–69.
36. Cobb, *The Idea of the Brain*, 236.
37. Naomi I. Eisenberger, Matthew D. Lieberman, and Kipling D. Williams, "Does Rejection Hurt? An fMRI Study of Social Exclusion," *Science* 302, no. 5643 (2003): 290–92.
38. C. Nathan DeWall et al., "Acetaminophen Reduces Social Pain: Behavioral and Neural Evidence," *Psychological Science* 21, no. 7 (2010): 931–37.
39. David Chalmers, "The Hard Problem of Consciousness," in *The Blackwell Companion to Consciousness*, ed. Max Velmans, Susan Schneider, and Jeffrey Gray (Oxford, Blackwell Publishers, 2007): 225–35.

2. Consciousness

1. F. Gooding, "From Its Myriad Tips," *London Review of Books* 43, no. 10 (2021): 68–69.
2. Jeremy Bentham, *The Collected Works of Jeremy Bentham: An Introduction to the Principles of Morals and Legislation* (London: Clarendon Press, 1996).
3. John Locke, *An Essay Concerning Human Understanding*, ed. Paul H. Nidditch (Oxford: Clarendon Press, 1988), 389. For discussion, see Sydney Shoemaker, "The Inverted Spectrum," *Journal of Philosophy* 79, no. 7 (1982): 357–81.
4. Marcel Proust, *Swann's Way*, ed. Paul Negri (Newton Abbot, UK: David & Charles, 2012), 72.
5. Thomas Nagel, "What Is It Like to Be a Bat?," *Philosophical Review* 83, no. 4 (1974): 435–50.
6. William James, *The Principles of Psychology* (New York: Henry Holt, 1908), 488.
7. Alison Gopnik, *The Philosophical Baby: What Children's Minds Tell Us About Truth, Love, and the Meaning of Life* (New York: Random House, 2009).
8. For discussion, see Paul Bloom, "What's Inside a Big Baby Head?," *Slate*, August 9, 2009, https://slate.com/culture/2009/08/alison-gopnik-s-the-philosophical-baby.html.
9. There is a nice exploration of this problem in the science fiction book by Andy Weir, *Project Hail Mary: A Novel* (New York: Ballantine Books, 2021).
10. For discussion, see Jason Schukraft, "Does Critical Flicker-Fusion Frequency Track the Subjective Experience of Time?," *Rethinking Priorities*, August 3, 2020, https://rethinkpriorities.org/publications/does-critical-flicker-fusion-frequency-track-the-subjective-experience-of-time.
11. Frank Tong et al., "Binocular Rivalry and Visual Awareness in Human Extrastriate Cortex," *Neuron* 21, no. 4 (1998): 753–59.

12. Adrian M. Owen et al., "Detecting Awareness in the Vegetative State," *Science* 313, no. 5792 (2006): 1402.

13. James Somers, "The Science of Mind Reading," *New Yorker*, December 6, 2021.

14. Bernard J. Baars, "In the Theatre of Consciousness. Global Workspace Theory, a Rigorous Scientific Theory of Consciousness," *Journal of Consciousness Studies* 4, no. 4 (1997): 292–309.

15. Lucia Melloni, Liad Mudrik, Michael Pitts, and Christof Koch, "Making the Hard Problem of Consciousness Easier," *Science* 372, no. 6545 (2021): 911–12.

16. For an accessible review, see Paul M. Churchland, *Matter and Consciousness* (Cambridge, MA: MIT Press, 2013).

17. John R. Searle, "Minds, Brains, and Programs," *Behavioral and Brain Sciences* 3, no. 3 (1980): 417–24.

18. Eric Mandelbaum, "Everything and More: The Prospects of Whole Brain Emulation," *Journal of Philosophy*, in press.

19. David Chalmers, "The Hard Problem of Consciousness," in *The Blackwell Companion to Consciousness*, ed. Max Velmans, Susan Schneider, and Jeffrey Gray (Oxford, Blackwell Publishers, 2007): 225–35, 203.

20. Ned Block, "On a Confusion About a Function of Consciousness," *Behavioral and Brain Sciences* 18, no. 2 (1995): 227–47.

21. Edgar Rubin, *Synsoplevede Figurer: Studier i Psykologisk Analyse* (Copenhagen: Gyldendal, Nordisk Forlag, 1915).

22. Barry Arons, "A Review of the Cocktail Party Effect," *Journal of the American Voice I/O Society* 12, no. 7 (1992): 35–50.

23. Stevan L. Nielsen and Irwin G. Sarason, "Emotion, Personality, and Selective Attention," *Journal of Personality and Social Psychology* 41, no. 5 (1981): 945–60.

24. Daniel J. Simons and Christopher F. Chabris, "Gorillas in Our Midst: Sustained Inattentional Blindness for Dynamic Events," *Perception* 28, no. 9 (1999): 1059–74.

25. David L. Strayer, Frank A. Drews, and William A. Johnston, "Cell Phone-Induced Failures of Visual Attention During Simulated Driving," *Journal of Experimental Psychology: Applied* 9, no. 1 (2003): 23–32; D. A. Redelmeier and R. J. Tibshirani, "Association Between Cellular-Telephone Calls and Motor Vehicle Collisions," *New England Journal of Medicine* 336 (1997): 453–58.

26. Edward Gilman Slingerland, *Trying Not to Try: Ancient China, Modern Science, and the Power of Spontaneity* (New York: Broadway Books, 2014).

27. James, *The Principles of Psychology*, 122.

28. Daniel H. Pink, *When: The Scientific Secrets of Perfect Timing* (New York: Riverhead Books, 2019).

29. Maria Konnikova, *Mastermind: How to Think Like Sherlock Holmes*, epigraph (New York: Viking Penguin, 2013).

30. Jerry Fodor, *The Modularity of Mind* (Cambridge, MA: MIT Press, 1983).

31. J. Ridley Stroop, "Studies of Interference in Serial Verbal Reactions," *Journal of Experimental Psychology* 18, no. 6 (1935): 643.

32. Matthew A. Killingsworth and Daniel T. Gilbert, "A Wandering Mind Is an Un-

happy Mind," *Science* 330, no. 6006 (2010): 932.

33. Thomas Gilovich, Victoria Husted Medvec, and Kenneth Savitsky, "The Spotlight Effect in Social Judgment: An Egocentric Bias in Estimates of the Salience of One's Own Actions and Appearance," *Journal of Personality and Social Psychology* 78, no. 2 (2000): 211–22.

34. Thomas Gilovich and Victoria Husted Medvec, "The Experience of Regret: What, When, and Why," *Psychological Review* 102, no. 2 (1995): 379–95.

35. Gustave Le Bon, *The Crowd: A Study of the Popular Mind* (Mineola, NY: Dover Publications, 2002), 8.

36. Ruud Hortensius and Beatrice De Gelder, "From Empathy to Apathy: The Bystander Effect Revisited," *Current Directions in Psychological Science* 27, no. 4 (2018): 249–56.

37. Edward Slingerland, *Drunk: How We Sipped, Danced, and Stumbled Our Way to Civilization* (New York: Little, Brown Spark, 2021), 290.

38. Roy F. Baumeister, "Masochism as Escape from Self," *Journal of Sex Research* 25, no. 1 (1988): 28–59.

39. Pat Califia, "Doing It Together: Gay Men, Lesbians and Sex," *Advocate* 7 (1983): 24–27. For more discussion, see Paul Bloom, *The Sweet Spot: The Pleasures of Suffering and the Search for Meaning* (New York: Ecco/HarperCollins, 2021).

40. Robert Trivers, *The Folly of Fools: The Logic of Deceit and Self-Deception in Human Life* (New York: Basic Books, 2011).

41. See Steven Pinker, *How the Mind Works* (New York: W. W. Norton, 1997).

42. Rory Sutherland, *Alchemy: The Surprising Power of Ideas That Don't Make Sense* (New York: Random House, 2019).

43. Sutherland, *Alchemy*, 17.

44. Heidi Moawad, MD, "Ondine's Curse: Causes, Symptoms, and Treatment," Neurology Live, April 17, 2018, https://www.neurologylive.com/view/ondines-curse -causes-symptoms-and-treatment.

3. Freud and the Unconscious

1. Peter D. Kramer, *Freud: Inventor of the Modern Mind* (New York: HarperCollins, 2009), 2. Much of what follows is based on this excellent book.

2. Kramer, *Freud*, 1.

3. Kramer, *Freud*, 2.

4. George Prochnik, "The Curious Conundrum of Freud's Persistent Influence," *New York Times*, August 14, 2017. The book is Frederick Crews, *Freud: The Making of an Illusion* (New York: Henry Holt, 2017).

5. Kramer, *Freud*, 52.

6. Sigmund Freud, *Dora: An Analysis of a Case of Hysteria*, ed. Philip Rieff (New York: Touchstone Books, 1963), 69.

7. Sigmund Freud and James Strachey, *The Interpretation of Dreams*, vol. 4 (New York: Gramercy Books, 1996).

8. Cara C. MacInnis and Gordon Hodson, "Is Homophobia Associated with an Implicit Same-Sex Attraction?," *Journal of Sex Research* 50, no. 8 (2013): 777–85.

9. Karl Popper, *The Logic of Scientific Discovery* (New York: Basic Books, 1959).

10. For discussion, see Michael D. Gordon, "The Quest to Tell Science from Pseudoscience," *Boston Review*, March 23, 2021.

11. Sigmund Freud, *The Interpretation of Dreams*, ed. Abraham Brill (New York: Macmillan, 1918), 249.

12. Frederick Crews, "The Verdict on Freud," *Psychological Science* 7, no. 2 (1996): 63–68.

13. Seth Stephens-Davidowitz, *Everybody Lies: Big Data, New Data, and What the Internet Can Tell Us About Who We Really Are* (New York: HarperCollins, 2017).

14. Joseph Heath, "Absent-Mindedness as Dominance Behaviour," *In Due Course* (Blog), September 5, 2017, http://induecourse.ca/absent-mindedness-as-dominance-behaviour/.

15. Sigmund Freud, *The Basic Writings of Sigmund Freud*, ed. and trans. Abraham Brill (New York: Modern Library), 109.

16. Prochnik, "The Curious Conundrum of Freud's Persistent Influence."

4. The Skinnerian Revolution

1. John Watson, *Behaviorism*, ed. Gregory Kimble (New Brunswick, NJ: Transaction Publishers, 1998), 82.

2. Cited by Duane P. Schultz and Sydney Ellen Schultz, *A History of Modern Psychology* (Boston: Cengage Learning, 2015).

3. Michael Specter, "Drool," *New Yorker*, November 24, 2014.

4. Virginia Woolf, *Mrs. Dalloway* (New York: Warbler Classics, 2020), 1.

5. John B. Watson, "Psychology as the Behaviorist Views It," *Psychological Review* 20, no. 2 (1913): 158–77.

6. Geir Overskeid, "Looking for Skinner and Finding Freud," *American Psychologist* 62, no. 6 (2007): 590–95.

7. Overskeid, "Looking for Skinner and Finding Freud."

8. Specter, "Drool."

9. John B. Watson and Rosalie Rayner, "Conditioned Emotional Reactions," *Journal of Experimental Psychology* 3, no. 1 (1920): 1–14.

10. Hall P. Beck, Sharman Levinson, and Gary Irons, "The Evidence Supports Douglas Merritte as Little Albert," *American Psychologist* (2010): 301–3. For a critical discussion, see Russell A. Powell, "Little Albert Still Missing," *American Psychologist* (2010): 299–300.

11. Laura Harcourt, "Creating the Coffee Break: Marketing and the Manipulation of Demand," Isenberg Marketing, January 28, 2016, https://isenbergmarketing.wordpress.com/2016/01/28/creating-the-coffee-break-marketing-and-the-manipulation-of-demand/. But there are other origin stories as well: see "Who Invented the Coffee Break," *Coffee for Less* (Blog), December 18, 2017, https://www.coffeeforless.com/blogs/coffee-for-less-blog/who-invented-the-coffee-break.

12. Joseph Wolpe et al., "The Current Status of Systematic Desensitization," *American Journal of Psychiatry* 130, no. 9 (1973): 961–65.

13. Hidehiro Watanabe and Makoto Mizunami, "Pavlov's Cockroach: Classical Conditioning of Salivation in an Insect," *PLoS One* 2, no. 6 (2007): e529.

14. Edward Lee Thorndike, *The Fundamentals of Learning* (New York: AMS Press, 1971).

15. Specter, "Drool."

16. B. F. Skinner, "Selection by Consequences," *Behavioral and Brain Sciences* 7, no. 4 (1984): 477–81.

17. B. F. Skinner, "'Superstition' in the Pigeon," *Journal of Experimental Psychology* 38, no. 2 (1948): 168–72.

18. Keller Breland and Marian Breland, "The Misbehavior of Organisms," *American Psychologist* 16, no. 11 (1961): 681–84.

19. John Garcia, Donald J. Kimeldorf, and Robert A. Koelling, "Conditioned Aversion to Saccharin Resulting from Exposure to Gamma Radiation," *Science* 122, no. 3160 (1955): 157–58.

20. Edward C. Tolman, "Cognitive Maps in Rats and Men," *Psychological Review* 55, no. 4 (1948): 189–208.

21. For review, see Lynn Nadel and Howard Eichenbaum, "Introduction to the Special Issue on Place Cells," *Hippocampus* 9, no. 4 (1999): 341–45.

22. B. F. Skinner, *Verbal Behavior* (Acton, MA: Echo Points Book & Media, 2008), 456.

23. Noam Avram Chomsky, "A Review of Skinner's Verbal Behavior," *Language* 35, no. 1 (1959): 26–58.

24. Kenneth MacCorquodale, "On Chomsky's Review of Skinner's Verbal Behavior," *Journal of the Experimental Analysis of Behavior* 13, no. 1 (1970): 83–99.

25. Chomsky, "A Review of Skinner's Verbal Behavior."

26. Chomsky, "A Review of Skinner's Verbal Behavior," 26–58.

27. Skinner, *Verbal Behavior*, 458.

5. Piaget's Project

1. Avshalom Caspi, "The Child Is Father of the Man: Personality Continuities from Childhood to Adulthood," *Journal of Personality and Social Psychology* 78, no. 1 (2000): 158–72.

2. William James, *The Principles of Psychology* (New York: Henry Holt, 1908), 488.

3. Cited by Philippe Rochat, *The Infant's World* (Cambridge, MA: Harvard University Press, 2009).

4. John Locke, *An Essay Concerning Human Understanding* (London: Tegg and Son, 1836), 51.

5. James J. Gross et al., *Interactive Psychology: People in Perspective* (New York: W. W. Norton, 2020), study unit 11.1.

6. Susan Curtiss, *Genie: A Psycholinguistic Study of a Modern-Day Wild Child* (Cambridge, MA: Academic Press, 2014).

7. David Eagleman, *Livewired: The Inside Story of the Ever-Changing Brain* (Toronto: Doubleday Canada, 2020), 26.

8. Gross et al., *Interactive Psychology*, study unit 11.6.

9. W. E. Dixon, *Twenty Studies That Revolutionized Child Psychology* (Saddle River, NJ: Pearson, 2015), 13.

10. Cited by Richard Kohler, *Jean Piaget* (London: Bloomsbury Publishing, 2014), 72.

11. George Berkeley, *A Treatise Concerning the Principles of Human Knowledge* (Philadelphia: J. B. Lippincott, 1881).

12. Rheta De Vries, "Constancy of Generic Identity in the Years Three to Six," *Monographs of the Society for Research in Child Development* 34, no. 3 (1969): iii–67.

13. Jeffrey Jenson Arnett, *Adolescence and Emerging Adulthood: A Cultural Approach* (Upper Saddle River, NJ: Prentice Hall, 2010), 89.

14. J. Roy Hopkins, "The Enduring Influence of Jean Piaget," *Psychology Today*, December 1, 2011, https://www.psychologicalscience.org/observer/jean-piaget.

15. Dixon, *Twenty Studies That Revolutionized Child Psychology*, 12.

16. For a review, see Elizabeth S. Spelke, *What Babies Know: Core Knowledge and Composition*, vol. 1 (Oxford: Oxford University Press, 2022).

17. Koleen McCrink and Karen Wynn, "Large-Number Addition and Subtraction by 9-Month-Old Infants," *Psychological Science* 15, no. 11 (2004): 776–81. For an opposing view, see Sami R. Yousif and Frank C. Keil, "Area, Not Number, Dominates Estimates of Visual Quantities," *Scientific Reports* 10, no. 1 (2020): 1–13.

18. R. Baillargeon, E. S. Spelke, and S. Wasserman, "Object Permanence in Five-Month-Old Infants," *Cognition* 20, no. 3 (1985): 191–208.

19. Karen Wynn, "Addition and Subtraction by Human Infants," *Nature* 358, no. 6389 (1992): 749–50.

20. Elizabeth Spelke, "Initial Knowledge: Six Suggestions," *Cognition* 50, no. 1–3 (1994): 431–45.

21. This research is summarized in Susan A. Gelman, *The Essential Child: Origins of Essentialism in Everyday Thought* (New York: Oxford University Press, 2003).

22. Sheila J. Walker, "Culture, Domain Specificity and Conceptual Change: Natural Kind and Artifact Concepts," *British Journal of Developmental Psychology* 17, no. 2 (1999): 203–19.

23. For review of these studies, see Susan A. Gelman, "Learning from Others: Children's Construction of Concepts," *Annual Review of Psychology* 60 (2009): 115–40.

24. Frank Keil, *Concepts, Kinds, and Cognitive Development* (Cambridge, MA: MIT Press, 1989).

25. Meredith Meyer et al., "My Heart Made Me Do It: Children's Essentialist Beliefs About Heart Transplants," *Cognitive Science* 41, no. 6 (2017): 1694–1712.

26. Mark H. Johnson et al., "Newborns' Preferential Tracking of Face-Like Stimuli and Its Subsequent Decline," *Cognition* 40, no. 1–2 (1991): 1–19.

27. Teresa Farroni et al., "Eye Contact Detection in Humans from Birth," *Proceedings of the National Academy of Sciences* 99, no. 14 (2002): 9602–5.

28. Lauren B. Adamson and Janet E. Frick, "The Still Face: A History of a Shared Experimental Paradigm," *Infancy* 4, no. 4 (2003): 451–73.

29. Amanda L. Woodward, "Infants Selectively Encode the Goal Object of an Actor's Reach," *Cognition* 69, no. 1 (1998): 1–34. See also G. Gergely and G. Csibra, "Teleological Reasoning in Infancy: The Naïve Theory of Rational Action," *Trends in Cognitive Sciences* 7, no. 7 (2003): 287–92.

30. Valerie Kuhlmeier, Karen Wynn, and Paul Bloom, "Attribution of Dispositional States by 12-Month-Olds," *Psychological Science* 14, no. 5 (2003): 402–8.

31. J. Kiley Hamlin, Karen Wynn, and Paul Bloom, "Social Evaluation by Preverbal Infants," *Nature* 450, no. 7169 (2007): 557–59.

32. Michael Tomasello, "Uniquely Primate, Uniquely Human," *Developmental Science* 1, no. 1 (1998): 1–16.

33. All examples from Karen Bartsch and Henry M. Wellman, *Children Talk About the Mind* (Oxford: Oxford University Press, 1995).

34. The first "false belief" experiment was done by Heinz Wimmer and Josef Perner, "Beliefs About Beliefs: Representation and Constraining Function of Wrong Beliefs in Young Children's Understanding of Deception," *Cognition* 13, no. 1 (1983): 103–28. The specific scenario described here is from Simon Baron-Cohen, Alan M. Leslie, and Uta Frith, "Does the Autistic Child Have a 'Theory of Mind'?," *Cognition* 21, no. 1 (1985): 37–46.

35. David Premack and Guy Woodruff, "Does the Chimpanzee Have a Theory of Mind?," *Behavioral and Brain Sciences* 1, no. 4 (1978): 515–26.

36. For example, C. Krachun, J. Call, and M. Tomasello, "A New Change-of-Contents False Belief Test: Children and Chimpanzees Compared," *International Journal of Comparative Psychology* 23, no. 2 (2010): 145–65.

37. Kristine H. Onishi and Renée Baillargeon, "Do 15-Month-Old Infants Understand False Beliefs?," *Science* 308, no. 5719 (2005): 255–58.

38. Paul Bloom and Tamsin German, "Two Reasons to Abandon the False Belief Task as a Test of Theory of Mind," *Cognition* 77, no. 1 (2000): B25–B31.

39. Susan A. J. Birch and Paul Bloom, "The Curse of Knowledge in Reasoning About False Beliefs," *Psychological Science* 18, no. 5 (2007): 382–86.

40. Birch and Bloom, "The Curse of Knowledge in Reasoning About False Beliefs."

41. Rebecca Saxe, "The New Puzzle of Theory of Mind Development," in *Navigating the Social World: What Infants, Children, and Other Species Can Teach Us*, eds. Mahzarin Banaji and Susan Gelman (New York: Oxford University Press, 2013), 107–12.

42. Susan Carey, *Conceptual Change in Childhood* (Cambridge, MA: MIT Press, 1985); Susan Carey, *The Origin of Concepts* (New York: Oxford University Press, 2009). See also Annette Karmiloff-Smith and Bärbel Inhelder, "If You Want to Get Ahead, Get a Theory," *Cognition* 3, no. 3 (1974): 195–212.

43. Thomas S. Kuhn, *The Structure of Scientific Revolutions* (Chicago: University of Chicago Press, 2021).

44. Alison Gopnik, "Explanation as Orgasm," *Minds and Machines* 8, no. 1 (1998): 101–18.

45. A. E. Stahl and L. Feigenson, "Observing the Unexpected Enhances Infants' Learning and Exploration," *Science* 348, no. 6230 (2015): 91–94.

6. The Ape That Speaks

1. David Crystal, *How Language Works: How Babies Babble, Words Change Meaning, and Languages Live or Die* (New York: Avery/Penguin, 2007), 1.

2. For discussion, see Steven Pinker, *The Language Instinct: How the Mind Creates Language* (New York: HarperCollins, 1994).

3. Charles Darwin, *The Descent of Man, and Selection in Relation to Sex* (New York: Appleton and Company, 1871), 53.

4. Dorothy V. M. Bishop, "What Causes Specific Language Impairment in Children?," *Current Directions in Psychological Science* 15, no. 5 (2006): 217–21.

5. Steven Pinker and Paul Bloom, "Natural Language and Natural Selection," *Behavioral and Brain Sciences* 13, no. 4 (1990): 707–27.

6. Philip Lieberman, "The Evolution of Human Speech: Its Anatomical and Neural Bases," *Current Anthropology* 48, no. 1 (2007): 39–66.

7. Much of the discussion here draws on Pinker, *The Language Instinct*.

8. Crystal, *How Language Works*.

9. Paul Bloom, *How Children Learn the Meanings of Words* (Cambridge, MA: MIT Press, 2000).

10. Paul Bloom, "Myths of Word Learning," in *Weaving a Lexicon*, eds. Geoffrey Hall and Sandra Waxman (Cambridge, MA: MIT Press, 2004), 205–24.

11. Pinker, *The Language Instinct*, 120.

12. "Do You Speak Corona? A Guide to COVID-19 Slang," *Economist*, April 8, 2020, https://www.economist.com/1843/2020/04/08/do-you-speak-corona-a-guide-to-COVID-19-slang.

13. Youka Nagase, "There's a Japanese Word for Drinking Online with Friends: On-Nomi," *Time Out*, March 25, 2020, https://www.timeout.com/tokyo/news/theres-a-japanese-word-for-drinking-online-with-friends-on-nomi-032620.

14. Noam Chomsky, *Syntactic Structures* (The Hague/Paris: Mouton de Gruyter, 1957), 15.

15. Pinker, *The Language Instinct*, 77.

16. Rachel L. Harris and T. Tarchak, "Showers and Pants Are So 2019," *New York Times*, September 11, 2020, https://www.nytimes.com/2020/09/21/opinion/coronavirus-six-word-memoirs.html.

17. For example, Paul Smolensky, "On the Proper Treatment of Connectionism," *Behavioral and Brain Sciences* 11, no. 1 (1988): 1–74.

18. For example, Adam H. Marblestone, Greg Wayne, and Konrad P. Kording, "Toward an Integration of Deep Learning and Neuroscience," *Frontiers in Computational Neuroscience* (2016): 1–41.

19. Steven Pinker, *Words and Rules: The Ingredients of Language* (New York: Basic Books, 2015).

20. Steven Pinker and Alan Prince, "The Nature of Human Concepts: Evidence from an Unusual Source," in *The Nature of Concepts*, ed. Philip Van Loocke (London and New York: Routledge, 1999), 20–63.

21. See, for example, Herbert S. Terrace, *Why Chimpanzees Can't Learn Language and Only Humans Can* (New York: Columbia University Press, 2019).

22. Juliane Kaminski, Josep Call, and Julia Fischer, "Word Learning in a Domestic Dog: Evidence for 'Fast Mapping,'" *Science* 304, no. 5677 (2004): 1682–83; for commentary, see Paul Bloom, "Can a Dog Learn a Word?," *Science* 304, no. 5677 (2004): 1605–6.

23. Paul Bloom and Lori Markson, "Capacities Underlying Word Learning," *Trends in Cognitive Sciences* 2, no. 2 (1998): 67–73.

24. John Macnamara, *Names for Things: A Study of Human Learning* (Cambridge, MA: MIT Press, 1982); Bloom, *How Children Learn the Meanings of Words*.

25. Anthony J. DeCasper and William P. Fifer, "Of Human Bonding: Newborns Prefer Their Mothers' Voices," *Science* 208, no. 4448 (1980): 1174–76.

26. Jacques Mehler et al., "A Precursor of Language Acquisition in Young Infants," *Cognition* 29, no. 2 (1988): 143–78.

27. Christine Moon, Robin Panneton Cooper, and William P. Fifer, "Two-Day-Olds Prefer Their Native Language," *Infant Behavior and Development* 16, no. 4 (1993): 495–500.

28. Janet F. Werker et al., "Developmental Aspects of Cross-Language Speech Perception," *Child Development* (1981): 349–55.

29. Laura Ann Petitto and Paula F. Marentette, "Babbling in the Manual Mode: Evidence for the Ontogeny of Language," *Science* 251, no. 5000 (1991): 1493–96.

30. Elika Bergelson and Daniel Swingley, "At 6–9 Months, Human Infants Know the Meanings of Many Common Nouns," *Proceedings of the National Academy of Sciences* 109, no. 9 (2012): 3253–58.

31. Bloom, *How Children Learn the Meanings of Words*.

32. Michael C. Frank et al., *Variability and Consistency in Early Language Learning: The Wordbank Project* (Cambridge, MA: MIT Press, 2021), 57–58.

33. Kathy Hirsh-Pasek and Roberta Michnick Golinkoff, "The Intermodal Preferential Looking Paradigm: A Window onto Emerging Language Comprehension," in *Methods for Assessing Children's Syntax*, ed. Dana McDaniel, Cecile McKee, and Helen Smith Cairns (Cambridge, MA: MIT Press, 1996), 105–24.

34. Yael Gertner, Cynthia Fisher, and Julie Eisengart, "Learning Words and Rules: Abstract Knowledge of Word Order in Early Sentence Comprehension," *Psychological Science* 17, no. 8 (2006): 684–91.

35. Pinker, *The Language Instinct*, 271.

36. Jean Berko, "The Child's Learning of English Morphology," *Word* 14, no. 2–3 (1958): 150–77.

37. Pinker, *Words and Rules*.

38. Elissa L. Newport, Daphne Bavelier, and Helen J. Neville, "Critical Thinking About Critical Periods: Perspectives on a Critical Period for Language Acquisition," in *Language, Brain and Cognitive Development: Essays in Honor of Jacques Mehler*, ed. Emmanuel Dupoux (Cambridge, MA: MIT Press, 2001), 481–502.

39. Joshua K. Hartshorne, Joshua B. Tenenbaum, and Steven Pinker, "A Critical Period for Second Language Acquisition: Evidence from 2/3 Million English Speakers," *Cognition* 177 (2018): 263–77.

40. Bishop, "What Causes Specific Language Impairment in Children?"

41. Alejandrina Cristia et al., "Child-Directed Speech Is Infrequent in a Forager-Farmer Population: A Time Allocation Study," *Child Development* 90, no. 3 (2019): 759–73. For a popular treatment, see Dana G. Smith, "Parents in a Remote Amazon Village Barely Talk to Their Babies—and the Kids Are Fine," *Scientific American*, December 5, 2017, https://www.scientificamerican.com/article/parents-in-a-remote-amazon-village-barely-talk-to-their-babies-mdash-and-the-kids-are-fine/.

42. Noam Chomsky, *On Language* (New York: New Press, 1998), 9–10.

43. Jenny R. Saffran, Richard N. Aslin, and Elissa L. Newport, "Statistical Learning by 8-Month-Old Infants," *Science* 274, no. 5294 (1996): 1926–28.

44. Bloom, *How Children Learn the Meanings of Words*.

45. John Locke, *An Essay Concerning Human Understanding* (Cleveland, OH: Meridian Books, 1964), 108.

46. For discussion, see Bloom, *How Children Learn the Meanings of Words*.

47. Willard Van Orman Quine, *Word and Object* (Cambridge, MA: MIT Press, 1960), 25.

48. For instance, Gregory L. Murphy and Edward E. Smith, "Basic-Level Superiority in Picture Categorization," *Journal of Verbal Learning and Verbal Behavior* 21, no. 1 (1982): 1–20.

49. Roger W. Brown, "Linguistic Determinism and the Part of Speech," *Journal of Abnormal and Social Psychology* 55, no. 1 (1957): 1–5.

50. Nancy Katz, Erica Baker, and John Macnamara, "What's in a Name? A Study of How Children Learn Common and Proper Names," *Child Development* (1974): 469–73.

51. Lila Gleitman, "The Structural Sources of Verb Meanings," *Language Acquisition* 1, no. 1 (1990): 3–55.

52. Helen Keller, *The Story of My Life*, vol. 1 (Alexandria, Egypt: Library of Alexandria, 2004).

53. Leonard Talmy, "Fictive Motion in Language and 'Ception,'" *Language and Space* 21 (1996): 1–276.

54. John McWhorter, *Nine Nasty Words: English in the Gutter: Then, Now, and Forever* (New York: Avery/Penguin Books, 2021).

55. George Lakoff and Mark Johnson, "Conceptual Metaphor in Everyday Language," *Journal of Philosophy* 77, no. 8 (1980), 453–86.

56. Noam Avram Chomsky, *Language and Mind* (Cambridge, UK: Cambridge University Press, 2006), xv.

57. Peter Grice, "Logic and Conversation," in *Speech Acts*, eds. Peter Cole and Jerry Morgan (Leiden: Brill, 1975), 41–58.

58. H. H. Clark and D. H. Schunk, "Polite Responses to Polite Requests," *Cognition* 8 (1980): 111–43.

59. Benjamin Lee Whorf, *Language, Thought, and Reality: Selected Writings of Benjamin Lee Whorf*, ed. John Caroll (Cambridge, MA: MIT Press, 1956), 212. For a critical discussion, see Steven Pinker, *The Stuff of Thought: Language as a Window into Human Nature* (New York: Viking Penguin, 2007).

60. Gregory L. Murphy, "On Metaphoric Representation," *Cognition* 60, no. 2 (1996): 173–204.

61. For review and discussion, see Pinker, *The Stuff of Thought*; Gary Lupyan et al., "Effects of Language on Visual Perception," *Trends in Cognitive Sciences* 24, no. 11 (2020): 930–44; John H. McWhorter, *The Language Hoax: Why the World Looks the Same in Any Language* (New York: Oxford University Press, 2014).

62. Jonathan Winawer et al., "Russian Blues Reveal Effects of Language on Color Discrimination," *Proceedings of the National Academy of Sciences* 104, no. 19 (2007): 7780–85.

63. Winawer et al., "Russian Blues Reveal Effects of Language on Color Discrimination."

64. McWhorter, *The Language Hoax*, 149.

65. Stanislas Dehaene, *The Number Sense: How the Mind Creates Mathematics* (New York: Oxford University Press, 1997), 91–92.

66. Karen Wynn, "Addition and Subtraction by Human Infants," *Nature* 358, no. 6389 (1992): 749–50.

67. For discussion, see Elizabeth S. Spelke, *What Babies Know: Core Knowledge and Composition*, vol. 1 (Oxford: Oxford University Press, 2022).

68. Jill G. de Villiers and Peter A. de Villiers, "The Role of Language in Theory of Mind Development," *Topics in Language Disorders* 34, no. 4 (2014): 313–28.

69. Jennie E. Pyers and Ann Senghas, "Language Promotes False-Belief Understanding: Evidence from Learners of a New Sign Language," *Psychological Science* 20, no. 7 (2009): 805–12.

70. De Villiers and de Villiers, "The Role of Language in Theory of Mind Development."

71. Daniel C. Dennett, *Kinds of Minds: Toward an Understanding of Consciousness* (New York: Basic Books, 2008), 17.

7. The World in Your Head

1. William James, *The Principles of Psychology* (New York: Henry Holt, 1908), 403.

2. George Berkeley, *A Treatise Concerning the Principles of Human Knowledge* (Philadelphia: J. B. Lippincott & Company, 1881).

3. John Horgan, "Do Our Questions Create the World," *Scientific American* (Blog), June 6, 2018, https://blogs.scientificamerican.com/cross-check/do-our-questions -create-the-world/.

4. James Boswell, *The Life of Samuel Johnson*, ed. Christopher Hibbert (New York: Penguin Books, 1986), 102.

5. Thomas Reid, *Essays on the Intellectual Powers of Man* (New York: Cambridge University Press, 2011), 45.

6. Cited by Steven Pinker, *Rationality: What It Is, Why It Seems Scarce, Why It Matters* (New York: Viking Penguin, 2021), 298.

7. For example, Stacey Aston et al., "Exploring the Determinants of Color Perception Using #Thedress and Its Variants: The Role of Spatio-Chromatic Context, Chromatic Illumination, and Material–Light Interaction," *Perception* 49, no. 11 (2020): 1235–51; Karl R. Gegenfurtner, Marina Bloj, and Matteo Toscani, "The Many Colours of 'the Dress,'" *Current Biology* 25, no. 13 (2015): R543–R544.

8. Jerry A. Fodor, *The Modularity of Mind* (Cambridge, MA: MIT Press, 1983), 54.

9. Stanislas Dehaene, *Consciousness and the Brain: Deciphering How the Brain Codes Our Thoughts* (New York: Viking Penguin, 2014), 60.

10. Gösta Ekman, "Weber's Law and Related Functions," *Journal of Psychology* 47, no. 2 (1959): 343–52.

11. Steven Barthelme and Frederick Barthelme, *Double Down: Reflections on Gambling and Loss* (Boston: Houghton Mifflin Harcourt, 2001), 25.

12. James J. Gross et al., *Interactive Psychology: People in Perspective* (New York: W. W. Norton, 2020), study unit 4.1.

13. Reed Johnson, "The Mystery of S., the Man with an Impossible Memory," *New Yorker*, August 12, 2017.

14. Annu Singh, "Computer Vision: From a Summer Intern Project to Redefining AI Future," Medium, August 6, 2019, https://medium.com/analytics-vidhya/computer -vision-from-a-summer-intern-project-to-redefining-ai-future-5dc87fdc9f72.

15. Leda Cosmides and John Tooby, "Beyond Intuition and Instinct Blindness: Toward an Evolutionarily Rigorous Cognitive Science," *Cognition* 50, no. 1–3 (1994): 41–77.

16. Kurt Koffka, *Principles of Gestalt Psychology* (Abingdon, Oxon, UK: Routledge, 2013).

17. Philip J. Kellman and Elizabeth S. Spelke, "Perception of Partly Occluded Objects in Infancy," *Cognitive Psychology* 15, no. 4 (1983): 483–524.

18. Gaetano Kanizsa, "Margini Quasi-Percettivi in Campi con Stimolazione Omogenea," *Rivista di Psicologia* 49, no. 1 (1955): 7–30.

19. Richard M. Warren, "Perceptual Restoration of Missing Speech Sounds," *Science* 167, no. 3917 (1970): 392–93.

20. Scott Alexander, "Mysticism and Pattern Matching," Slate Star Codex, August 8, 2015, https://slatestarcodex.com/2015/08/28/mysticism-and-pattern-matching/.

21. For discussion, see C. Firestone and B. J. Scholl, "Cognition Does Not Affect Perception: Evaluating the Evidence for 'Top-Down' Effects," *Behavioral and Brain Sciences* 39 (2016): 1–77.

22. John Locke, *An Essay Concerning Human Understanding*, ed. Paul H. Nidditch (Oxford: Clarendon Press, 1979).

23. For discussion, see Christina Starmans and Paul Bloom, "Nothing Personal: What Psychologists Get Wrong About Identity," *Trends in Cognitive Sciences* 22, no. 7 (2018): 566–68; M. Finlay and C. Starmans, "Not the Same Same: Distinguishing Between Similarity and Identity in Judgments of Change," *Cognition* 218 (2022): 104953.

24. Richard C. Atkinson and Richard M. Shiffrin, "Human Memory: A Proposed System and Its Control Processes," in *The Psychology of Learning and Motivation*, vol. 2, eds. Kenneth Spence and Janet Taylor Spence (Cambridge, MA: Academic Press, 1968), 89–195.

25. Fergus I. M. Craik and Robert S. Lockhart, "Levels of Processing: A Framework for Memory Research," *Journal of Verbal Learning and Verbal Behavior* 11, no. 6 (1972): 671–84.

26. George A. Miller, "The Magical Number Seven, Plus or Minus Two: Some Limits on Our Capacity for Processing Information," *Psychological Review* 63, no. 2 (1956): 81–97.

27. Steven J. Luck and Edward K. Vogel, "Visual Working Memory Capacity: From Psychophysics and Neurobiology to Individual Differences," *Trends in Cognitive Sciences* 17, no. 8 (2013): 391–400.

28. K. Anders Ericsson, "Superior Working Memory in Experts," in *The Cambridge Handbook of Expertise and Expert Performance*, eds. K. Anders Ericsson et al. (Cambridge, UK: Cambridge University Press, 2018), 696–713.

29. Paul Reber, "What Is the Memory Capacity of the Human Brain?," *Scientific American*, May 1, 2010, https://www.scientificamerican.com/article/what-is-the-memory-capacity/.

30. Fergus I. M. Craik and Endel Tulving, "Depth of Processing and the Retention of Words in Episodic Memory," *Journal of Experimental Psychology: General* 104, no. 3 (1975): 268–94.

31. Joshua Foer, *Moonwalking with Einstein: The Art and Science of Remembering Everything* (New York: Penguin Books, 2012), 33.

32. Masaki Nishida and Matthew P. Walker, "Daytime Naps, Motor Memory Consolidation and Regionally Specific Sleep Spindles," *PLoS One* 2, no. 4 (2007): e341.

33. Marcel Proust, *Swann's Way*, trans. C. K. Scott Moncrieff, in *Remembrance of Things Past* (London: Chatto & Windus, 1922), 52.

34. Donald W. Goodwin et al., "Alcohol and Recall: State-Dependent Effects in Man," *Science* 163, no. 3873 (1969): 1358–60.

35. Penelope A. Lewis and Hugo D. Critchley, "Mood-Dependent Memory," *Trends in Cognitive Sciences* 7, no. 10 (2003): 431–33.

36. Michael Connelly, *Blood Work* (New York: Little, Brown and Company, 1997), 186.

37. Jean C. Augustinack et al., "HM's Contributions to Neuroscience: A Review and Autopsy Studies," *Hippocampus* 24, no. 11 (2014): 1267–86. For a popular summary, see Donald G. MacKay, *Remembering: What 50 Years of Research with Famous*

Amnesia Patient HM Can Teach Us About Memory and How It Works (Buffalo, NY: Prometheus Books, 2019).

38. Francis Eustache, Béatrice Desgranges, and Pierre Messerli, "Edouard Claparède et la Mémoire Humaine," *Revue Neurologique (Paris)* 152, no. 10 (1996): 602–10.

39. Sigmund Freud, "Three Essays on the Theory of Sexuality," in *The Standard Edition of the Complete Psychological Works of Sigmund Freud,* ed. James Strachey (London: Hogarth Press, 1953), 174.

40. JoNell A. Usher and Ulric Neisser, "Childhood Amnesia and the Beginnings of Memory for Four Early Life Events," *Journal of Experimental Psychology: General* 122, no. 2 (1993): 155.

41. Carole Peterson, Qi Wang, and Yubo Hou, "'When I Was Little': Childhood Recollections in Chinese and European Canadian Grade School Children," *Child Development* 80, no. 2 (2009): 506–18.

42. Cited by Elizabeth Loftus and Katherine Ketcham, *Witness for the Defense: The Accused, the Eyewitness, and the Expert Who Puts Memory on Trial* (New York: St. Martin's Press, 1991), 19.

43. Henry L. Roediger and Kathleen B. McDermott, "Creating False Memories: Remembering Words Not Presented in Lists," *Journal of Experimental Psychology: Learning, Memory, and Cognition* 21, no. 4 (1995): 803–14.

44. Gordon H. Bower, John B. Black, and Terrence J. Turner, "Scripts in Memory for Text," *Cognitive Psychology* 11, no. 2 (1979): 177–220.

45. Larry L. Jacoby et al., "Becoming Famous Overnight: Limits on the Ability to Avoid Unconscious Influences of the Past," *Journal of Personality and Social Psychology* 56, no. 3 (1989): 326.

46. Elizabeth F. Loftus, David G. Miller, and Helen J. Burns, "Semantic Integration of Verbal Information into a Visual Memory," *Journal of Experimental Psychology: Human Learning and Memory* 4, no. 1 (1978): 19–31.

47. Elizabeth F. Loftus, "Leading Questions and the Eyewitness Report," *Cognitive Psychology* 7, no. 4 (1975): 560–72.

48. For review, see Elizabeth F. Loftus, "Eyewitness Science and the Legal System," *Annual Review of Law and Social Science* 14, no. 1 (2018): 1–10.

49. Brandon L. Garrett, *Convicting the Innocent: Where Criminal Prosecutions Go Wrong* (Cambridge, MA: Harvard University Press, 2011), 93–159.

50. Rachel Aviv, "Remembering the Murder You Didn't Commit," *New Yorker,* June 12, 2017, https://www.newyorker.com/magazine/2017/06/19/remembering -the-murder-you-didnt-commit.

8. The Rational Animal

1. The paragraph here is modified from Paul Bloom, "The War on Reason," *Atlantic,* March 15, 2014).

2. A. Alfred Taubman, *Threshold Resistance: The Extraordinary Career of a Luxury Retailing Pioneer* (New York: HarperCollins, 2007), 62–64.

3. "List of Cognitive Biases," Wikipedia, https://en.wikipedia.org/wiki/List_of_cog nitive_biases.

4. Matthew S. McGlone and Jessica Tofighbakhsh, "Birds of a Feather Flock Conjointly (?): Rhyme as Reason in Aphorisms," *Psychological Science* 11, no. 5 (2000): 424–28.

5. Amos Tversky and Daniel Kahneman, "Availability: A Heuristic for Judging Frequency and Probability," *Cognitive Psychology* 5, no. 2 (1973): 207–32.

6. Amos Tversky and Daniel Kahneman, "Extensional Versus Intuitive Reasoning: The Conjunction Fallacy in Probability Judgment," *Psychological Review* 90, no. 4 (1983): 293–315.

7. Tversky and Kahneman, "Extensional Versus Intuitive Reasoning."

8. Daniel Kahneman and Amos Tversky, "Evidential Impact of Base Rates," in *Judgment Under Uncertainty: Heuristics and Biases*, eds. Daniel Kahneman, Paul Slovic, and Amos Tversky (Cambridge, UK: Cambridge University Press, 1985), 153–60.

9. Thanks to Chaz Firestone for this example.

10. Amos Tversky and Daniel Kahneman, "The Framing of Decisions and the Psychology of Choice," in *Behavioral Decision Making*, ed. George Wright (Boston: Springer, 1985), 25–41.

11. For discussion, see Rory Sutherland, *Alchemy: The Surprising Power of Ideas That Don't Make Sense* (New York: Random House, 2019).

12. Tversky and Kahneman, "The Framing of Decisions and the Psychology of Choice." Note, however, that José Luis Bermúdez, "Rational Framing Effects: A Multidisciplinary Case," *Behavioral and Brain Sciences* (2022): 1–67, argues that there are some cases where framing effects can be rational.

13. Eldar Shafir, Itamar Simonson, and Amos Tversky, "Reason-Based Choice," *Cognition* 49, no. 1–2 (1993): 11–36.

14. Peter C. Wason, "Reasoning About a Rule," *Quarterly Journal of Experimental Psychology* 20, no. 3 (1968): 273–81.

15. Wason, "Reasoning About a Rule."

16. John Theodore Macnamara, *A Border Dispute: The Place of Logic in Psychology* (Cambridge, MA: MIT Press, 1986).

17. L. Cosmides, "The Logic of Social Exchange: Has Natural Selection Shaped How Humans Reason? Studies with the Wason Selection Task," *Cognition* 31, no. 3 (1989): 187–276.

18. Daniel Kahneman, *Thinking, Fast and Slow* (New York: Farrar, Straus and Giroux, 2011).

19. Shane Frederick, "Cognitive Reflection and Decision Making," *Journal of Economic Perspectives* 19, no. 4 (2005): 25–42.

20. Keela S. Thomson and Daniel M. Oppenheimer, "Investigating an Alternate Form of the Cognitive Reflection Test," *Judgment & Decision Making* 11, no. 1 (2016).

21. Gordon Pennycook and David G. Rand, "Lazy, Not Biased: Susceptibility to Partisan Fake News Is Better Explained by Lack of Reasoning Than by Motivated Reasoning," *Cognition* 188 (2019): 39–50.

22. Amitai Shenhav, David G. Rand, and Joshua D. Greene, "Divine Intuition: Cognitive Style Influences Belief in God," *Journal of Experimental Psychology: General* 141, no. 3 (2012): 423.

23. Gordon Pennycook et al., "Fighting COVID-19 Misinformation on Social Media: Experimental Evidence for a Scalable Accuracy-Nudge Intervention," *Psychological Science* 31, no. 7 (2020): 770–80.

24. David Leonhardt, "COVID's Partisan Errors," *New York Times*, March 18, 2021, https://www.nytimes.com/2021/03/18/briefing/atlanta-shootings-kamala-harris -tax-deadline-2021.html.

25. Geoffrey L. Cohen, "Party over Policy: The Dominating Impact of Group Influence on Political Beliefs," *Journal of Personality and Social Psychology* 85, no. 5 (2003): 808–22.

26. Peter H. Ditto et al., "At Least Bias Is Bipartisan: A Meta-Analytic Comparison of Partisan Bias in Liberals and Conservatives," *Perspectives on Psychological Science* 14, no. 2 (2019): 273–91.

27. Keith E. Stanovich, Richard F. West, and Maggie E. Toplak, "Myside Bias, Rational Thinking, and Intelligence," *Current Directions in Psychological Science* 22, no. 4 (2013): 259–64.

28. Julia Galef, *The Scout Mindset: Why Some People See Things Clearly and Others Don't* (New York: Portfolio/Penguin, 2021), 7–8.

29. Jay J. Van Bavel et al., "How Social Media Shapes Polarization," *Trends in Cognitive Sciences* 25, no. 11 (2021): 913–16; Jonathan Haidt, *The Righteous Mind: Why Good People Are Divided by Politics and Religion* (New York: Vintage Books, 2012).

30. Peter Nauroth et al., "Social Identity Threat Motivates Science-Discrediting Online Comments," *PLoS One* 10, no. 2 (2015): e0117476.

31. Kiara Minto et al., "A Social Identity Approach to Understanding Responses to Child Sexual Abuse Allegations," *PLoS One* 11, no. 4 (2016): e0153205.

32. Max H. Bazerman and Don A. Moore, *Judgment in Managerial Decision Making* (Hoboken, NJ: John Wiley & Sons, 2012).

33. David Byler and Yan Wu, "Opinion: Will You Fall into the Conspiracy Theory Rabbit Hole? Take Our Quiz and Find Out," *Washington Post*, October 6, 2021, https://www.washingtonpost.com/opinions/interactive/2021/conspiracy-theory -quiz/.

34. Hugo Mercier, *Not Born Yesterday* (Princeton, NJ: Princeton University Press, 2020).

35. Steven Pinker, *Rationality: What It Is, Why It Seems Scarce, Why It Matters* (New York: Viking Penguin, 2021), 300.

36. Pinker, *Rationality*.

37. Garrett Hardin, "The Tragedy of the Commons: The Population Problem Has No Technical Solution; It Requires a Fundamental Extension in Morality," *Science* 162, no. 3859 (1968): 1243–48.

9. Hearts and Minds

1. Edward L. Thorndike, "Valuations of Certain Pains, Deprivations, and Frustrations," *Pedagogical Seminary and Journal of Genetic Psychology* 51, no. 2 (1937): 227–39.

2. Quoted from Bridgette Martin Hard et al., *Psychology: People in Perspective*, 1st ed. (New York: W. W. Norton, 2020).

3. For instance, Daniel Yon, Cecilia Heyes, and Clare Press, "Beliefs and Desires in the Predictive Brain," *Nature Communications* 11, no. 1 (2020): 1–4. For a popular overview, see Andy Clark, *Surfing Uncertainty: Prediction, Action, and the Embodied Mind* (New York: Oxford University Press, 2015).

4. Zekun Sun and Chaz Firestone, "The Dark Room Problem," *Trends in Cognitive Sciences* 24, no. 5 (2020): 346–48.

5. William James, *The Principles of Psychology* (New York: Henry Holt, 1908).

6. James, *The Principles of Psychology*, 293.

7. For a good review of the theory of natural selection, see Richard Dawkins, *The Blind Watchmaker: Why the Evidence of Evolution Reveals a Universe Without Design* (New York: W. W. Norton, 1996).

8. George Christopher Williams, *Adaptation and Natural Selection: A Critique of Some Current Evolutionary Thought* (Princeton, NJ: Princeton University Press, 2018).

9. Stephen Jay Gould, Richard Lewontin, and Christopher M. Anderson, "A Philosophical Critique of the Arguments Presented in *The Spandrels of San Marco and the Panglossian Paradigm: A Critique of the Adaptationist Programme*," *Proceedings of the Royal Society B: Biological Sciences* 205 (1979): 581–89.

10. Jac T. M. Davis et al., "Cultural Components of Sex Differences in Color Preference," *Child Development* 92, no. 4 (2021): 1574–89.

11. George John Romanes and Charles Darwin, *Mental Evolution in Animals: With a Posthumous Essay on Instinct by Charles Darwin* (London: Kegan Paul, Trench, 1883), 108.

12. William James, *The Principles of Psychology*, vol. 1 (Mineola, NY: Dover Publications, 1950), 382.

13. Paul Ekman and Dacher Keltner, "Universal Facial Expressions of Emotion," in *Nonverbal Communication: Where Nature Meets Culture*, eds. Ullica Segerstrale and Peter Molnar (Oxfordshire, UK: Routledge, 1997), 27–46.

14. David Matsumoto and Bob Willingham, "Spontaneous Facial Expressions of Emotion of Congenitally and Noncongenitally Blind Individuals," *Journal of Personality and Social Psychology* 96, no. 1 (2009): 1–10.

15. Lisa Feldman Barrett, "Are Emotions Natural Kinds?," *Perspectives on Psychological Science* 1, no. 1 (2006): 28–58; Lisa Feldman Barrett, *How Emotions Are Made: The Secret Life of the Brain* (London: Pan Macmillan, 2017).

16. Saba Safdar et al., "Variations of Emotional Display Rules Within and Across Cultures: A Comparison Between Canada, USA, and Japan," *Canadian Journal*

of Behavioural Science/Revue Canadienne des Sciences du Comportement 41, no. 1 (2009): 1–31.

17. Much of the following is drawn, with minor modifications, from Paul Bloom, *The Sweet Spot: The Pleasures of Suffering and the Search for Meaning* (New York: Ecco/HarperCollins, 2021).

18. Barbara L. Fredrickson and Robert W. Levenson, "Positive Emotions Speed Recovery from the Cardiovascular Sequelae of Negative Emotions," *Cognition & Emotion* 12, no. 2 (1998): 191–220.

19. Susan M. Hughes and Shevon E. Nicholson, "Sex Differences in the Assessment of Pain Versus Sexual Pleasure Facial Expressions," *Journal of Social, Evolutionary, and Cultural Psychology* 2, no. 4 (2008): 289–98.

20. Oriana R. Aragón et al., "Dimorphous Expressions of Positive Emotion: Displays of Both Care and Aggression in Response to Cute Stimuli," *Psychological Science* 26, no. 3 (2015): 259–73.

21. Some of what follows is drawn, with minor modifications, from Paul Bloom, *Descartes' Baby: How the Science of Child Development Explains What Makes Us Human* (New York: Random House, 2005).

22. Steven Pinker, *How the Mind Works* (New York: W. W. Norton, 1997), 372; see also Robert H. Frank, *Passions Within Reason: The Strategic Role of the Emotions* (New York: W. W. Norton, 1988).

23. Richard Hanley, *Is Data Human? The Metaphysics of Star Trek* (Hampshire, UK: Boxtree, 1998).

24. Oliver Sacks, "The Last Hippie," *New York Review of Books*, March 26, 1992, 53–62.

25. Antonio R. Damasio, *Descartes' Error: Emotion, Reason and the Human Brain* (New York: Penguin Books, 1994), 36.

26. James, *The Principles of Psychology*, vol. 1 (Mineola, NY: Dover Publications, 1950), 308.

27. For a summary, see Joseph LeDoux, *Anxious: The Modern Mind in the Age of Anxiety* (New York: Simon and Schuster, 2015).

28. Frank Herbert, *Dune* (New York: Berkley Books, 1990), 12.

29. William W. Eaton, O. Joseph Bienvenu, and Beyon Miloyan, "Specific Phobias," *Lancet Psychiatry* 5, no. 8 (2018): 678–86.

30. Stefanie Hoehl et al., "Itsy Bitsy Spider . . . : Infants React with Increased Arousal to Spiders and Snakes," *Frontiers in Psychology* 8 (2017): 1710. See also Vanessa LoBue and Judy S. DeLoache, "Detecting the Snake in the Grass: Attention to Fear-Relevant Stimuli by Adults and Young Children," *Psychological Science* 19, no. 3 (2008): 284–89.

31. Example from David A. Pizarro and Paul Bloom, "The Intelligence of the Moral Intuitions: A Comment on Haidt (2001)," *Psychological Review* 110 (2003): 193–96.

32. Some of what follows is drawn, with minor modifications, from Paul Bloom, *Just Babies: The Origins of Good and Evil* (New York: Broadway Books, 2013).

33. For reviews, see Paul Rozin, Jonathan Haidt, and Clark R. McCauley, "Disgust:

The Body and Soul Emotion," in *Handbook of Emotions*, 3rd ed., eds. Michael Lewis, Jeannette M. Haviland-Jones, and Lisa F. Barrett (New York: Guilford Press, 1999), 429, 757–776; for a different perspective, see William Ian Miller, *The Anatomy of Disgust* (Cambridge, MA: Harvard University Press, 1998).

34. Jonathan Haidt, Clark McCauley, and Paul Rozin, "Individual Differences in Sensitivity to Disgust: A Scale Sampling Seven Domains of Disgust Elicitors," *Personality and Individual Differences* 16, no. 5 (1994): 701–13. For a modified version, see Bunmi O. Olatunji et al., "The Disgust Scale: Item Analysis, Factor Structure, and Suggestions for Refinement," *Psychological Assessment* 19, no. 3 (2007): 281.

35. Sigmund Freud, *Civilization and Its Discontents* (New York: W. W. Norton, 2010), 52.

36. Paul Rozin et al., "The Child's Conception of Food: Differentiation of Categories of Rejected Substances in the 16 Months to 5 Year Age Range," *Appetite* 7, no. 2 (1986): 141–51.

37. Daniel M. T. Fessler, Serena J. Eng, and C. David Navarrete, "Elevated Disgust Sensitivity in the First Trimester of Pregnancy: Evidence Supporting the Compensatory Prophylaxis Hypothesis," *Evolution and Human Behavior* 26, no. 4 (2005): 344–51.

38. Bruno Wicker et al., "Both of Us Disgusted in My Insula: The Common Neural Basis of Seeing and Feeling Disgust," *Neuron* 40, no. 3 (2003): 655–64; Paul Wright et al., "Disgust and the Insula: fMRI Responses to Pictures of Mutilation and Contamination," *Neuroreport* 15, no. 15 (2004): 2347–51.

39. Val Curtis, Robert Aunger, and Tamer Rabie, "Evidence That Disgust Evolved to Protect from Risk of Disease," *Proceedings of the Royal Society of London, Series B: Biological Sciences* 271, suppl. 4 (2004): S131–S133.

40. For a good overview, see David M. Buss, *The Evolution of Desire: Strategies of Human Mating* (New York: Basic Books, 2016).

41. Peter Salovey, lecture, Yale Courses, Introduction to Psychology, YouTube, https://www.youtube.com/watch?v=kZoBgX8rScg.

42. Donn Byrne, "An Overview (and Underview) of Research and Theory Within the Attraction Paradigm," *Journal of Social and Personal Relationships* 14, no. 3 (1997): 417–31.

43. Jill M. Mateo, "Kin-Recognition Abilities and Nepotism as a Function of Sociality," *Proceedings of the Royal Society of London, Series B: Biological Sciences* 269, no. 1492 (2002): 721–27.

44. Edward Westermarck, *The History of Human Marriage*, 3 vols., 5th ed. (New York: Allerton Book Co., 1922).

45. Debra Lieberman, John Tooby, and Leda Cosmides, "The Architecture of Human Kin Detection," *Nature* 445, no. 7129 (2007): 727–31.

46. Some of what follows is drawn, with minor modifications, from Paul Bloom, *How Pleasure Works: The New Science of Why We Like What We Like* (New York: Random House, 2010).

47. The analogy here is made by a fictional neuroscientist in Ted Chiang, "Liking

What You See: A Documentary," in *Stories of Your Life and Others* (New York: Orb Books, 2003).

48. Gillian Rhodes, "The Evolutionary Psychology of Facial Beauty," *Annual Review of Psychology* 57 (2006): 199–226.

49. Alan Slater et al., "Newborn Infants Prefer Attractive Faces," *Infant Behavior and Development* 21, no. 2 (1998): 345–54.

50. Judith H. Langlois and Lori A. Roggman, "Attractive Faces Are Only Average," *Psychological Science* 1, no. 2 (1990): 115–21.

51. Thomas R. Alley and Michael R. Cunningham, "Article Commentary: Averaged Faces Are Attractive, but Very Attractive Faces Are Not Average," *Psychological Science* 2, no. 2 (1991): 123–25.

52. David M. Buss, "Sex Differences in Human Mate Preferences: Evolutionary Hypotheses Tested in 37 Cultures," *Behavioral and Brain Sciences* 12, no. 1 (1989): 1–14.

53. Kathryn V. Walter et al., "Sex Differences in Mate Preferences Across 45 Countries: A Large-Scale Replication," *Psychological Science* 31, no. 4 (2020): 408–23.

54. Christian Rudder, *Dataclysm: Who We Are (When We Think No One's Looking)* (New York: Crown Publishers, 2014).

55. The discussion that follows is drawn, with minor modifications, from Bloom, *How Pleasure Works*.

56. Robert L. Trivers, "Parental Investment and Sexual Selection," in *Sexual Selection and the Descent of Man*, ed. Bernard G. Campbell (Chicago: Aldine, 1972), 136–79. Trivers, I should add, is one of the more interesting characters in the field. The blurb for his autobiography starts: "Unlike other renowned scientists, Robert Trivers has spent time behind bars, drove a getaway car for Huey P. Newton, and founded an armed group in Jamaica to protect gay men from mob violence."

57. Steve Stewart-Williams, *The Ape That Understood the Universe: How the Mind and Culture Evolve* (Cambridge, UK: Cambridge University Press, 2018), 72–74.

58. David P. Schmitt, "Universal Sex Differences in the Desire for Sexual Variety: Tests from 52 Nations, 6 Continents, and 13 Islands," *Journal of Personality and Social Psychology* 85, no. 1 (2003): 85. For different perspectives on how to make sense of these cultural differences, see Alice H. Eagly and Wendy Wood, "The Origins of Sex Differences in Human Behavior: Evolved Dispositions Versus Social Roles," *American Psychologist* 54, no. 6 (1999); Steven W. Gangestad, Martie G. Haselton, and David M. Buss, "Evolutionary Foundations of Cultural Variation: Evoked Culture and Mate Preferences," *Psychological Inquiry* 17, no. 2 (2006): 75–95.

59. Stewart-Williams, *The Ape That Understood the Universe*, 76.

60. Buss, "Sex Differences in Human Mate Preferences: Evolutionary Hypotheses Tested in 37 Cultures."

61. For discussion, see Bloom, *Just Babies*.

62. Stanley Milgram, Leon Mann, and Susan Harter, "The Lost-Letter Technique: A Tool of Social Research," *Public Opinion Quarterly* 29, no. 3 (1965): 437–38.

63. Joseph Henrich, Steven J. Heine, and Ara Norenzayan, "The Weirdest People in the World?," *Behavioral and Brain Sciences* 33, no. 2–3 (2010): 61–83.

64. Bloom, *Just Babies.*

65. Felix Warneken and Michael Tomasello, "Altruistic Helping in Human Infants and Young Chimpanzees," *Science* 311, no. 5765 (2006): 1301–3.

66. Sarah Blaffer Hrdy, *Mothers and Others: The Evolutionary Origins of Mutual Understanding* (Cambridge, MA: Harvard University Press, 2009).

67. Adam Smith, *The Theory of Moral Sentiments* (Los Angeles: Logos Books, 2018), 83.

68. Robert L. Trivers, "The Evolution of Reciprocal Altruism," *Quarterly Review of Biology* 46, no. 1 (1971): 35–57.

69. Frans De Waal, "One for All," *Scientific American* 311, no. 3 (2014): 68–71.

70. For discussion, see Justin W. Martin et al., "When Do We Punish People Who Don't?," *Cognition* 193 (2019): 104040.

71. For discussion, see Paul Bloom, "Did God Make These Babies Moral? Intelligent Design's Oldest Attack on Evolution Is as Popular as Ever," *New Republic* 13 (2014).

72. Francis Collins, *The Language of God: A Scientist Presents Evidence for Belief* (New York: Free Press, 2008).

73. From Peter Singer, *The Expanding Circle: Ethics, Evolution, and Moral Progress* (Princeton, NJ: Princeton University Press, 2011), 136. The rest of the paragraph is from, with minor modifications, Bloom, *Just Babies.*

74. For a popular exposition of this argument, see Kevin Simler and Robin Hanson, *The Elephant in the Brain: Hidden Motives in Everyday Life* (New York: Oxford University Press, 2017).

75. Jillian J. Jordan et al., "Third-Party Punishment as a Costly Signal of Trustworthiness," *Nature* 530, no. 7591 (2016): 473–76.

76. Randolph M. Nesse, *Good Reasons for Bad Feelings: Insights from the Frontier of Evolutionary Psychiatry* (New York: Dutton, 2019), 162.

10. A Brief Note on a Crisis

1. John C. Turner et al., "Self and Collective: Cognition and Social Context," *Personality and Social Psychology Bulletin* 20, no. 5 (1994): 454–63.

2. For instance, Jay J. Bavel et al., "Using Social and Behavioural Science to Support COVID-19 Pandemic Response," *Nature Human Behaviour* 4, no. 5 (2020): 460–71.

3. For a popular treatment, see Jesse Singal, *The Quick Fix: Why Fad Psychology Can't Cure Our Social Ills* (New York: Farrar, Straus and Giroux, 2021). For my own take, see Paul Bloom, "Afterword: Crisis? What Crisis?," in *Psychological Science Under Scrutiny: Recent Challenges and Proposed Solutions,* ed. Scott O. Lillenfeld and Irwin D. Waldman (Chichester: Wiley-Blackwell, 2017), 349–55.

4. Open Science Collaboration, "Estimating the Reproducibility of Psychological Science," *Science* 349, no. 6251 (2015). For critical remarks, see Daniel T. Gilbert et al., "A Response to the Reply to Our Technical Comment on 'Estimating the Reproducibility of Psychological Science'" (2016), https:/gking.harvard.edu/files /gking/filesgkpw_response_to_osc_rebutal.pdf.

5. For discussion, see Brian A. Nosek et al., "Replicability, Robustness, and Reproducibility in Psychological Science," *Annual Review of Psychology* 73, no. 1 (2022): 719–48.

6. Michael Inzlicht, "The Replication Crisis Is Not Over," *Getting Better* (Blog), http://michaelinzlicht.com/getting-better.

7. Joseph P. Simmons, Leif D. Nelson, and Uri Simonsohn, "False-Positive Citations," *Perspectives on Psychological Science* 13, no. 2 (2018): 255–59.

8. C. Glenn Begley and Lee M. Ellis, "Raise Standards for Preclinical Cancer Research," *Nature* 483, no. 7391 (2012): 531–33.

9. Diederik Stapel, "Faking Science: A True Story of Academic Fraud," trans. Nicholas J. L. Brown, 2016, http://nick.brown.free.fr/stapel/FakingScience-20161115.pdf.

10. Joseph Henrich, Steven J. Heine, and Ara Norenzayan, "The Weirdest People in the World?," *Behavioral and Brain Sciences* 33, no. 2 (2010): 61–83; see also Jeffrey Jensen Arnett, "The Neglected 95%: A Challenge to Psychology's Philosophy of Science," *American Psychologist* 64, no. 6 (2009): 571–74.

11. See Bloom, "Afterword: Crisis? What Crisis?"

12. Joseph Henrich, *The WEIRDest People in the World: How the West Became Psychologically Peculiar and Particularly Prosperous* (New York: Farrar, Straus and Giroux, 2020), 21.

11. Social Butterflies

1. Thomas Gilovich, Victoria Husted Medvec, and Kenneth Savitsky, "The Spotlight Effect in Social Judgment: An Egocentric Bias in Estimates of the Salience of One's Own Actions and Appearance," *Journal of Personality and Social Psychology* 78, no. 2 (2000): 211–22.

2. Cited by David G. Myers, *The Inflated Self* (New York: Seabury Press, 1980).

3. Irene Scopelliti et al., "Bias Blind Spot: Structure, Measurement, and Consequences," *Management Science* 61, no. 10 (2015): 2468–86.

4. Christopher F. Chabris and Daniel J. Simons, *The Invisible Gorilla: And Other Ways Our Intuitions Deceive Us* (New York: Harmony Books, 2010).

5. For instance, Peter E. De Michele, Bruce Gansneder, and Gloria B. Solomon, "Success and Failure Attributions of Wrestlers: Further Evidence of the Self-Serving Bias," *Journal of Sport Behavior* 21, no. 3 (1998): 242–55; James Shepperd, Wendi Malone, and Kate Sweeny, "Exploring Causes of the Self-Serving Bias," *Social and Personality Psychology Compass* 2, no. 2 (2008): 895–908.

6. Fiery Cushman, "Rationalization Is Rational," *Behavioral and Brain Sciences* 43, no. e28 (2019): 1–59.

7. Leon Festinger and James M. Carlsmith, "Cognitive Consequences of Forced Compliance," *Journal of Abnormal and Social Psychology* 58, no. 2 (1959): 203–10.

8. Jack W. Brehm, "Postdecision Changes in the Desirability of Alternatives," *Journal of Abnormal and Social Psychology* 52, no. 3 (1956): 384–89; Louisa C. Egan, Laurie R. Santos, and Paul Bloom, "The Origins of Cognitive Dissonance: Evidence from Children and Monkeys," *Psychological Science* 18, no. 11 (2007): 978–83. But

see M. Keith Chen and Jane L. Risen, "How Choice Affects and Reflects Preferences: Revisiting the Free-Choice Paradigm," *Journal of Personality and Social Psychology* 99, no. 4 (2010): 573–94, for a critical discussion.

9. Tali Sharot, Cristina M. Velasquez, and Raymond J. Dolan, "Do Decisions Shape Preference? Evidence from Blind Choice," *Psychological Science* 21, no. 9 (2010): 1231–35; Louisa C. Egan, Paul Bloom, and Laurie R. Santos, "Choice-Induced Preferences in the Absence of Choice: Evidence from a Blind Two Choice Paradigm with Young Children and Capuchin Monkeys," *Journal of Experimental Social Psychology* 46, no. 1 (2010): 204–7.

10. Nalini Ambady, Frank J. Bernieri, and Jennifer A. Richeson, "Toward a Histology of Social Behavior: Judgmental Accuracy from Thin Slices of the Behavioral Stream," in *Advances in Experimental Social Psychology*, vol. 32, ed. Mark P. Zanna (Cambridge, MA: Academic Press, 2000), 201–271.

11. For review, see Max Weisbuch and Nalini Ambady, "Thin-Slice Vision," *Science of Social Vision* (2011): 228–47.

12. Michal Kosinski, "Facial Recognition Technology Can Expose Political Orientation from Naturalistic Facial Images," *Scientific Reports* 11, no. 1 (2021): 1–7.

13. Lee Ross, "From the Fundamental Attribution Error to the Truly Fundamental Attribution Error and Beyond: My Research Journey," *Perspectives on Psychological Science* 13, no. 6 (2018): 750–69. For some qualifications, see Bertram F. Malle, "The Actor-Observer Asymmetry in Attribution: A (Surprising) Meta-Analysis," *Psychological Bulletin* 132, no. 6 (2006): 895.

14. Lee D. Ross, Teresa M. Amabile, and Julia L. Steinmetz, "Social Roles, Social Control, and Biases in Social-Perception Processes," *Journal of Personality and Social Psychology* 35, no. 7 (1977): 485–94.

15. Paul Rozin, "Paul Rozin: Time Management," *Research Digest*, October 4, 2009, http://bps-research-digest.blogspot.com/2009/10/paul-rozin-time-management .html.

16. Martie G. Haselton and David M. Buss, "Error Management Theory: A New Perspective on Biases in Cross-Sex Mind Reading," *Journal of Personality and Social Psychology* 78, no. 1 (2000): 81–91.

17. Martie G. Haselton and Daniel Nettle, "The Paranoid Optimist: An Integrative Evolutionary Model of Cognitive Biases," *Personality and Social Psychology Review* 10, no. 1 (2006): 47–66.

18. Julia Galef, *The Scout Mindset: Why Some People See Things Clearly and Others Don't* (New York: Portfolio/Penguin, 2021), 33.

19. Randolph M. Nesse, *Good Reasons for Bad Feelings: Insights from the Frontier of Evolutionary Psychiatry* (New York: Dutton, 2019).

20. Nesse, *Good Reasons for Bad Feelings: Insights from the Frontier of Evolutionary Psychiatry*, 73.

21. Cited by Daniel T. Gilbert, "Ordinary Personology," *Handbook of Social Psychology* 2 (1998): 89–150.

22. Clifford Geertz, "'From the Native's Point of View': On the Nature of Anthropo-

logical Understanding," *Bulletin of the American Academy of Arts and Sciences* (1974): 26–45.

23. Joseph Henrich, *The WEIRDest People in the World: How the West Became Psychologically Peculiar and Particularly Prosperous* (New York: Farrar, Straus and Giroux, 2020), 21–22.

24. Joseph Henrich, Steven J. Heine, and Ara Norenzayan, "The Weirdest People in the World?," *Behavioral and Brain Sciences* 33, no. 2 (2010): 61–83; Henrich, *The WEIRDest People in the World.*

25. Henrich, *The WEIRDest People in the World*, 33.

26. Discussed in M. Popova, "The Psychology of Conformity," *Atlantic*, January 17, 2012, https://www.theatlantic.com/health/archive/2012/01/the-psychology-of-conformity/251371/.

27. Robert Cialdini, "Don't Throw In the Towel: Use Social Influence Research," *APS Observer* 18 (2005): 33–39.

28. Solomon E. Asch, "An Experimental Investigation of Group Influence," in *Symposium on Preventive and Social Psychiatry* (Washington, DC: Walter Reed Army Institute of Research, 1958), 17.

29. Ian Parker, "Obedience," *Granta* 71, no. 4 (2000): 99–125.

30. Stanley Milgram, "Behavioral Study of Obedience," *Journal of Abnormal and Social Psychology* 67, no. 4 (1963): 371–78.

31. A. L. Forest et al., "Turbulent Times, Rocky Relationships: Relational Consequences of Experiencing Physical Instability," *Psychological Science* 26, no. 8 (2015): 1261–71.

32. Erik G. Helzer and David A. Pizarro, "Dirty Liberals! Reminders of Physical Cleanliness Influence Moral and Political Attitudes," *Psychological Science* 22, no. 4 (2011): 517–22.

33. Y. Inbar, D. A. Pizarro, and P. Bloom, "Disgusting Smells Cause Decreased Liking of Gay Men," *Emotion* 12, no. 1 (2012): 23–27.

34. Joshua M. Ackerman, Christopher C. Nocera, and John A. Bargh, "Incidental Haptic Sensations Influence Social Judgments and Decisions," *Science* 328, no. 5986 (2010): 1712–15.

35. Ackerman, Nocera, and Bargh, "Incidental Haptic Sensations Influence Social Judgments and Decisions."

36. Dana R. Carney, Amy J. C. Cuddy, and Andy J. Yap, "Power Posing: Brief Nonverbal Displays Affect Neuroendocrine Levels and Risk Tolerance," *Psychological Science* 21, no. 10 (2010): 1363–68.

37. Kathleen D. Vohs, "Money Priming Can Change People's Thoughts, Feelings, Motivations, and Behaviors: An Update on 10 Years of Experiments," *Journal of Experimental Psychology: General* 144, no. 4 (2015): e86.

38. S. Christian Wheeler, Jonah Berger, and Marc Meredith, "Can Where People Vote Influence How They Vote? The Influence of Polling Location Type on Voting Behavior," Stanford University Graduate School of Business Research Paper 1926, 2006.

39. John A. Bargh and Idit Shalev, "The Substitutability of Physical and Social Warmth in Daily Life," *Emotion* 12, no. 1 (2012): 154–62.

40. John A. Bargh, Mark Chen, and Lara Burrows, "Automaticity of Social Behavior: Direct Effects of Trait Construct and Stereotype Activation on Action," *Journal of Personality and Social Psychology* 71, no. 2 (1996): 230.

41. Chen-Bo Zhong and Katie Liljenquist, "Washing Away Your Sins: Threatened Morality and Physical Cleansing," *Science* 313, no. 5792 (2006): 1451–52.

42. Diederik A. Stapel and Siegwart Lindenberg, "Coping with Chaos: How Disordered Contexts Promote Stereotyping and Discrimination," *Science* 332, no. 6026 (2011): 251–53.

43. John A. Bargh and Tanya L. Chartrand, "The Unbearable Automaticity of Being," *American Psychologist* 54, no. 7 (1999): 462–79.

44. Robert S. Wyer Jr., *The Automaticity of Everyday Life: Advances in Social Cognition*, vol. x (Hove, UK: Psychology Press, 2014).

45. Lawrence E. Williams, Julie Y. Huang, and John A. Bargh, "The Scaffolded Mind: Higher Mental Processes Are Grounded in Early Experience of the Physical World," *European Journal of Social Psychology* 39, no. 7 (2009): 1257–67.

46. Zhong and Liljenquist, "Washing Away Your Sins: Threatened Morality and Physical Cleansing." For discussion, see Spike W. S. Lee and Norbert Schwarz, "Grounded Procedures: A Proximate Mechanism for the Psychology of Cleansing and Other Physical Actions," *Behavioral and Brain Sciences* 44 (2021): 1–78.

47. Claude Mathias Messner and Adrian Gadient-Brügger, "Nazis by Kraut: A Playful Application of Moral Self-Licensing," *Psychology* 6, no. 09 (2015): 1144–49.

48. Ian Ayres, Mahzarin Banaji, and Christine Jolls, "Race Effects on eBay," *RAND Journal of Economics* 46, no. 4 (2015): 891–917.

49. Marc W. Patry, "Attractive but Guilty: Deliberation and the Physical Attractiveness Bias," *Psychological Reports* 102, no. 3 (2008): 727–33.

50. Jesse Singal, *The Quick Fix: Why Fad Psychology Can't Cure Our Social Ills* (New York: Farrar, Straus and Giroux, 2021), 9.

51. Lee and Schwarz, "Grounded Procedures: A Proximate Mechanism for the Psychology of Cleansing and Other Physical Actions."

52. Amy J. C. Cuddy, S. Jack Schultz, and Nathan E. Fosse, "P-Curving a More Comprehensive Body of Research on Postural Feedback Reveals Clear Evidential Value for Power-Posing Effects: Reply to Simmons and Simonsohn (2017)," *Psychological Science* 29, no. 4 (2018): 656–66.

53. Joseph Henrich and Francisco J. Gil-White, "The Evolution of Prestige: Freely Conferred Deference as a Mechanism for Enhancing the Benefits of Cultural Transmission," *Evolution and Human Behavior* 22, no. 3 (2001): 165–96.

12. Is Everyone a Little Bit Racist?

1. Masha Gessen, "We Need to Change the Terms of the Debate on Trans Kids," *New Yorker*, January 13, 2021, https://www.newyorker.com/news/our-columnists/we-need-to-change-the-terms-of-the-debate-on-trans-kids.

2. David M. Messick and Diane M. Mackie, "Intergroup Relations," *A Psychology* 40 (1989): 45–81.

3. Shelley E. Taylor et al., "Categorical and Contextual Bases of Person Memory and Stereotyping," *Journal of Personality and Social Psychology* 36, no. 7 (1978): 778–93.

4. Jim Sidanius and Felicia Pratto, *Social Dominance: An Intergroup Theory of Social Hierarchy and Oppression* (New York: Cambridge University Press, 2001).

5. Robert Kurzban, John Tooby, and Leda Cosmides, "Can Race Be Erased? Coalitional Computation and Social Categorization," *Proceedings of the National Academy of Sciences* 98, no. 26 (2001): 15387–92.

6. Michael A. Woodley of Menie et al., "A Meta-Analysis of the 'Erasing Race' Effect in the United States and Some Theoretical Considerations," *Frontiers in Psychology* 11 (2020): 1635.

7. Gregory Murphy, *The Big Book of Concepts* (Cambridge, MA: MIT Press, 2004).

8. Murphy, *The Big Book of Concepts*, 1.

9. Lee Jussim et al., "The Unbearable Accuracy of Stereotypes," *Handbook of Prejudice, Stereotyping, and Discrimination* 199 (2009): 227.

10. Qinggong Li et al., "Susceptibility to Being Lured Away by a Stranger: A Real-World Field Test of Selective Trust in Early Childhood," *Psychological Science* 31, no. 12 (2020): 1488–96.

11. John Sides, "Democrats Are Gay, Republicans Are Rich: Our Stereotypes of Political Parties Are Amazingly Wrong," *Washington Post*, May 13, 2016, https://www.washingtonpost.com/news/monkey-cage/wp/2016/05/23/democrats-are-gay-republicans-are-rich-our-stereotypes-of-political-parties-are-amazingly-wrong/.

12. For discussion of essentialism, see Susan A. Gelman, *The Essential Child: Origins of Essentialism in Everyday Thought* (New York: Oxford University Press, 2003); Paul Bloom, *How Pleasure Works: The New Science of Why We Like What We Like* (New York: Random House, 2010).

13. Alexander Noyes and Frank C. Keil, "Asymmetric Mixtures: Common Conceptual Priorities for Social and Chemical Kinds," *Psychological Science* 29, no. 7 (2018): 1094–1103.

14. Andrei Cimpian and Erika Salomon, "The Inherence Heuristic: An Intuitive Means of Making Sense of the World, and a Potential Precursor to Psychological Essentialism," *Behavioral and Brain Sciences* 37, no. 5 (2014): 461–80.

15. All these lines of research are explored in more detail in Paul Bloom, *Just Babies: The Origins of Good and Evil* (New York: Broadway Books, 2013).

16. Muzafer Sherif et al., *The Robbers Cave Experiment: Intergroup Conflict and Cooperation* (Norman: University of Oklahoma Book Exchange, 1961); Gina Perry, *The Lost Boys: Inside Muzafer Sherif's Robbers Cave Experiment* (Pontiac, MI: Scribe Publishing Co., 2018).

17. David Shariatmadari, "A Real-Life Lord of the Flies: The Troubling Legacy of the Robbers Cave Experiment," *Guardian*, April 16, 2016. See also Perry, *The Lost Boys*.

18. I have a similar feeling about the well-known Stanford Prison Experiment, which,

despite its fame, involves too much playacting on the part of the participants to be a good psychological study. For discussion, see Thibault Le Texier, "Debunking the Stanford Prison Experiment," *American Psychologist* 74, no. 7 (2019): 823–39.

19. Henri Tajfel et al., "Social Categorization and Intergroup Behaviour," *European Journal of Social Psychology* 1, no. 2 (1971): 149–78. See also Brian Mullen, Rupert Brown, and Colleen Smith, "Ingroup Bias as a Function of Salience, Relevance, and Status: An Integration," *European Journal of Social Psychology* 22, no. 2 (1992): 103–22.

20. For review, see Yarrow Dunham, "Mere Membership," *Trends in Cognitive Sciences* 22, no. 9 (2018): 780–93.

21. Dunham, "Mere Membership."

22. For review of all studies described in this section, see Katherine D. Kinzler, "Language as a Social Cue," *Annual Review of Psychology* 72 (2021): 241–64.

23. For example, Yair Bar-Haim et al., "Nature and Nurture in Own-Race Face Processing," *Psychological Science* 17, no. 2 (2006): 159–63; David J. Kelly et al., "Cross-Race Preferences for Same-Race Faces Extend Beyond the African Versus Caucasian Contrast in 3-Month-Old Infants," *Infancy* 11, no. 1 (2007): 87–95.

24. Katherine D. Kinzler, Kristin Shutts, and Joshua Correll, "Priorities in Social Categories," *European Journal of Social Psychology* 40, no. 4 (2010): 581–92.

25. Katherine D. Kinzler et al., "Accent Trumps Race in Guiding Children's Social Preferences," *Social Cognition* 27, no. 4 (2009): 623–34.

26. Deborah Belle, "'I Can't Operate, That Boy Is My Son!': Gender Schemas and a Classic Riddle," *Sex Roles* 85, no. 3 (2021): 161–71.

27. Ian Ayres, Mahzarin Banaji, and Christine Jolls, "Race Effects on eBay," *RAND Journal of Economics* 46, no. 4 (2015): 891–917.

28. For discussion, see Mahzarin R. Banaji and Anthony G. Greenwald, *Blindspot: Hidden Biases of Good People* (New York: Bantam Books, 2013).

29. First introduced by Anthony G. Greenwald, Debbie E. McGhee, and Jordan L. K. Schwartz, "Measuring Individual Differences in Implicit Cognition: The Implicit Association Test," *Journal of Personality and Social Psychology* 74, no. 6 (1998): 1464–80; the conceptual basis for this work was introduced in Anthony G. Greenwald and Mahzarin R. Banaji, "Implicit Social Cognition: Attitudes, Self-Esteem, and Stereotypes," *Psychological Review* 102, no. 1 (1995): 4–24.

30. Project Implicit, https://implicit.harvard.edu/implicit/.

31. Banaji and Greenwald, *Blindspot.*

32. Tessa E. S. Charlesworth and Mahzarin R. Banaji, "Patterns of Implicit and Explicit Attitudes: I. Long-Term Change and Stability from 2007 to 2016," *Psychological Science* 30, no. 2 (2019): 174–92.

33. Sarah Patton Moberg, Maria Krysan, and Deanna Christianson, "Racial Attitudes in America," *Public Opinion Quarterly* 83, no. 2 (2019): 450–71.

34. Shanto Iyengar and Sean J. Westwood, "Fear and Loathing Across Party Lines: New Evidence on Group Polarization," *American Journal of Political Science* 59, no. 3 (2015): 690–707.

35. Bertram Gawronski et al., "Temporal Stability of Implicit and Explicit Measures: A Longitudinal Analysis," *Personality and Social Psychology Bulletin* 43, no. 3 (2017): 300–312.

36. Frederick L. Oswald et al., "Predicting Ethnic and Racial Discrimination: A Meta-Analysis of IAT Criterion Studies," *Journal of Personality and Social Psychology* 105, no. 2 (2013): 171–92.

37. Eric Luis Uhlmann, David A. Pizarro, and Paul Bloom, "Varieties of Social Cognition," *Journal for the Theory of Social Behaviour* 38, no. 3 (2008): 293–322.

38. Eric Luis Uhlmann, Victoria L. Brescoll, and Elizabeth Levy Paluck, "Are Members of Low Status Groups Perceived as Bad, or Badly Off? Egalitarian Negative Associations and Automatic Prejudice," *Journal of Experimental Social Psychology* 42, no. 4 (2006): 491–99.

39. B. Keith Payne and Jason W. Hannay, "Implicit Bias Reflects Systemic Racism," *Trends in Cognitive Sciences* 25, no. 11 (2021): 927–36.

40. Brian A. Nosek et al., "National Differences in Gender-Science Stereotypes Predict National Sex Differences in Science and Math Achievement," *Proceedings of the National Academy of Sciences* 106, no. 26 (2009): 10593–597.

41. Jordan B. Leitner et al., "Racial Bias Is Associated with Ingroup Death Rate for Blacks and Whites: Insights from Project Implicit," *Social Science & Medicine* 170 (2016): 220–27.

42. The discussion that follows is modified from Bloom, *Just Babies*.

13. Uniquely You

1. Noam Chomsky, *On Language* (New York: New Press, 1998), 61–62.

2. B. Schwartz et al., "Maximizing Versus Satisficing: Happiness Is a Matter of Choice," *Journal of Personality and Social Psychology* 83, no. 5 (2002): 1178–97.

3. Walter Mischel, Yuichi Shoda, and Rodolfo Mendoza-Denton, "Situation-Behavior Profiles as a Locus of Consistency in Personality," *Current Directions in Psychological Science* 11, no. 2 (2002): 50–54.

4. Paul Bloom, *Against Empathy: The Case for Rational Compassion* (New York: Ecco/HarperCollins, 2017); Roy F. Baumeister, *Evil: Inside Human Cruelty and Violence* (New York: Henry Holt, 1996).

5. Ian Parker, "Obedience," *Granta* 71, no. 4 (2000): 99–125.

6. Damion Searls, *The Inkblots: Hermann Rorschach, His Iconic Test, and the Power of Seeing* (New York: Broadway Books, 2017).

7. Howard N. Garb et al., "Roots of the Rorschach Controversy," *Clinical Psychology Review* 25, no. 1 (2005): 97–118.

8. Joseph Stromberg and Estellen Caswell, "Why the Myers-Briggs Test Is Totally Meaningless," *Vox*, October 8, 2015, https://www.vox.com/2014/7/15/5881947/myers-briggs-personality-test-meaningless.

9. Gordon W. Allport, *Pattern and Growth in Personality* (New York: Holt, Reinhart & Winston, 1961).

10. Oliver P. John and Richard W. Robins, "Gordon Allport," in *Fifty Years of Personality Psychology*, ed. Kenneth H. Craik, Robert Hogan, and Raymond N. Wolfe (New York: Springer, 1993), 215–36.

11. For discussion, see Rachel L. C. Mitchell and Veena Kumari, "Hans Eysenck's Interface Between the Brain and Personality: Modern Evidence on the Cognitive Neuroscience of Personality," *Personality and Individual Differences* 103 (2016): 74–81.

12. Raymond B. Cattell, *Personality and Motivation Structure and Measurement* (New York: World Book Co.: 1957).

13. Lewis R. Goldberg, "An Alternative 'Description of Personality': The Big-Five Factor Structure," *Journal of Personality and Social Psychology* 59, no. 6 (1990): 1216–29.

14. Michael C. Ashton, Kibeom Lee, and Reinout E. De Vries, "The HEXACO Honesty-Humility, Agreeableness, and Emotionality Factors: A Review of Research and Theory," *Personality and Social Psychology Review* 18, no. 2 (2014): 139–52.

15. Christiane Nieß and Hannes Zacher, "Openness to Experience as a Predictor and Outcome of Upward Job Changes into Managerial and Professional Positions," *PLoS One* 10, no. 6 (2015): e0131115.

16. Samuel D. Gosling et al., "A Room with a Cue: Personality Judgments Based on Offices and Bedrooms," *Journal of Personality and Social Psychology* 82, no. 3 (2002): 379–98.

17. James C. Tate and Britton L. Shelton, "Personality Correlates of Tattooing and Body Piercing in a College Sample: The Kids Are Alright," *Personality and Individual Differences* 45, no. 4 (2008): 281–85.

18. David P. Schmitt, "The Big Five Related to Risky Sexual Behaviour Across 10 World Regions: Differential Personality Associations of Sexual Promiscuity and Relationship Infidelity," *European Journal of Personality* 18, no. 4 (2004): 301–19.

19. Erik E. Noftle and Richard W. Robins, "Personality Predictors of Academic Outcomes: Big Five Correlates of GPA and SAT Scores," *Journal of Personality and Social Psychology* 93, no. 1 (2007): 116–30.

20. S. E. Hampson and H. S. Friedman, "Personality and Health: A Lifespan Perspective," in *Handbook of Personality: Theory and Research*, ed. O. P. John, R. W. Robins, and L. A. Pervin (New York: Guilford Press, 2008), 770–94.

21. Cameron Anderson et al., "Who Attains Social Status? Effects of Personality and Physical Attractiveness in Social Groups," *Journal of Personality and Social Psychology* 81, no. 1 (2001): 116–32.

22. Terri D. Fisher and James K. McNulty, "Neuroticism and Marital Satisfaction: The Mediating Role Played by the Sexual Relationship," *Journal of Family Psychology* 22, no. 1 (2008): 112–22.

23. Steve Stewart-Williams, *The Ape That Understood the Universe: How the Mind and Culture Evolve* (New York: Cambridge University Press, 2018), 84.

24. Timothy A. Judge and Daniel M. Cable, "The Effect of Physical Height on

Workplace Success and Income: Preliminary Test of a Theoretical Model," *Journal of Applied Psychology* 89, no. 3 (2004): 428–41.

25. Wiebke Bleidorn et al., "Personality Maturation Around the World: A Cross-Cultural Examination of Social-Investment Theory," *Psychological Science* 24, no. 12 (2013): 2530–40.

26. Jonathan Rauch, *The Happiness Curve: Why Life Gets Better After 50* (New York: Thomas Dunne Books, 2018).

27. Robert R. McCrae and Antonio Terracciano, "Personality Profiles of Cultures: Aggregate Personality Traits," *Journal of Personality and Social Psychology* 89, no. 3 (2005): 407–25.

28. Michael Muthukrishna, Joseph Henrich, and Edward Slingerland, "Psychology as a Historical Science," *Annual Review of Psychology* 72 (2021): 717–49.

29. Dana R. Carney et al., "The Secret Lives of Liberals and Conservatives: Personality Profiles, Interaction Styles, and the Things They Leave Behind," *Political Psychology* 29, no. 6 (2008): 807–40.

30. Zhiguo Luo et al., "Gender Identification of Human Cortical 3-D Morphology Using Hierarchical Sparsity," *Frontiers in Human Neuroscience* 13 (2019): 29.

31. David P. Schmitt et al., "Why Can't a Man Be More Like a Woman? Sex Differences in Big Five Personality Traits Across 55 Cultures," *Journal of Personality and Social Psychology* 94, no. 1 (2008): 168–82.

32. Tim Kaiser, Marco Del Giudice, and Tom Booth, "Global Sex Differences in Personality: Replication with an Open Online Dataset," *Journal of Personality* 88, no. 3 (2020): 415–49.

33. Scotty Barry Kaufmann, "Taking Sex Differences in Personality Seriously," scottbarrykaufmann.com, December 16, 2019, https://scottbarrykaufman.com/taking-sex-differences-in-personality-seriously./

34. Quoted by Stuart Ritchie, *Intelligence: All That Matters* (London: John Murray, 2015).

35. Steven Pinker, *The Blank Slate: The Modern Denial of Human Nature* (New York: Viking Penguin, 2003).

36. Michael Lewis, *The Undoing Project: A Friendship That Changed the World* (London: Allen Lane, 2016), 96.

37. Fredrik DeBoer, *The Cult of Smart: How Our Broken Education System Perpetuates Social Injustice* (New York: All Points Books, 2020), 27.

38. Much of the discussion that follows draws on Ritchie, *Intelligence: All That Matters*.

39. Heiner Rindermann, *Cognitive Capitalism: Human Capital and the Wellbeing of Nations* (Cambridge, UK: Cambridge University Press, 2018).

40. Ritchie, *Intelligence: All That Matters*.

41. Ian J. Deary, Alison Pattie, and John M. Starr, "The Stability of Intelligence from Age 11 to Age 90 Years: The Lothian Birth Cohort of 1921," *Psychological Science* 24, no. 12 (2013): 2361–68.

42. Linda S. Gottfredson, "Mainstream Science on Intelligence: An Editorial with 52 Signatories, History, and Bibliography," *Intelligence* 24, no. 1 (1997): 13–23.

43. Daniel L. Schacter, Daniel T. Gilbert, and Daniel M. Wegner, *Psychology*, 2nd ed. (New York: W. W. Norton, 2020), 394.

44. Katie M. Williams et al., "Phenotypic and Genotypic Correlation Between Myopia and Intelligence," *Scientific Reports* 7, no. 1 (2017): 1–8.

45. Harrison J. Kell, David Lubinski, and Camilla P. Benbow, "Who Rises to the Top? Early Indicators," *Psychological Science* 24, no. 5 (2013): 648–59.

46. David Lubinski, Camilla P. Benbow, and Harrison J. Kell, "Life Paths and Accomplishments of Mathematically Precocious Males and Females Four Decades Later," *Psychological Science* 25, no. 12 (2014): 2217–32.

47. Angela L. Duckworth et al., "Grit: Perseverance and Passion for Long-Term Goals," *Journal of Personality and Social Psychology* 92, no. 6 (2007): 1087–1101; Angela Duckworth, *Grit: The Power of Passion and Perseverance* (New York: Scribner, 2016), 90.

48. For review, see Steven Pinker, *The Better Angels of Our Nature: Why Violence Has Declined* (New York: Viking Penguin, 2012).

49. Roy F. Baumeister, *Evil: Inside Human Cruelty and Violence* (New York: Henry Holt, 1996).

50. Abigail A. Marsh et al., "Neural and Cognitive Characteristics of Extraordinary Altruists," *Proceedings of the National Academy of Sciences* 111, no. 42 (2014): 15036–41.

51. Adam Smith, *The Theory of Moral Sentiments* (Los Angeles: Logos Books, 2018), 165.

52. Naomi P. Friedman, Marie T. Banich, and Matthew C. Keller, "Twin Studies to GWAS: There and Back Again," *Trends in Cognitive Sciences* 25, no. 10 (2021): 855–69.

53. Much of the discussion that follows is based on Kathryn Paige Harden, *The Genetic Lottery: Why DNA Matters for Social Equality* (Princeton, NJ: Princeton University Press, 2021).

54. Eric Turkheimer, "Three Laws of Behavior Genetics and What They Mean," *Current Directions in Psychological Science* 9, no. 5 (2000): 160–64.

55. Emily A. Willoughby et al., "Free Will, Determinism, and Intuitive Judgments About the Heritability of Behavior," *Behavior Genetics* 49, no. 2 (2019): 136–53.

56. Richard C. Lewontin, "Race and Intelligence," *Bulletin of the Atomic Scientists* 26, no. 3 (1970): 2–8.

57. James R. Flynn, *What Is Intelligence?: Beyond the Flynn Effect* (Cambridge, UK: Cambridge University Press, 2007).

58. For discussion, see Pinker, *The Better Angels of Our Nature.*

59. Bernt Bratsberg and Ole Rogeberg, "Flynn Effect and Its Reversal Are Both Environmentally Caused," *Proceedings of the National Academy of Sciences* 115, no. 26 (2018): 6674–78.

60. Elliot M. Tucker-Drob and Timothy C. Bates, "Large Cross-National Differences in Gene × Socioeconomic Status Interaction on Intelligence," *Psychological Science* 27, no. 2 (2016): 138–49.

61. Richard E. Nisbett, "Intelligence: New Findings and Theoretical Developments," *American Psychologist* 67, no. 2 (2012): 130.

62. For discussion, see Harden, *The Genetic Lottery*.

63. Christopher Jencks, *Inequality: A Reassessment of the Effect of Family and Schooling in America* (New York: Basic Books, 1972).

64. Eric Turkheimer, "Three Laws of Behavior Genetics and What They Mean," *Current Directions in Psychological Science* 9, no. 5 (2000): 160–64.

65. For an exploration of this idea, see Judith Rich Harris, *The Nurture Assumption: Why Children Turn Out the Way They Do* (New York: Simon and Schuster, 2011).

66. For one example among many, Zvi Strassberg et al., "Spanking in the Home and Children's Subsequent Aggression Toward Kindergarten Peers," *Development and Psychopathology* 6, no. 3 (1994): 445–61.

67. K. Paige Harden, "'Reports of My Death Were Greatly Exaggerated': Behavior Genetics in the Postgenomic Era," *Annual Review of Psychology* 72 (2021): 37–60.

68. Christopher F. Chabris et al., "The Fourth Law of Behavior Genetics," *Current Directions in Psychological Science* 24, no. 4 (2015): 304–12.

69. Ming Qian, "The Effects of Iodine on Intelligence in Children: A Meta-Analysis of Studies Conducted in China," *Asia Pacific Journal of Clinical Nutrition* 14, no. 1 (2005): 32–42; Serve Heidari et al., "Correlation Between Lead Exposure and Cognitive Function in 12-Year-Old Children: A Systematic Review and Meta-Analysis," *Environmental Science and Pollution Research* 28, no. 32 (2021): 43064–73.

70. DeBoer, *The Cult of Smart: How Our Broken Education System Perpetuates Social Injustice*.

71. Harden, *The Genetic Lottery: Why DNA Matters for Social Equality*.

72. For similar concerns, see Daniel Markovits, *The Meritocracy Trap* (New York: Penguin Press, 2019).

14. Suffering Minds

1. Christopher J. Ferguson and John Colwell, "Lack of Consensus Among Scholars on the Issue of Video Game 'Addiction,'" *Psychology of Popular Media* 9, no. 3 (2020): 359–66.

2. Rory C. Reid and Martin P. Kafka, "Controversies About Hypersexual Disorder and the DSM-5," *Current Sexual Health Reports* 6, no. 4 (2014): 259–64.

3. Thomas S. Szasz, *The Myth of Mental Illness: Foundations of a Theory of Personal Conduct* (New York: HarperCollins, 2011); see also Thomas Szasz, "The Myth of Mental Illness: 50 Years Later," *Psychiatrist* 35, no. 5 (2011): 179–82.

4. Jerome Groopman, "The Troubled History of Psychiatry," *New Yorker*, May 20, 2019.

5. For instance, Roy Richards Grinker, "Being Trans Is Not a Mental Disorder," *New York Times*, December 7, 2018, https://www.nytimes.com/2018/12/06/opinion/trans-gender-dysphoria-mental-disorder.html.

6. Johann Hari, *Lost Connections: Why You're Depressed and How to Find Hope* (London: Bloomsbury Publishing, 2019).

7. Ronald Pies, "The Bereavement Exclusion and DSM-5: An Update and Commentary," *Innovations in Clinical Neuroscience* 11, no. 7–8 (2014): 19–22.

8. Hari, *Lost Connections*, 51.

9. For instance, Steve Silberman, *NeuroTribes: The Legacy of Autism and the Future of Neurodiversity* (New York: Avery/Penguin, 2015).

10. Tyler Cowen and Daniel Gross, *Talent: How to Identify Energizers, Creatives, and Winners Around the World* (New York: St. Martin's Press, 2022), 158.

11. Oliver Sacks, *An Anthropologist on Mars: Seven Paradoxical Tales* (New York: Vintage Books, 1996), 246.

12. Getinet Ayano, "Schizophrenia: A Concise Overview of Etiology, Epidemiology Diagnosis and Management: Review of Literatures," *Journal of Schizophrenia Research* 3, no. 2 (2016): 2–7.

13. Thomas Munk Laursen, "Life Expectancy Among Persons with Schizophrenia or Bipolar Affective Disorder," *Schizophrenia Research* 131, no. 1–3 (2011): 101–4.

14. Kathryn M. Abel, Richard Drake, and Jill M. Goldstein, "Sex Differences in Schizophrenia," *International Review of Psychiatry* 22, no. 5 (2010): 417–28.

15. For an accessible summary, see Deanna M. Barch, "Schizophrenia Spectrum Disorders," in *Noba Textbook Series: Psychology*, eds. Robert Biswas-Diener and Ed Diener (Champaign, IL: DEF Publishers, 2020), http://noba.to/5d98nsy4.

16. Katherine H. Karlsgodt, Daqiang Sun, and Tyrone D. Cannon, "Structural and Functional Brain Abnormalities in Schizophrenia," *Current Directions in Psychological Science* 19, no. 4 (2010): 226–31.

17. Karlsgodt, Sun, and Cannon, "Structural and Functional Brain Abnormalities in Schizophrenia."

18. Elena Ivleva, Gunvant Thaker, and Carol A. Tamminga, "Comparing Genes and Phenomenology in the Major Psychoses: Schizophrenia and Bipolar 1 Disorder," *Schizophrenia Bulletin* 34, no. 4 (2008): 734–42.

19. M. S. Farrell et al., "Evaluating Historical Candidate Genes for Schizophrenia," *Molecular Psychiatry* 20, no. 5 (2015): 555–62.

20. Christopher F. Chabris et al., "The Fourth Law of Behavior Genetics," *Current Directions in Psychological Science* 24, no. 4 (2015): 304–12.

21. Sarah Bendall et al., "Childhood Trauma and Psychotic Disorders: A Systematic, Critical Review of the Evidence," *Schizophrenia Bulletin* 34, no. 3 (2008): 568–79.

22. Shirli Werner, Dolores Malaspina, and Jonathan Rabinowitz, "Socioeconomic Status at Birth Is Associated with Risk of Schizophrenia: Population-Based Multilevel Study," *Schizophrenia Bulletin* 33, no. 6 (2007): 1373–78.

23. Trisha A. Jenkins, "Perinatal Complications and Schizophrenia: Involvement of the Immune System," *Frontiers in Neuroscience* 7 (2013): 1–9.

24. Hannah Gardener, Donna Spiegelman, and Stephen L. Buka, "Perinatal and Neonatal Risk Factors for Autism: A Comprehensive Meta-Analysis," *Pediatrics* 128, no. 2 (2011): 344–55.

25. Hai-yin Jiang et al., "Maternal Infection During Pregnancy and Risk of Autism Spectrum Disorders: A Systematic Review and Meta-Analysis," *Brain, Behavior, and Immunity* 58 (2016): 165–72.

26. Ousseny Zerbo et al., "Month of Conception and Risk of Autism," *Epidemiology (Cambridge, Mass.)* 22, no. 4 (2011): 469.

27. Elaine F. Walker, Tammy Savoie, and Dana Davis, "Neuromotor Precursors of Schizophrenia," *Schizophrenia Bulletin* 20, no. 3 (1994): 441–51.

28. Quoted in "James Joyce and His Daughter Lucia (The Subtle Border Between Madness and Genius)," Faena Aleph, https://www.faena.com/aleph/james-joyce -and-his-daughter-lucia-the-subtle-border-between-madness-and-genius.

29. For an accessible summary, see Anda Gershon and Renee Thompson, "Mood Disorders," in *Noba Textbook Series: Psychology*.

30. Andrew Solomon, "Anatomy of Melancholy," *New Yorker*, January 4, 1998.

31. Deborah S. Hasin et al., "Epidemiology of Adult DSM-5 Major Depressive Disorder and Its Specifiers in the United States," *Journal of the American Medical Association Psychiatry* 75, no. 4 (2018): 336–46.

32. Aislinne Freeman et al., "The Role of Socio-Economic Status in Depression: Results from the COURAGE (Aging Survey in Europe)," *BMC Public Health* 16, no. 1 (2016): 1–8.

33. Matthew Cobb, *The Idea of the Brain: The Past and Future of Neuroscience* (London: Profile Books, 2020), 304.

34. Cristy Phillips, "Brain-Derived Neurotrophic Factor, Depression, and Physical Activity: Making the Neuroplastic Connection," *Neural Plasticity*, (2017): article 7260130.

35. Susan Nolen-Hoeksema, Blair E. Wisco, and Sonja Lyubomirsky, "Rethinking Rumination," *Perspectives on Psychological Science* 3, no. 5 (2008): 400–424.

36. Susan Nolen-Hoeksema, "Sex Differences in Unipolar Depression: Evidence and Theory," *Psychological Bulletin* 101, no. 2 (1987): 259–22.

37. Scott Siskind, "Depression," Lorien Psychiatry, 2022, https://lorienpsych.com /2021/06/05/depression/.

38. Andrew Solomon, *The Noonday Demon* (New York: Simon and Schuster, 2014), 129.

39. Alastair G. Cardno and Michael J. Owen, "Genetic Relationships Between Schizophrenia, Bipolar Disorder, and Schizoaffective Disorder," *Schizophrenia Bulletin* 40, no. 3 (2014): 504–15.

40. Kay Redfield Jamison, *Touched with Fire* (New York: Simon and Schuster, 1996).

41. For an accessible summary, see David H. Barlow and Kristen K. Ellard, "Anxiety and Related Disorders," in *Noba Textbook Series: Psychology*.

42. Ronald C. Kessler et al., "Lifetime Prevalence and Age-of-Onset Distributions of DSM-IV Disorders in the National Comorbidity Survey Replication," *Archives of General Psychiatry* 62, no. 6 (2005): 593–602.

43. Kessler et al., "Lifetime Prevalence and Age-of-Onset Distributions of DSM-IV Disorders in the National Comorbidity Survey Replication."

44. Peter Muris and H. Harald Merckelbach, "Specific Phobia: Phenomenology, Epidemiology, and Etiology," in *Intensive One-Session Treatment of Specific Phobias*, eds. Thompson E. Davis III, Thomas H. Ollendick, and Lars-Göran Öst (New York: Springer, 2012), 3–18.

45. Emily J. Fawcett, Hilary Power, and Jonathan M. Fawcett, "Women Are at Greater Risk of OCD Than Men: A Meta-Analytic Review of OCD Prevalence Worldwide," *Journal of Clinical Psychiatry* 81, no. 4 (2020): 13075.

46. Judith L. Rapoport, *The Boy Who Couldn't Stop Washing: The Experience and Treatment of Obsessive-Compulsive Disorder* (New York: Signet, 1991), 43–44.

47. Hari, *Lost Connections*, 15.

48. For an accessible review, see Delena van Heugten-van der Kloet, "Dissociative Disorders," in *Noba Textbook Series: Psychology*.

49. Steven Jay Lynn et al., "Dissociation and Dissociative Disorders Reconsidered: Beyond Sociocognitive and Trauma Models Toward a Transtheoretical Framework," *Annual Review of Clinical Psychology* 18 (2022): 259–89.

50. Paulette Marie Gillig, "Dissociative Identity Disorder: A Controversial Diagnosis," *Psychiatry (Edgmont)* 6, no. 3 (2009): 24–29.

51. Steven Jay Lynn et al., "Dissociation and Its Disorders: Competing Models, Future Directions, and a Way Forward," *Clinical Psychology Review* 73 (2019): 101755.

52. Dalena van Heugten-van der Kloet et al., "Imagining the Impossible Before Breakfast: The Relation Between Creativity, Dissociation, and Sleep," *Frontiers in Psychology* 6 (2015): 324.

53. For an accessible review, see Hannah Boettcher, Stefan Hofmann, and Jade Wu, "Therapeutic Orientations," in *Noba Textbook Series: Psychology*.

54. For a general discussion, see Emily A. Holmes et al., "The Lancet Psychiatry Commission on Psychological Treatments Research in Tomorrow's Science," *Lancet Psychiatry* 5, no. 3 (2018): 237–86.

55. Randolph M. Nesse, *Good Reasons for Bad Feelings: Insights from the Frontier of Evolutionary Psychiatry* (New York: Dutton, 2019), 20.

56. Nesse, *Good Reasons for Bad Feelings*, 6.

57. For discussion, see Bruce E. Wampold et al., "A Meta-Analysis of Outcome Studies Comparing Bona Fide Psychotherapies: Empirically, 'All Must Have Prizes,'" *Psychological Bulletin* 122, no. 3 (1997): 203–15; Rick Budd and Ian Hughes, "The Dodo Bird Verdict—Controversial, Inevitable and Important: A Commentary on 30 Years of Meta-Analyses," *Clinical Psychology & Psychotherapy: An International Journal of Theory & Practice* 16, no. 6 (2009): 510–22.

58. Saul Rosenzweig, "Some Implicit Common Factors in Diverse Methods of Psychotherapy," *American Journal of Orthopsychiatry* 6, no. 3 (1936): 412–15.

59. Louis G. Castonguay and Larry E. Beutler, eds., *Principles of Therapeutic Change That Work* (New York: Oxford University Press, 2005).

60. Groopman, "The Troubled History of Psychiatry."

61. Maria Konnikova, "The New Criteria for Mental Disorders," *New Yorker*, May 8, 2013.

62. Adam Rogers, "Star Neuroscientist Tom Insel Leaves the Google-Spawned Verily for . . . a Startup?," *Wired*, May 2017.

63. Groopman, "The Troubled History of Psychiatry."

64. Mary O'Hara and Pamela Duncan, "Why 'Big Pharma' Stopped Searching for the Next Prozac," *Guardian*, January 27, 2016, https://www.theguardian.com /society/2016/jan/27/prozac-next-psychiatric-wonder-drug-research-medicine -mental-illness.

65. Nesse, *Good Reasons for Bad Feelings*.

66. Robert A. Power et al., "Polygenic Risk Scores for Schizophrenia and Bipolar Disorder Predict Creativity," *Nature Neuroscience* 18, no. 7 (2015): 953–55.

67. For review, see Stephane A. De Brito et al., "Psychopathy," *Nature Reviews Disease Primers* 7, no. 1 (2021): 1–21.

68. R. D. Hare, *The Hare PCL-R*, 2nd ed. (Toronto: Multi-Health Systems, 2003).

69. Ellison M. Cale and Scott O. Lilienfeld, "Sex Differences in Psychopathy and Antisocial Personality Disorder: A Review and Integration," *Clinical Psychology Review* 22, no. 8 (2002): 1179–207.

70. K. Dutton, *The Wisdom of Psychopaths: What Saints, Spies, and Serial Killers Can Teach Us About Success* (Toronto: Anchor Canada, 2012).

71. Ana Sanz-García et al., "Prevalence of Psychopathy in the General Adult Population: A Systematic Review and Meta-Analysis," *Frontiers in Psychology* (2021): 3278.

72. Nick Haslam et al., "Dimensions over Categories: A Meta-Analysis of Taxometric Research," *Psychological Medicine* 50, no. 9 (2020): 1418–32.

15. The Good Life

1. For discussion, see Scott Barry Kaufman, *Transcend: The New Science of Self-Actualization* (New York: Tarcher/Perigee, 2021).

2. M. E. Seligman and M. Csikszentmihalyi, "Positive Psychology: An Introduction," *American Psychologist* 55, no. 1 (2000): 5.

3. For instance, Dan Gilbert, "The Surprising Science of Happiness," TED Video, 20:52, 2004, https://www.ted.com/talks/dan_gilbert_the_surprising_science_of _happiness.

4. For critical discussion, see Barbara Ehrenreich, *Bright-Sided: How Positive Thinking Is Undermining America* (New York: Metropolitan Books, 2009); Jesse Singal, *The Quick Fix: Why Fad Psychology Can't Cure Our Social Ills* (New York: Farrar, Straus and Giroux, 2021).

5. Steven Pinker, *Enlightenment Now: The Case for Reason, Science, Humanism, and Progress* (New York: Penguin Books, 2018).

6. Tyler Cowen, *Stubborn Attachments* (San Francisco: Stripe Press, 2018), 17.

7. Paul Bloom, *The Sweet Spot: The Pleasures of Suffering and the Search for Meaning* (New York: Ecco/HarperCollins, 2021).

8. Ed Diener, "Subjective Well-Being: The Science of Happiness and a Proposal for a National Index," *American Psychologist* 55, no. 1 (2000): 34–43.

9. Daniel Kahneman, Peter P. Wakker, and Rakesh Sarin, "Back to Bentham? Explorations of Experienced Utility," *Quarterly Journal of Economics* 112, no. 2 (1997): 375–406.

10. Johannes C. Eichstaedt et al., "Lifestyle and Wellbeing: Exploring Behavioral and Demographic Covariates in a Large US Sample," *International Journal of Wellbeing* 10, no. 4 (2020): 87–112.

11. Bloom, *The Sweet Spot*.

12. Ed Diener et al., "Happiest People Revisited," *Perspectives on Psychological Science* 13, no. 2 (2018): 176–84.

13. Lara B. Aknin et al., "Happiness and Prosocial Behavior: An Evaluation of the Evidence," in *World Happiness Report*, eds. John F. Helliwell, Richard Layard, and Jeffrey D. Sachs (New York: Sustainable Development Solutions Network, 2019), https://worldhappiness.report/ed/2019/happiness-and-prosocial-behavior-an -evaluation-of-the-evidence/.

14. Aknin et al., "Happiness and Prosocial Behavior: An Evaluation of the Evidence."

15. Ricky N. Lawton, Iulian Gramatki, Will Watt, and Daniel Fujiwara, "Does Volunteering Make Us Happier, or Are Happier People More Likely to Volunteer? Addressing the Problem of Reverse Causality When Estimating the Wellbeing Impacts of Volunteering," *Journal of Happiness Studies* 22, no. 2 (2021): 599–624.

16. Aknin et al., "Happiness and Prosocial Behavior: An Evaluation of the Evidence."

17. David Lykken and Auke Tellegen, "Happiness Is a Stochastic Phenomenon," *Psychological Science* 7, no. 3 (1996): 186–89.

18. Sonja Lyubomirsky, *The How of Happiness: A Scientific Approach to Getting the Life You Want* (New York: Penguin Books, 2008).

19. Daniel Gilbert, *Stumbling on Happiness* (New York: Vintage Books, 2009).

20. Elizabeth W. Dunn, Timothy D. Wilson, and Daniel T. Gilbert, "Location, Location, Location: The Misprediction of Satisfaction in Housing Lotteries," *Personality and Social Psychology Bulletin* 29, no. 11 (2003): 1421–32.

21. Paul Dolan and Robert Metcalfe, "'Oops . . . I Did It Again': Repeated Focusing Effects in Reports of Happiness," *Journal of Economic Psychology* 31, no. 4 (2010): 732–37.

22. Timothy D. Wilson, Jay Meyers, and Daniel T. Gilbert, "'How Happy Was I, Anyway?': A Retrospective Impact Bias," *Social Cognition* 21, no. 6 (2003): 421–46.

23. Daniel T. Gilbert et al., "Immune Neglect: A Source of Durability Bias in Affective Forecasting," *Journal of Personality and Social Psychology* 75, no. 3 (1998): 617–38.

24. Daniel Kahneman and Richard H. Thaler, "Anomalies: Utility Maximization and Experienced Utility," *Journal of Economic Perspectives* 20, no. 1 (2006): 221–24.

25. Gilbert et al., "Immune Neglect: A Source of Durability Bias in Affective Forecasting," 617–638, references removed from quote.

26. Ed Diener et al., "Findings All Psychologists Should Know from the New Science on Subjective Well-Being," *Canadian Psychology/Psychologie Canadienne* 58, no. 2 (2017): 87–104. To look at the most recent data on happiness across countries, see https://worldhappiness.report.

27. John F. Helliwell et al., "World Happiness Report 2021," https://happiness-report
.s3.amazonaws.com/2021/WHR+21.pdf.

28. Anna Wierzbicka, "'Happiness' in Cross-Linguistic & Cross-Cultural Perspective," *Daedalus* 133, no. 2 (2004): 34–43. The discussion that follows is drawn, with
minor modifications, from Bloom, *The Sweet Spot*.

29. Helliwell et al., "World Happiness Report 2021"; John F. Helliwell, Richard Layard, Jeffrey D. Sachs, and Jan-Emmanuel De Neve, "World Happiness Report
2021: Happiness, Trust, and Deaths Under COVID-19," https://worldhappiness
.report/ed/2021/happiness-trust-and-deaths-under-COVID-19/.

30. Diener et al., "Findings All Psychologists Should Know from the New Science on
Subjective Well-Being."

31. Summarized in "World Happiness Report 2018," at https://worldhappiness
.report/ed/2018/.

32. For instance, Ed Diener and Robert Biswas-Diener, "Will Money Increase Subjective Well-Being?," *Social Indicators Research* 57, no. 2 (2002): 119–69.

33. Andrew T. Jebb et al., "Happiness, Income Satiation and Turning Points Around
the World," *Nature Human Behaviour* 2, no. 1 (2018): 33–38.

34. "Life Experiences and Income Inequality in the United States," Robert Wood
Johnson Foundation, January 9, 2020, https://www.rwjf.org/en/library/research
/2019/12/life-experiences-and-income-inequality-in-the-united-states.html;
Christopher Ingraham, "The 1% Are Much More Satisfied with Their Lives Than
Everyone Else, Survey Finds," *Washington Post*, January 9, 2020, https://www
.washingtonpost.com/business/2020/01/09/1-are-much-more-satisfied-with
-their-lives-than-everyone-else-survey-finds/. For similar results with life satisfaction, see Matthew A. Killingsworth, "Experienced Well-Being Rises with Income,
Even Above $75,000 per Year," *PNAS* 118, no. 4 (January 18, 2021), https://www
.pnas.org/content/118/4/e2016976118.short.

35. Grant E. Donnelly et al., "The Amount and Source of Millionaires' Wealth (Moderately) Predict Their Happiness," *Personality and Social Psychology Bulletin* 44,
no. 5 (2018): 684–99.

36. Philip Brickman, Dan Coates, and Ronnie Janoff-Bulman, "Lottery Winners and
Accident Victims: Is Happiness Relative?," *Journal of Personality and Social Psychology* 36, no. 8 (1978): 917–27.

37. Jonathan Gardner and Andrew J. Oswald, "Money and Mental Wellbeing: A Longitudinal Study of Medium-Sized Lottery Wins," *Journal of Health Economics* 26,
no. 1 (2007): 49–60.

38. Jonathan Gardner and Andrew Oswald, "Does Money Buy Happiness? A Longitudinal Study Using Data on Windfalls," working paper, Warwick University,
2001, https://users.nber.org/~confer/2001/midmf01/oswald.pdf.

39. Ed Diener, Richard E. Lucas, and Shigehiro Oishi, "Advances and Open Questions in the Science of Subjective Well-Being," *Collabra: Psychology* 4, no. 1
(2018).

40. Arthur A. Stone et al., "A Snapshot of the Age Distribution of Psychological Well-

Being in the United States," *Proceedings of the National Academy of Sciences* 107, no. 22 (2010): 9985–90. For discussion, see Jonathan Rauch, *The Happiness Curve: Why Life Gets Better After 50* (New York: Thomas Dunne Books, 2018).

41. Angus Deaton and Arthur A. Stone, "Two Happiness Puzzles," *American Economic Review* 103, no. 3 (2013): 591–97.

42. Cassondra Batz and Louis Tay, "Gender Differences in Subjective Well-Being," in *Handbook of Well-Being* , ed. Ed Diener, Shigehiro Oishi, and Louis Tay (Salt Lake City, UT: DEF Publishers, 2018).

43. David G. Myers and Ed Diener, "The Scientific Pursuit of Happiness," *Perspectives on Psychological Science* 13, no. 2 (2018): 218–25.

44. Sonja Lyubomirsky, Laura King, and Ed Diener, "The Benefits of Frequent Positive Affect: Does Happiness Lead to Success?," *Psychological Bulletin* 131, no. 6 (2005): 803–55.

45. Maike Luhmann et al., "Subjective Well-Being and Adaptation to Life Events: A Meta-Analysis," *Journal of Personality and Social Psychology* 102, no. 3 (2012): 592–615.

46. Kristen Schultz Lee and Hiroshi Ono, "Marriage, Cohabitation, and Happiness: A Cross-National Analysis of 27 Countries," *Journal of Marriage and Family* 74, no. 5 (2012): 953–72.

47. The discussion that follows is drawn, with minor modifications, from Bloom, *The Sweet Spot.*

48. Daniel Kahneman et al., "A Survey Method for Characterizing Daily Life Experience: The Day Reconstruction Method," *Science* 306, no. 5702 (2004): 1776–80.

49. Richard E. Lucas et al., "Reexamining Adaptation and the Set Point Model of Happiness: Reactions to Changes in Marital Status," *Journal of Personality and Social Psychology* 84, no. 3 (2003): 527–39; Luhmann et al., "Subjective Well-Being and Adaptation to Life Events: A Meta-Analysis."

50. Jean M. Twenge, W. Keith Campbell, and Craig A. Foster, "Parenthood and Marital Satisfaction: A Meta-Analytic Review," *Journal of Marriage and Family* 65, no. 3 (2003): 574–83.

51. Chuck Leddy, "Money, Marriage, Kids," *Harvard Gazette*, February 21, 2013, https://news.harvard.edu/gazette/story/2013/02/money-marriage-kids.

52. Jennifer Senior, *All Joy and No Fun: The Paradox of Modern Parenthood* (New York: Ecco/HarperCollins, 2014), 49.

53. S. Katherine Nelson et al., "In Defense of Parenthood: Children Are Associated with More Joy Than Misery," *Psychological Science* 24, no. 1 (2013): 3–10.

54. Jennifer Glass, Robin W. Simon, and Matthew A. Andersson, "Parenthood and Happiness: Effects of Work-Family Reconciliation Policies in 22 OECD Countries," *American Journal of Sociology* 122, no. 3 (2016): 886–929.

55. Senior, *All Joy and No Fun*, 256–57.

56. Angus Deaton and Arthur A. Stone, "Evaluative and Hedonic Wellbeing Among Those With and Without Children at Home," *Proceedings of the National Academy of Sciences* 111, no. 4 (2014): 1328–33.

57. Zadie Smith, "Joy," *New York Review of Books*, January 10, 2013, 4.

58. Pinker, *Enlightenment Now*, 267.

59. Peter D. Kramer, *Listening to Prozac: A Psychiatrist Explores Antidepressant Drugs and the Remaking of the Self* (New York: Penguin Books, 1994).

60. Brett Q. Ford et al., "Desperately Seeking Happiness: Valuing Happiness Is Associated with Symptoms and Diagnosis of Depression," *Journal of Social and Clinical Psychology* 33, no. 10 (2014): 890–905.

61. B. Ford and I. Mauss, "The Paradoxical Effects of Pursuing Positive Emotion," in *Positive Emotion: Integrating the Light Sides and Dark Sides*, eds. J. Gruber and J. T. Moskowitz (Oxford: Oxford University Press, 2014): 363–82.

62. Ford and Mauss, "The Paradoxical Effects of Pursuing Positive Emotion."

63. Helga Dittmar et al., "The Relationship Between Materialism and Personal Well-Being: A Meta-Analysis," *Journal of Personality and Social Psychology* 107, no. 5 (2014): 879–924. The discussion that follows is drawn, with minor modifications, from Bloom, *The Sweet Spot*.

64. Brett Q. Ford et al., "Culture Shapes Whether the Pursuit of Happiness Predicts Higher or Lower Well-Being," *Journal of Experimental Psychology: General* 144, no. 6 (2015): 1053–62.

65. Shigehiro Oishi and Erin C. Westgate, "A Psychologically Rich Life: Beyond Happiness and Meaning," *Psychological Review* (2021). Advance online publication.

66. Daniel Kahneman et al., "When More Pain Is Preferred to Less: Adding a Better End," *Psychological Science* 4, no. 6 (1993): 401–5.

67. Simon Kemp, Christopher D. B. Burt, and Laura Furneaux, "A Test of the Peak-End Rule with Extended Autobiographical Events," *Memory & Cognition* 36, no. 1 (2008): 132–38. The discussion that follows is drawn, with minor modifications, from Bloom, *The Sweet Spot*.

68. Kahneman et al., "When More Pain Is Preferred to Less: Adding a Better End."

69. Donald A. Redelmeier, Joel Katz, and Daniel Kahneman, "Memories of Colonoscopy: A Randomized Trial," *Pain* 104, no. 1–2 (2003): 187–94.

Index

above-average effect, 263–64
access consciousness, 43–45, 51–52
Adelson, Edward, 169
Adler, Alfred, 57
adoption studies, 328–30
adrenaline, 229
affective forecasting, 373–74
age
   happiness and, 318, 369, 375,
     378
   implicit bias test and, 298–99,
     301
   personality and, 318
   as social category, 285, 286, 288,
     289, 290–91, 295, 365
agnosia, 27–28
AI systems, 22, 37, 137, 166–67,
   282
Alexander, Richard, 251
Alexander, Scott, 173
Allport, Gordon, 313
altruism, 245, 247, 251, 326, 372
American Psychiatric Association,
   344

amygdala, 229
anal stage, 60–61
anterior insular cortex, 232
anterograde amnesia, 184–85
anxiety, 270–71, 366
anxiety disorders, 346, 354–55,
   359, 360, 363, 364, 367
appearance-reality confusion,
   108–9
Aristotle, 15, 46, 136, 191–92
artificial selection, 222
Asch, Solomon, 275, 311
Astonishing Hypothesis, 10
asymmetric feedback, 268
attention
   babies' social understanding
     and, 118
   fear and, 229
   habituation and, 111–12
   William James on, 160
   physical world and, 160, 161
   unconscious attention, 45, 46
   willful control of, 39, 47
Auden, W. H., 47

autism, 58–59, 345–46, 350, 366
axons, 18, 19
Ayers, Nathaniel, 348

babbling, 140, 141
Bacon, Francis, 270
Banaji, Mahzarin, 298, 299
Barrow, John, 1, 4
Barthelme, Frederick, 165
Barthelme, Steven, 165
base-rate neglect, 195, 278
behavioral economics, 191
behavioral psychology
    on associations, 135, 279–80
    challenges to, 85–86
    classical conditioning and,
        75–77, 79, 80, 83, 88
    on consciousness, 35, 74
    on differences between species,
        72–73, 84
    empiricism and, 100
    generalization and, 77–78
    on language, 125
    on learning, 72, 75, 77, 80, 84,
        85–86, 88, 93
    on mental representations,
        73–74, 84, 86, 89, 93, 109,
        160–61
    operant conditioning and, 80–
        84, 85, 86, 87, 88, 103
    on phobias, 78, 79, 93
    reinforced trials and, 77
    schools of, 72
    self-reinforcement and, 91–92,
        219–20
    B. F. Skinner and, 81–83,
        89–92, 125
    social priming and, 279
    theories of, 71–72
    unreinforced trials and, 77

behavior genetics, 328–30, 334,
    338, 350, 373
behavior therapy, 358
Benedict XVI (pope), 181
Bentham, Jeremy, 37, 249
Bergman, Peter, 268
Berkeley, George, 105, 161
bêtes machines (beast machines), 11
Beyoncé, 309
biases
    availability bias, 193–95, 201
    confirmation bias and, 199–200,
        206
    consciousness of, 52
    cultural influences and, 271
    effect of, 2, 191, 281
    implicit bias test, 283, 298–301
    language bias, 296
    overriding of, 303
    positivity bias, 268, 269–70, 271
    rationality and, 192–96, 199–
        200, 201, 206, 212
    "rhyme as reason" bias, 192–93
    self-enhancement bias, 264–65,
        267, 268
    social biases, 263–68, 274,
        297–99
    unconscious effects and, 281,
        301
Biden, Joseph, 205
Binet, Alfred, 322
binocular disparity, 172
biological adaptations, 79
bipolar disorder, 351, 354, 363
Bisson, Terry, 16–17
blind auditions, 303
Block, Ned, 43–44
Boddy, Lynne, 36
Boswell, James, 161
brain

association and, 279
asymmetry of, 28
attending to numbers and, 110,
    112, 157
computer analogy and, 22–23,
    33, 35, 179
conceptions of, 22, 23
conscious experience and, 2,
    15–16, 32–33, 42, 43, 49
contralateral organization of, 28
fear and, 229
fMRI of, 27, 32, 33, 40–41
goal of, 217–18, 223
listening to speech and, 110
memory capacity of, 179
mind-brain relationship, 31, 32,
    161, 385
natural experiments and, 27
parts of, 23–26, *26*, 32
pattern discovery and, 136
plasticity in, 352–53
right-left differences in, 28–29
sensory areas of, 42
as sentient meat, 16–17, 33
split-brain cases, 29–30
subcortical structures of, 25
thought and, 10, 14, 15, 260, 386
brain damage, effects of, 8, 9, 24,
    27, 184, 228
brain scanners, 14
brain tumors, 9–10
Breland, Keller, 85
Breland, Marian, 85
Breuer, Josef, 64
British Empiricists, 100, 103, 105,
    135, 279
Broca, Paul, 24
Broca's area, 24, 128
Brosnan, Pierce, 136
Bush, George W., 209, 373

café wall illusion, 162, *162*
Candid Camera (television show),
    274
Carey, Susan, 122–23
Caspi, Avshalom, 98–99
Cattell, Raymond, 314
central hyperventilation syndrome,
    54
cerebellum, 25
Chalmers, David, 33, 43
Charcot, Jean-Martin, 63
Charlesworth, Tessa, 299
child effect, 335
children. *See also* developmental
    psychology
adults compared to, 2, 99–100
babies' social understanding,
    116–19, 150, 260
child-as-scientist view, 123–24
consciousness and, 39, 47, 49,
    99–100
curse of knowledge and,
    121
differences in thinking and,
    122–24
effects of parenting and, 335–37
essentialism and, 114–16, 123
false-belief task and, 119–22
Sigmund Freud's developmental
    theory, 59–62
goal expectations and, 117
happiness and, 379–81
kindness and, 244
language learning of, 138,
    139–42, 260
looking-time patterns of, 117,
    120–21, 192
mental life of babies, 110–18,
    157
perception of, 164

children (*continued*)
  principles of physical world and,
    113–15, 123
  radical conceptual change and,
    122–23
  responses to babies, 226–27
  self-conscious emotions and, 49
  stereotypes and, 289
  theory of mind and, 116–21,
    148, 157
chimpanzees, 118–20, 138, 158
choice, conceptions of, 3, 4
Chomsky, Noam
  on brain as mental organ, 23
  on individual differences,
    307
  on language, 90, 152–53
  on language learning, 144–45
  on B. F. Skinner, 90–92, 219
  on syntax, 133
circadian rhythm, 101
Claparède, Édouard, 185
classical conditioning
  behavioral psychology and,
    75–77, 79, 80, 83, 88
  generalization and, 77–78
  learning and, 80, 85–86, 88
  multiple trials and, 78, 79, 85
  operant conditioning combined
    with, 83
  Ivan Pavlov and, 75–77, 88, 93,
    135
  stages of, 76–77
  systematic desensitization and,
    79–80
clinical psychology, 49, 341
Clinton, Hillary, 205, 208
coalitions, 286
Cobb, Matthew, 32
cocktail party effect, 45, 47

cognitive-behavioral therapy
    (CBT), 64, 358
cognitive dissonance, 265–66, 374
cognitive maps, 86–89
cognitive psychology
  on consciousness, 35–36, 45, 49
  on knowledge, 109
  reaction-time studies of, 109
  on stereotypes, 286–87
Cognitive Reflection Test, 204
cognitive therapy, 358
Collins, Francis, 248
color, perception of, 162, 167
communication systems, of animal
    species, 138
compatibility principle, 182–83
compliance, 277–78
computational systems, 42–43
computers
  computer analogy for brain,
    22–23, 33, 35, 179
  internal representations and,
    86
computer simulations, 12
concepts, function of, 287–88
concrete operational stage, 106–7
conformity, 274–78
Confucius, 46, 249
connectionism, 135–36
Connelly, Michael, 183–84,
    186–87
consciousness
  access consciousness and, 43–45,
    51–52
  behavioral psychology on, 35, 74
  as biological phenomenon, 42,
    43
  brain and, 2, 15–16, 32–33, 42,
    43, 49
  children and, 39, 47, 49, 99–100

computational systems and, 42–43

conception of, 4, 14, 15, 32–33, 37–38, 279

differences in experiences and, 37–39

disorders of, 49

expertise and, 178

first-person nature of, 50–51

Global Workspace Theory and, 42

habits and, 46–47, 221

inebriation and, 51

limits of, 44–46

mind-wandering and, 49

neural expression of, 40–41

obliteration of, 53–54

pain and, 51

passage of time and, 39–40

phenomenal consciousness and, 43–44

science of the mind and, 36–37, 160

social psychology on, 49, 74–75

spotlight effect and, 49–50

working memory and, 177

conservation, Jean Piaget on, 106, 107, 108

cooperation, 244

Corkin, Suzanne, 185

corpus callosum, 29–30

cortex
electrical activity in, 40
higher-order functions and, 27
motor and sensory maps of, 26–27
as part of brain, 25
parts of, 26, *26*

Cosmides, Leda, 166

countertransference, 64

COVID-19
diagnosis of, 366
false claims about, 204
politicized beliefs about, 205–6
six-word memoirs, 135
social psychology and, 255
words coined during, 132

Cowen, Tyler, 345–46, 369

Crews, Frederick, 56

Crick, Francis, 10

critical flicker-fusion frequency (CFF), 40

Crosby, Bing, 33

Crystal, David, 125–26

cultural influences
biases and, 271
brain and, 28
childhood memories and, 184
emotions and, 226
essentialism and, 115
Sigmund Freud on, 56
happiness and, 375
individualism and, 271–73
individual variation within, 240–41
kindness and, 243–44
language and, 128, 130, 140, 143, 152, 155, 158
learning and, 101
motivations and, 223
personality and, 318–19
Jean Piaget on, 107
sexuality and, 236, 238, 239, 240
social categories and, 286
thought and, 152

Curtis, Valerie, 232–33

Dalai Lama, 205

Damasio, Antonio, 228

Darwin, Charles
  on brain, 10, 21
  on disgust, 232
  Francis Galton and, 322
  on language, 128
  on natural selection, 222
  Jean Piaget compared to, 107
  Alfred Russel Wallace and, 248
Darwinian perspective
  differential reproductive success
    and, 82
  kindness and, 242
  mental representations and,
    87–88
  morality and, 248
  on sexuality, 234–35, 241, 247
DeBoer, Fredrik, 321–22, 339
decision making, 30, 308
deductive logic, 200–201
deep work, 47
defense mechanisms, 62–63, 74
Dehaene, Stanislas, 157, 163
Delphi (AI system), 137–38
DeLuise, Dom, 180–81
dendrites, 17–18, 19
Dennett, Daniel, 119, 158
depersonalization/derealization
    disorder, 356
depression
  diagnosis of, 344–45, 346, 364,
    366
  environmental influences and,
    352
  genetics and, 352
  prevalence of, 352, 354, 363, 378
  symptoms of, 351–53, 355
  treatment of, 346, 359
depth of processing, 179–80
de Saussure, Ferdinand, 129
Descartes, René
  on consciousness, 37
  on dualism, 10–15, 21, 31, 33,
    161
  on pituitary gland, 25
  on thinking, 73
destiny, developmental psychology
    and, 98–99
developmental psychology
  on child effect, 335
  on consciousness, 39, 47, 49
  destiny and, 98–99
  differences in thinking in adults
    and children, 122–24
  on essentialism, 114–16, 123
  Sigmund Freud on, 59–62, 104
  language learning and, 140
  on learning, 100–104
  low-tech nature of, 97
  nature-nurture interaction and,
    98–99, 100, 101
  Jean Piaget on, 102, 103, 104–8,
    109, 114, 119, 120, 124
  on radical cognitive change,
    122–24
  stability of characteristics over
    time, 98–99
  theory of mind and, 116–21,
    148, 157
De Vries, Rheta, 106, 108–9
Diagnostic and Statistical Manual
    of Mental Disorders, 342, 343,
    355, 356, 361, 364
dichotomies, 302–3
Dick, Philip K., 162
disgust, 3, 231–33, 235–36
displacement, 63
dissociative amnesia, 356
dissociative disorders, 356–57
dissociative identity disorder
    (DID), 348, 356, 357–58

Dodo Bird effect, 360

dogs, language learning of, 138–39, 141, 158

dopamine, 349

dreams, 62, 64, 66, 67

dualism, 10–15, 21, 25, 31, 33, 161

Duckworth, Angela, 326

Dunham, Yarrow, 295

Eagleman, David, 101

ecstasy (drug), 362

EEG (electroencephalogram), 40

ego, 59–60, 74

egocentrism, 105–6, 119

Einstein, Albert, 55–56, 181

Eisenberger, Naomi, 32

EKG, 110

Electra complex, 61–62

electroconvulsive therapy (ECT), 359

Elizabeth Stuart, Queen of Bohemia, 14

emotions
  brain activity and, 30
  cultural norms and, 226
  disgust, 3, 231–33, 235–36, 241
  evolution of, 225, 317
  facial expressions and, 225, 231, 232
  fear and, 3, 228–30, 241, 261
  lust, 233–34, 241
  morality and, 242
  motivations and, 217, 227–28, 231, 234–35, 241–42
  physiological responses and, 225–26
  self-conscious emotions, 49
  theories of, 217, 225, 226–28

empathy, 39

empiricism, 100, 103, 122

environmental influences
  empiricism and, 100
  Flynn effect and, 331–32, 338–39
  on happiness, 375–76
  heritability and, 332–36, 338
  individual differences and, 309, 327–28, 329, 330, 331, 337
  language learning and, 140, 145
  memory and, 182–83
  natural selection and, 222
  parenting and, 335–37
  schizophrenia and, 350, 352
  social priming and, 279, 280–82
  stability of characteristics over time, 98–99

Epicurus, 368

epilepsy, 29, 185

Epstein, Jeffrey, 208

ERP, 110

essentialism, 114–16, 123, 291

eugenics, 322

evolution. *See also* natural selection
  disgust and, 232, 235–36
  of emotions, 225, 317
  fear and, 229–30
  happiness and, 381–82
  instincts and, 221
  kindness and, 246
  for language, 128
  morality and, 225, 242
  motivations and, 223–25, 250
  nature-nurture interaction and, 101
  Jean Piaget and, 104
  sexuality and, 234–35, 240, 320
  social categories and, 286

evolutionarily stable strategies, 245–46

evolutionary psychology
  on learning, 85–86
  on morality, 251
  on pair bonding, 240
  on perception, 166–67
  on phobias, 79
experienced happiness, 370
experimenter demands, 108
Eysenck, Hans, 313–14

facial-recognition algorithm,
  266–67
false alarms, 270
falsifiability, 65–67, 107, 199, 220
familiarity effect, 296
fear, 3, 228–30, 261. *See also* phobias
Festinger, Leon, 374
figure/ground perception, 44, *44*,
  52
Firestone, Chaz, 219–20
Fishburne, Laurence, 285
Flanagan, Owen, 10
Flaubert, Gustave, 310
Flynn, James, 331–32, 338
fMRI, 27, 32, 33, 40–41, 110
fNIRS (fNIRS—functional near-
  infrared spectroscopy), 110
Fodor, Jerry, 31–32, 163, 164
Foer, Joshua, 180–81
formal operational stage, 107
Foxx, Jamie, 348
Francis (pope), 205
Frederick, Shane, 204
free association, 64
free riders, 245–46, 247, 250
Freud, Anna, 62
Freud, Sigmund
  on absent-mindedness, 68–69
  Gordon Allport and, 313
  on amnesia, 185
  on centrality of sexuality, 69
  on consciousness, 52, 279
  on defense mechanisms, 62–63,
    193
  developmental theory of,
    59–62, 104
  on disgust, 231
  on dreams, 62, 64, 66, 67
  in history of psychology, 68,
    192
  on human nature, 2, 56, 107
  Jean Piaget and, 102–3
  on pleasure principle, 59
  on preconscious, 178
  psychoanalysis and, 56, 61, 66,
    71–72, 74–75, 103, 109, 192,
    358
  on psychotherapy, 64, 66
  B. F. Skinner compared to, 71,
    72, 75, 92, 93, 107, 200
  theories of, 56–57, 64–67,
    71–72, 74, 75, 102, 374
  on trauma of toilet training,
    232
  on unconscious mind, 55, 56,
    57–59, 69, 71
Freudian slips, 67
frontal lobes, 8, 9, 24, 26, 121,
  228
fundamental attribution error,
  267–68, 271, 273, 309
fungi, 36

Gage, Phineas, 7–10, 14, 27, 121,
  184, 228
Galef, Julia, 206–7, 270
Gall, Franz Josef, 23–24
Galton, Francis, 322
Garcia, John, 85

Garland, Merrick, 283
Geertz, Clifford, 272
gender
  happiness and, 378
  implicit bias test and, 302
  individual differences and,
    307–8
  as social category, 284–85, 286,
    288, 289, 295, 297
gender dysphoria, 344
gender identity, 307–8, 344
generalized anxiety disorder,
    354–55
general positivity bias, 268
genetics
  behavior genetics, 328–30, 334,
    338, 350, 373
  group differences and, 330–31,
    332
  happiness and, 373–75
  heritability and, 327, 328, 330–
    36, 338–39
  individual differences and, 327,
    328–29, 330, 337–38, 339
  kin selection and, 242–43
  mental illnesses and, 346, 363
  natural selection and, 221–22
  polygenic traits, 338
  schizophrenia and, 349, 352
  stability of characteristics over
    time, 98–99
genital phase, 62
genome-wide complex trait analy-
    sis (GCTA), 329–30
Gessen, Masha, 284–85, 295
Gestalt psychology, 170–72
GI Bill of Rights, 75
Gilbert, Daniel, 373, 374, 380
Gilovich, Thomas, 49–50
Gleason, Jean Berko, 142

Gleitman, Lila, 139, 149–50
glial cells, 17, 25
Global Workspace Theory, 42
goals, theories of, 217–18
Gopnik, Alison, 39
Gore, Al, 373
Greenwald, Anthony, 298
grief, 344–45
Gross, Daniel, 345–46
group differences
  causes of, 317
  changes in, 317
  environmental influences and,
    331
  genetics and, 330–31, 332
  Gustave Le Bon on, 315–16
  meaning of, 316–17
  naturalistic fallacy and, 317–18
  in personality, 318–20
  physical differences, 317
group effect, 275–76
groups
  groupiness, 292
  mere membership and, 295
  research on, 283, 286, 287,
    293–97
  stereotypes of, 286–87, 302–3
  Us/Them psychology and, 292–
    93, 296, 303
groupthink, 207

habits, 46–47, 221
habituation, 111–12
Haeckel, Ernst, 103
Haldane, J. B. S., 242
Hamlin, Kiley, 117
handedness, 28
Hanks, Tom, 205
Hanley, Richard, 228
Hannay, Jason, 301–2

happiness
    affective forecasting, 373–74
    age and, 318, 369, 375, 378
    conditions of, 3, 369–72, 381
    decision making and, 308
    environmental influences on,
        375–76
    evolution and, 381–82
    health associated with, 378
    heritability of, 373–75
    improvements in, 382–84
    meaning and purpose, 370, 381,
        384
    memory and, 370, 384–85
    money and, 368, 371, 375,
        376–77, 383
    natural experiments and, 377
    parenthood and, 379–81
    positive psychology and, 367–68
    social relationships and, 371,
        377, 379–80, 383
    stability and, 372–73
    studies of, 370–72
    volunteering and, 372
Harden, Kathryn Paige, 335, 339
Hare, Robert, 364–65
Hare Psychopathy Checklist-
    Revised (PCL-R), 364–65
Hari, Johann, 345–46, 355
Harlow, John Martyn, 8
Haselton, Martie, 269
hazing, 266
heart, as source of sensation, 15–16
Henrich, Joseph, 259–60, 272–74
Heraclitus, 77
Herbert, Frank, 229
heritability, 327, 328, 330–36,
    338–39, 373
Hillel, 249
hippocampus, 25, 87, 180, 260

Hitchens, Christopher, 320–21
Homer, 12
Horney, Karen, 57
Hubble, Edwin, 1
human body, as machine, 11
human nature
    Sigmund Freud on, 2, 56, 107
    language and, 128
    psychology on, 2
Hume, David, 100, 125
hunger, 233–34
Hyman, Steven, 362
hypodescent, 291–92
hypophobia, 271
hypothalamus, 25
hypotheses, 65
hysteria, 63–64

Ichheiser, Gustav, 271
id, 59–60, 62, 63, 74
ignorance, 53
impartiality, 249–50
Implicit Associations Test (IAT),
    283, 298–302
inbreeding depression, 235
individual differences
    in decision making, 308
    descriptions of, 307
    environmental influences and,
        309, 327–28, 329, 330, 331,
        337
    genetics and, 327, 328–29, 330,
        337–38, 339
    intelligence and, 320–26, 327,
        339–40
    pathology and, 309
    personality and, 3, 309–13,
        326–27
    psychological tests and, 300–301
individual responsibility, 51

inhibited children, 98
innate differences, 72
Insel, Thomas, 361–62
instinct blindness, 166
instincts, theories of, 217, 220–22
intelligence
defining of, 324
differences in, 3
environmental influences and,
331–32, 333, 337–38
family environment and, 334
Flynn effect and, 331–32, 338
general intelligence, 323
heritability of, 333, 334, 337–38
individual differences in, 320–
26, 327, 339–40
as polygenic, 338
sentience and, 35–36
*International Classification of Dis-
eases*, 342, 364
Inzlicht, Michael, 256–57
IQ tests, 320–26, 328–29, 339

Jackson, Michael, 181
Jackson, Samuel L., 285
James, William
on attention, 160
on fear, 228–29
on habit, 46–47, 221
on mental life, 39, 99–100
on motivations, 220–21, 224–25
on perception, 164
on rational action, 20–21
Janet, Pierre, 63
Jencks, Christopher, 333–34, 339
Johnson, Samuel, 161
Joyce, James, 351, 363
Joyce, Lucia, 351, 363
judgments of others, 3, 266–67
Jung, Carl, 61–62, 312, 351

Kafka, Franz, 12
Kahneman, Daniel, 193, 196–97,
203, 321, 374, 379, 384–85
Kandinsky, Wassily, 294–96
Kanizsa Triangle illusion, 171–72,
172, 201
Kant, Immanuel, 2, 249
Keil, Frank, 116
Keillor, Garrison, 264
Keller, Helen, 151
ketamine, 362
kindness, 241, 242–47, 250, 326,
340, 372
King, Martin Luther, Jr., 49–50,
208
kin selection, 242
Kinzler, Katherine, 295, 296
Klee, Paul, 294–96
knowledge
cognitive psychology on, 109
empiricism and, 100
genetic epistemology and, 103
origins of, 2, 99, 103, 124
Konnikova, Maria, 47
Kramer, Peter, 55–56
Kuhlmeier, Valerie, 117
Kuhn, Thomas, 122–23

Lake Wobegon effect, 264, 268
language
access consciousness and, 51–52
associations and, 135–36
brain as computer and, 23
brain's processing of, 29
capacity of parts of brain and, 25
children's preferences for, 11,
296
Noam Chomsky on, 90, 152–53
cultural influences and, 128,
130, 140, 143, 152, 155, 158

language (*continued*)
defining of, 125–26
disorders of, 24–25
diversity of, 126–27
isolated impairments of, 143
larynx and, 128–29
metaphorical patterns in, 152
natural language, 126–27, 157
rules of, 106, 131, 132–37, 142, 144
as shaper of ideas, 154–55
sign languages, 126–27, 129, 130, 140, 157
taboo language, 152
teaching to nonhumans, 138–39
thought and, 2, 150–58
time and space in, 152
universality of, 127–28
vocabulary estimates and, 130–31
language acquisition, mechanisms of, 145–49
language learning
of apes, 139
associations and, 136
of babies, 139–40
of children, 138, 139–42, 260
critical period of, 142–43
of dogs, 138–39, 141, 158
environmental influences and, 140, 145
nativism and, 144–45
nature-nurture interaction and, 101
operant conditioning and, 143–44
of second languages, 142–43
sign language and, 140, 157
syntax and, 141–42, 143, 144, 145, 149–50

word meanings and, 147–49, 150
language processing, 172–73
larynx, 128–29
latency stage, 62
latent dream, 62
Law of Effect, 80–81, 83, 86, 110–11, 112, 215
learning
Bayesian learning and predictive processing, 125
behavioral psychology on, 72, 75, 77, 80, 84, 85–86, 88, 93
child-as-scientist view, 123
deep learning, 135
developmental psychology on, 100–104
disgust and, 233
evolutionary psychology on, 85–86
language and, 101
long-term memory and, 179–80
operant conditioning and, 80, 85, 86, 88, 101
Jean Piaget on, 103–4, 124
Le Bon, Gustave, 50, 315–16
Leibniz, Gottfried, 14, 125
Leonardo da Vinci, 23, 107
Lewis, Michael, 321
Lewontin, Richard C., 330–31, 338
Lieberman, Debra, 236
life span, 369
limbic system, 25
Liu, Lucy, 285
Lloyd, William Forster, 211–12
localization, 31–32
Locke, John
on differences in experience, 38
empiricism and, 100

on essentialism, 114
on knowledge, 109
on language, 125
on language learning, 147
on memory, 175
Loftus, Elizabeth, 188–89
lottery winners, 377
LSD, 362–63
lust, 233–34, 241

Macbeth effect, 280
McDonough, James, 276
Macnamara, John, 201, 203
McWhorter, John, 152, 156
maintenance rehearsal, 177
manifest dream, 62
Manilow, Barry, 49
Marley, Bob, 49
marriage, 379
Marx, Karl, 24
Maslow, Abraham, 367
materialism, 3, 10, 13, 32–33, 58
The Matrix (film), 12
Maximizers, 308
meaning and purpose, 370, 381, 384
medulla, 25
memory
    alterations of, 22
    anterograde amnesia and,
        184–85
    auditory sensory memory, 177
    autobiographical memory, 175
    brain activity and, 30
    brain injury and, 9, 14, 184
    of childhood, 185–86
    compatibility principle and,
        182–83
    consolidation of, 181
    depth of processing and, 179–
        80, 181

dissociative disorders and, 356
distortion of, 186, 188–89
expertise and, 178
happiness and, 370, 384–85
hedonic nature of experience
    and, 384
language and, 136
leading questions influencing,
    188–89
long-term memory, 177, 178–80
loss of, 175–76, 184
malleability of, 188
mental representations and, 159,
    160, 161
model of, 176, 176
perception and, 168, 187
procedural memory, 175, 185
retrieval cues and, 182–83
retrograde amnesia and, 184
rhyming and, 180, 181
semantic memory, 175
sensory memory, 176–77
visual sensory memory, 176
vivid images and, 180–81
working memory, 177–78, 179,
    190
memory confusion studies, 285,
    286
mental illnesses
    causes of, 3
    clinical psychology and, 341
    continuous view of, 366
    defining of, 341–43, 366
    diagnosis of, 361–62, 364,
        365–66, 386
    diathesis-stress model of, 350
    genetics and, 346, 363
    homosexuality as, 343, 344
    individual differences and, 309
    natural selection and, 363

mental illnesses (*continued*)
   personality disorders, 342
   treatment for, 3, 345, 361,
      362–63, 386
mental life
   of babies, 110–18, 157
   behavioral psychology and,
      73–74, 84, 86, 93
   mechanistic conception of, 3–4
   parts of, 23
   Jean Piaget on, 103–4, 120
mental representations
   behavioral psychology on,
      73–74, 84, 86, 89, 93, 109,
      160–61
   cognitive maps and, 86–87, 88
   Darwinian perspective on,
      87–88
   memory and, 159, 160, 161
   skepticism and, 161–62
   of social categories, 287, 297
Merritte, Douglas, 78
Milgram, Stanley, 243, 275–78,
   310–11
Mill, John Stuart, 100
Miller, George, 177–78
mind
   complexity of, 386
   John Locke on, 100
   mind-brain relationship, 31, 32,
      161, 385
   powers of, 278
   science of, 30–31, 36, 58, 125,
      161, 316, 386
   theory of mind, 116–21, 148,
      157
   unconscious mind, 45, 46, 52–
      53, 55, 56, 57–59, 69, 71
mind/body dualism, 10–12, 13, 21,
   25, 31, 33, 161

modernist prose, 73–74
Molaison, Henry, 185
Monti, Martin, 41
mood disorders, 355, 364, 367
moods, consciousness of, 52
morality
   biases and, 303
   conceptions of, 3, 4, 248
   consciousness and, 37
   emotions and, 242
   evolution and, 225, 242
   happiness and, 368
   impartiality and, 249–50
   intelligence and, 339–40
   mental illnesses and, 342, 366
   motivations and, 217, 225,
      242–44, 247, 248–51, 261
   physical purity and, 280,
      281–82
   research on, 2
   self-sacrificial moral acts,
      248–51
   stereotypes and, 289, 290
   superego and, 59–60
moral psychology, 2, 242, 247
morphemes, 131
morphology, 129–32, 135, 142,
   143, 153
motion parallax, 172
motivations
   complexity of, 217, 369
   emotions and, 217, 227–28, 231,
      234–35, 241–42
   evolution and, 223–25, 250
   homeostasis theories and, 217,
      218, 223
   identification of, 3
   kindness and, 241, 242–44, 245,
      246, 247, 250
   kin selection and, 242–43, 245

morality and, 217, 225, 242–44, 247, 248–51, 261

mutual aid and, 244–45

natural selection and, 223, 224, 244–45, 247, 248

pain/pleasure theory and, 218, 219, 220, 223–24

pluralism and, 369–70

predictive processing (PP) theory and, 217–20

psychological versus evolutionary motivations, 224–25

reputational theory and, 250–51

sexuality and, 217, 219, 224, 233–41, 247

social priming and, 279

Edward Thorndike on, 215–17, 219

Müller, Max, 130

multiple personality disorder, 356

multiple sclerosis, 18

Murphy, Gregory, 155, 287

myelin sheaths, 18, 121

Myers-Briggs Type Indicator, 312–13, 315

Nagel, Thomas, 39

naive realism, 162, 166

Nash, John, 348

National Institute of Mental Health, 361–62

nativism

Noam Chomsky and, 90, 144

essentialism and, 115

on fear, 230

language learning and, 144–45

on natural endowment of capacities, 100, 101, 122

social categories and, 292

naturalistic fallacy, 225, 317

natural selection

environmental influences and, 222

fear and, 261

genetics and, 221–22

kin selection and, 242, 245

language and, 129

mental illnesses and, 363

mental life and, 3

motivations and, 223, 224, 244–45, 247, 248

sexuality and, 233, 237, 241

skepticism and, 162

B. F. Skinner on, 82

nature-nurture interaction, 98–102, 103

Neese, Randolph, 359–60

negative framing, 197–98

negative reinforcement, 81

Neisser, Ulric, 72

Nesse, Randolph, 270–71, 360

Nettle, Daniel, 269

neurodiversity, 344, 345, 346, 366

neurons

assemblies of, 18–19

consciousness and, 42

function of, 19–20, 22, 30, 121, 165, 167, 302

parts of, 17–18, *17*

place cells and, 88

rational action and, 20–21

types of, 18

neuroscience

on biological basis on thought, 17, 22, 49

brain-imaging techniques of, 109

on consciousness, 49

dualism and, 14

on memory, 176

neuroscience (*continued*)
  on parts of brain, 23
  psychology compared to, 30–32
  on rats' brains, 87
neurotransmitters, 19–20, 349, 352
Nimoy, Leonard, 267–68
Nin, Anaïs, 162, 206
Nolen-Hoeksema, Susan, 353
nonshared environment, 328, 334
novelty, habituation and, 111–12

Obama, Barack, 92, 205
object permanence, 104, 107, 113, 164
obsessive-compulsive disorder, 355
occipital lobe, 26, 165
Oedipus complex, 61
OkCupid, 238
omnivores' dilemma, 233
Ondine's curse, 54
ontogeny, 103
operant conditioning
  language learning and, 143–44
  learning and, 80, 85, 86, 88, 101
  partial reinforcement effect in, 83–84, 88, 93, 260
  punishment and, 81, 81–82, 82, 83, 103, 143
  reinforcement in, 81–82, 83, 87, 91–92, 103, 143, 144
  shaping and, 81–82
optimism, 269, 386
oral stage, 60
organ transplants, 116

pain, 32, 51
pain/pleasure theory, 218, 219, 220, 223–24
parahippocampal gyrus, 41

parenting, effects of, 335–37
parietal lobe, 26, 42, 165, 260
Parker, Ian, 310
partial reinforcement effect, 83–84, 88, 93, 260
passage of time, subjective experience of, 39–40
Pauli, Wolfgang, 65
Pavlov, Ivan
  on behaviorism, 71
  classical conditioning and, 75–77, 88, 93, 135
  on differences between species, 73
  B. F. Skinner and, 81
Payne, Keith, 301–2
perception
  assumptions and, 168, 169, 172, 187, 189
  bottom-up information and, 168, 169, 172, 173–75
  brain as computer and, 23
  of color, 162, 167
  context as influence on, 168–69, 190
  disorders of, 27–28
  of distinct objects, 169–70
  explanation of, 14
  figure/ground perception, 44, *44*, 52
  fluctuations of, 167–68
  good continuation and, 171
  as graded, 18
  grouping rules, 170
  of physical world, 161
  proportions and, 164–65
  sensation distinguished from, 28, 160, 163–75, 187
  top-down information and, 168, 173–74

visual illusions and, 171–72, 189, 201, 278

visual perception studies on cats, 72

Perot, Ross, 38

persistent vegetative state, 41

personality

age and, 318

Gordon Allport on, 313

Big Five traits, 300, 314–15, 318–19, 330

Raymond Cattell on, 314

cultural influences and, 318–19

Hans Eysenck on, 313–14

family environment and, 334

group differences in, 318–20

heritability of, 334, 338, 375

HEXACO model, 315

individual differences and, 3, 309–13, 326–27

Myers-Briggs Type Indicator, 312–13, 315

self-control and, 326–27

personality disorders, 342

persuasion, 274–75

pessimism, 271

p-hacking, 256–57

phallic stage, 61–62

phenomenal consciousness, 43–44

Phillips, Emo, 21

phobias

behavioral psychology on, 78, 79, 93

evolutionary psychology on, 79

intense irrational fears as, 355

prevalence of specific phobias, 229–30

social phobias, 342

phonemes, 129, 139, 145

phonemic restoration effect, 173–74

phonology, 129, 140, 143, 153

phrenology, 23–24, 266

phylogeny, 103

physical things, limitations of, 10, 13

physics

psychophysics, 164

successive scientific worldview and, 122–23

Piaget, Jean

on accommodation, 104, 108

on assimilation, 103–4, 108

on developmental psychology, 102, 103, 104–8, 109, 114, 119, 120, 124

early life of, 102

essentialism, 115

Sigmund Freud and, 102–3

on genetic epistemology, 103

on learning, 103–4, 124

on memory, 186

methodology of, 108

on object permanence, 104, 107, 113, 164

on radical cognitive change, 122

on schemas, 103–4

stages of thinking, 104–7, 122

theoretical limitations of, 108

Three Mountains task and, 105

pineal gland, 15

Pinker, Steven, 13–14, 131, 135, 209, 227–28, 381–82

pituitary gland, 25

place cells, 87, 88

planning fallacy, 268–69, 278

Plato, 10, 23, 25, 100, 109

pleasure principle, 59

political differences
COVID-19 and, 205–6
group differences and, 319
rationality and, 205–6, 207,
210–11, 212
Popper, Karl, 64–65
positive framing, 197–98
positive psychology, 367–68
positive reinforcement, 81
positivity bias, 268, 269–70, 271
pragmatics, 153–54
predictive processing (PP) theory,
218, 219–20
preferential looking methods, 140,
141
prefrontal cortex, 228
preoperational stage, 105–6
primacy effect, 177
primary auditory area, 26
primary motor area, 26
primary somatosensory area, 26
primary visual area, 26
probability theory, 191, 193–99,
211, 260
problem solving, 80–81
procedural memory, 175, 185
Prochnik, George, 69
projection, 63
prosopagnosia, 27–28
Proust, Marcel, 38, 182
psilocybin, 362
psychoanalysis, 56, 61, 66, 71–72,
74–75, 103, 109, 358
psychodynamic therapy, 358
psychological tests
reliability of, 311–12, 315, 324
test-retest correlations, 300–301
validity of, 311, 312, 315, 324
psychology. *See also* behavioral psychology; clinical psychology;
cognitive psychology; developmental psychology; evolutionary psychology; positive
psychology; social psychology
attitudes toward, 386
humanist movement in, 367
neuroscience compared to, 30–32
psychopathy, 364–65, 366
psychophysics, 164
psychosis, 351, 354
Pugh, Emerson M., 21
punishment
motivations and, 242, 246, 247,
250
in operant conditioning, 81, 82,
83, 103, 143

QAnon, 205, 209, 210, 211
Quine, Willard V. O., 148

race
behavioral psychology on, 72
implicit bias test and, 299, 300,
302
as social category, 285–86, 289,
291, 296, 297–98
racial profiling, 290
racism
judgments based on race and,
283–84
scientific racism, 330
Ramón y Cajal, Santiago, 19
rational action, 20–21
rationality
availability bias and, 193–95,
201
base-rate neglect and, 195–96,
201, 202
biases and, 192–96, 199–200,
201, 206, 212

blind spots and, 191
confirmation bias and, 199–200,
   206
conspiracy theories and, 205,
   208–11
defining of, 210
failures of, 192, 201
fake news and, 204
framing effects and, 196–99, 207
of human beings, 2, 278, 282
politicized beliefs and, 205–6,
   207, 210–11, 212
probability theory and, 191,
   193–99, 211, 260
reasoning illusions and, 201–2
social priming and, 279
soldier mindset and, 206–7, 212
"System 1" versus "System 2"
   thinking, 203–5
tragedy of the commons and,
   211–12, 245
truth-seeking behavior and,
   210–11
rationalization, 62–63
Rawls, John, 93, 249
Rayner, Rosalie, 78–79
reaction formation, 63
reading
   as automatic, 47–48
   brain changed by, 28, 31
   Stroop Effect and, 48–49
Reagan, Ronald, 208
reality principle, 59
recency effect, 177
recursion, 134–35
regrets, 50
Reid, Thomas, 161–62
reinforcement, in operant condi-
      tioning, 81–82, 83, 87, 91–92,
      103, 143, 144

religion
   dualism and, 10
   happiness and, 368, 378
   language and, 152, 158
   morality and, 247, 249
   motivations and, 223
   parenting effects and, 335
   personality and, 318, 330
   as social category, 288
repression, 52, 64
responsibility, 4
retrograde amnesia, 184, 185
Ritchie, Stuart, 323
robots, 10–11
Rogers, Carl, 367
Romanes, George, 223–24
Ronson, Jon, 200
Rorschach, Hermann, 312
Rorschach Inkblot, 312, *312*
Rousseau, Jean-Jacques, 100
Rozin, Paul, 231–32, 233, 268–69
Rubin, Edgar, 44
Ruminative Response Scale, 353

Sacks, Oliver, 9, 27–28, 228, 346
Sapir, Edward, 154–55
Sapir-Whorf hypothesis, 154–55
Satisficers, 308
schizophrenia
   bipolar disorder linked with, 354
   causes of, 349, 363, 367
   environmental influences and,
      350, 352
   genetics and, 349, 352
   symptoms of, 347–52, 363
   treatment for, 346, 349, 359, 366
scientific racism, 330
scientific serendipity, 76
Seinfeld, Jerry, 50
self, 177

self-control, 326–27

self-focus
first-person nature of consciousness, 50–51
self-enhancement bias, 264–65,
267, 268
specialness and, 263–64

Senior, Jennifer, 380–81

sensation
heart as source of, 15–16
perception distinguished from,
28, 160, 163–75, 187

sensorimotor stage, 104, 106

sentient experience, 35–36, 43

serotonin, 352

sexism, 97

sexuality
attractiveness and, 236–37, 238
Sigmund Freud on, 69
individual differences and, 308,
309
lust and, 233–34, 241
motivations and, 217, 219, 224,
233–41, 247
pair bonding and, 240
personality and, 319
sex differences and, 237–41, 261,
319–20
sexual orientation and, 239

Shakespeare, William, 16

shaping, in operant conditioning,
81–82

shared environment, 327–29,
334–35

Sherif, Muzafer, 292–94, 296

Shirow, Masamune, 12

Shyamalan, M. Night, 356

sickle cell disease, 363

Simenon, George, 130

Simpson, O. J., 193

Singer, Peter, 249

skepticism, 161–62

Skinner, B. F.
on behaviorism, 72, 102
behavior therapy and, 358
Noam Chomsky on, 90–92,
219
on conscious experience, 35
on criminal justice system, 107
on environment, 100, 282
Sigmund Freud compared to, 71,
72, 75, 92, 93, 107, 200
on human nature, 2, 107
on learning, 101
operant conditioning and, 81–
83, 85, 88, 135, 143–44
on self-reinforcement, 91,
219–20
situational causes and, 279
on stimulus control, 89
on verbal behavior, 89–92, 125

sleep, 53, 181, 357–58

sleep-wake cycle, 101

Slingerland, Edward, 51

Smith, Adam, 245–46, 249,
326–27

Smith, Vernon, 346

Smith, Zadie, 381

social categories
biases and, 297–99
children's perceptions of, 295–96
essentialism and, 291–92
function of, 288–89, 290
mental representations of, 287,
297
research on, 284–85

social hierarchies, 286

social priming, 278–82

social psychology
on associations, 301–2

assumptions of, 273–74
on consciousness, 49, 74–75
group research and, 283, 286, 287, 293–95
novel methods and, 298–99
real-world interventions and, 255
replications of studies in, 255–59, 260, 281
on social priming, 278–82
Solomon, Andrew, 351–53
Sophocles, 61
spatial maps, 86–87, 88
Spelke, Elizabeth, 113–14
spotlight effect, 50, 263
Springsteen, Bruce, 102
Stapel, Diederik, 258, 281
Starmans, Christina, 15
Star Trek (television series), 227–28
Steno, Nicolaus, 24
Stephens-Davidowitz, Seth, 67
stereotypes
accuracy of, 289, 290
defense of, 287, 290–91
defining of, 286–87
of groups, 286–87, 302–3
morality and, 289, 290
Us/Them psychology and, 292
stimulus control, 89, 91
Stroop, John Ridley, 48
Stroop Effect, 48–49
subjective experience, of passage of time, 39–40
sublimation, 62
substance dualism, 15
Sullivan, Anne, 151
Sun, Zekun, 219–20
superego, 59–60, 62, 63, 74
surprise, in babies, 112–13

Sussman, Gerald, 166
Sutherland, Rory, 53
synapses, 19
synesthesia, 38, 165–66
syntax
of language, 129
language as influence on thought and, 153, 157
language learning and, 141–42, 143, 144, 145, 149–50
as system of rules, 132–38
syphilis infection, 361
systematic desensitization, 79–80
Szasz, Thomas, 343, 346, 361

taboo language, 152
Tajfel, Henri, 294, 295
talk therapy, 362
task demands, 108
teacher evaluations, 266
temporal experience
preoperational stage and, 105
subjectivity of, 39–40
temporal lobes, 9, 24–25, 26
Thaler, Richard, 374
therapeutic alliance, 360
therapy, forms of, 358–60, 362
Thorndike, Edward, 71–72, 80–81, 83, 88, 215–17, 219
thought
animal species and, 150
behavioral psychology on, 74, 93
biological basis of, 17, 22, 49
brain and, 10, 14, 15, 260, 386
depression and, 353
human's existence as thinking beings, 12, 73
language and, 2, 150–58
as manipulation of symbols, 136

thought (*continued*)
  physical seat of, 15
  Jean Piaget on stages of thinking, 104–7, 122
Thunberg, Greta, 346
TMS (transcranial magnetic stimulation), 27
Tolman, Edward, 86–88
Tooby, John, 166
tragedy of the commons, 211–12, 245
transcranial magnetic stimulation, 359
transference, 64
Trivers, Robert, 52, 238–39, 246
Trump, Barron, 58
Trump, Donald, 58, 205, 209–10, 309
Trump, Melania, 58
Turing, Alan, 22, 136
Turkheimer, Eric, 330, 334
Tversky, Amos, 193, 196–97, 321
twin studies, 329–30, 349, 350

Ullman, Tomer, 137
unconscious mind
  attention and, 45, 46
  deception and, 52–53
  Sigmund Freud on, 55, 56, 57–59, 69, 71
  repression and, 52
undercontrolled children, 98
Updike, John, 3–4
Us/Them psychology, 292–93, 296, 303

Vaillant, George, 374
van Gogh, Vincent, 354
Victoria (queen of England), 24

Wallace, Alfred Russel, 222, 248
Wallace, David Foster, 50
War of the Soups and Sparks, 19
Wason, Peter, 199–200
Wason Rule Discovery Task, 199–200, *199*
Wason Selection Task, 200, *200*, 201, 202, *203*
Watson, John, 71–72, 74, 75, 78–79, 100
Weber, Ernst, 164
Weber's Law, 164
WEIRD (Western Educated Industrial Rich Democracies) people, 259–60, 272–74
well-adjusted children, 98
Wells, H. G., 81
Wernicke, Carl, 24–25
Wernicke's area, 24–25, 128
Westermarck, Edward, 235–36
Whitehead, Alfred North, 89, 92
Whorf, Benjamin Lee, 154–56
Woolf, Virginia, 73–74, 354
Wordsworth, William, 98
World Happiness Report, 375
World Health Organization, 342
World Values Survey, 375
wu wei (effortless action), 46
Wynn, Karen, 117

Yuanming, Tao, 51

thought (*continued*)
  physical seat of, 15
  Jean Piaget on stages of think-
    ing, 104–7, 122
Thunberg, Greta, 346
TMS (transcranial magnetic stim-
  ulation), 27
Tolman, Edward, 86–88
Tooby, John, 166
tragedy of the commons, 211–12,
  245
transcranial magnetic stimulation,
  359
transference, 64
Trivers, Robert, 52, 238–39, 246
Trump, Barron, 58
Trump, Donald, 58, 205, 209–10,
  309
Trump, Melania, 58
Turing, Alan, 22, 136
Turkheimer, Eric, 330, 334
Tversky, Amos, 193, 196–97, 321
twin studies, 329–30, 349, 350

Ullman, Tomer, 137
unconscious mind
  attention and, 45, 46
  deception and, 52–53
  Sigmund Freud on, 55, 56,
    57–59, 69, 71
  repression and, 52
undercontrolled children, 98
Updike, John, 3–4
Us/Them psychology, 292–93, 296,
  303

Vaillant, George, 374
van Gogh, Vincent, 354
Victoria (queen of England), 24

Wallace, Alfred Russel, 222, 248
Wallace, David Foster, 50
War of the Soups and Sparks, 19
Wason, Peter, 199–200
Wason Rule Discovery Task,
  199–200, *199*
Wason Selection Task, 200, *200*,
  201, 202, *203*
Watson, John, 71–72, 74, 75,
  78–79, 100
Weber, Ernst, 164
Weber's Law, 164
WEIRD (Western Educated In-
  dustrial Rich Democracies)
  people, 259–60, 272–74
well-adjusted children, 98
Wells, H. G., 81
Wernicke, Carl, 24–25
Wernicke's area, 24–25, 128
Westermarck, Edward, 235–36
Whitehead, Alfred North, 89, 92
Whorf, Benjamin Lee, 154–56
Woolf, Virginia, 73–74, 354
Wordsworth, William, 98
World Happiness Report, 375
World Health Organization, 342
World Values Survey, 375
wu wei (effortless action), 46
Wynn, Karen, 117

Yuanming, Tao, 51

assumptions of, 273–74
on consciousness, 49, 74–75
group research and, 283, 286, 287, 293–95
novel methods and, 298–99
real-world interventions and, 255
replications of studies in, 255–59, 260, 281
on social priming, 278–82
Solomon, Andrew, 351–53
Sophocles, 61
spatial maps, 86–87, 88
Spelke, Elizabeth, 113–14
spotlight effect, 50, 263
Springsteen, Bruce, 102
Stapel, Diederik, 258, 281
Starmans, Christina, 15
Star Trek (television series), 227–28
Steno, Nicolaus, 24
Stephens-Davidowitz, Seth, 67
stereotypes
accuracy of, 289, 290
defense of, 287, 290–91
defining of, 286–87
of groups, 286–87, 302–3
morality and, 289, 290
Us/Them psychology and, 292
stimulus control, 89, 91
Stroop, John Ridley, 48
Stroop Effect, 48–49
subjective experience, of passage of time, 39–40
sublimation, 62
substance dualism, 15
Sullivan, Anne, 151
Sun, Zekun, 219–20
superego, 59–60, 62, 63, 74
surprise, in babies, 112–13

Sussman, Gerald, 166
Sutherland, Rory, 53
synapses, 19
synesthesia, 38, 165–66
syntax
of language, 129
language as influence on thought and, 153, 157
language learning and, 141–42, 143, 144, 145, 149–50
as system of rules, 132–38
syphilis infection, 361
systematic desensitization, 79–80
Szasz, Thomas, 343, 346, 361

taboo language, 152
Tajfel, Henri, 294, 295
talk therapy, 362
task demands, 108
teacher evaluations, 266
temporal experience
preoperational stage and, 105
subjectivity of, 39–40
temporal lobes, 9, 24–25, 26
Thaler, Richard, 374
therapeutic alliance, 360
therapy, forms of, 358–60, 362
Thorndike, Edward, 71–72, 80–81, 83, 88, 215–17, 219
thought
animal species and, 150
behavioral psychology on, 74, 93
biological basis of, 17, 22, 49
brain and, 10, 14, 15, 260, 386
depression and, 353
human's existence as thinking beings, 12, 73
language and, 2, 150–58
as manipulation of symbols, 136